SURFACE COAL MINING TECHNOLOGY

SURFACE COAL MINING TECHNOLOGY

Engineering and Environmental Aspects

Edited by R. Fung

Based on research prepared or sponsored by:

The Pennsylvania State University

U.S. Bureau of Mines

Central Research and Design Institute for Openpit Mining (Wroclaw, Poland)

NOYES DATA CORPORATION
Park Ridge, New Jersey, U.S.A.
1981

Library of Congress Catalog Card Number: 81-11036
ISBN: 0-8155-0866-2
ISSN: 0270-9155
Printed in the United States

Published in the United States of America by
Noyes Data Corporation
Noyes Building, Park Ridge, New Jersey 07656

Library of Congress Cataloging in Publication Data
Main entry under title:

Surface coal mining technology.

(Energy technology review; no. 71) (Pollution technology review; no. 83)
Bibliography: p.
Includes index.
1. Coal mines and mining. 2. Strip mining.
I. Fung, R. II. Series. III. Series: Pollution technology review; no. 83.
TN802.S87 622'.334 81-11036
ISBN 0-8155-0866-2 AACR2

Foreword

This book covers recommended methods, techniques, and alternatives for selecting and designing surface coal mining systems. It is based on a review and evaluation of methods reported in the literature and applied in the field.

While the high price of petroleum has resulted in conservation efforts, total energy demand will still increase substantially in the years ahead. This increase will try our abilities to discover, extract and beneficiate fuels in the necessary volumes. Coal will eventually replace considerable amounts of petroleum fuels.

Considering the difficulties experienced in mining without environmental damage, in finding acceptable locations for power plants, and in controlling stack emissions, there is a need for a thorough evaluation of options and cautious planning before a project is begun. This book, thus, will be a valuable tool for engineers and managers involved with planning for the expected growth in demand for coal and coal products.

Surface mine engineering aspects for the premining planning of surface coal mining operations in the eastern United States are addressed. Some facets of western and central United States surface coal mining are touched upon; however, large scale surface mining in the West is relatively new and, though production increases are expected to be rapid and great, the technology for eastern surface coal mining is better developed.

A comprehensive summary of the analyses required to lay out a surface mine, taking into consideration the limitations imposed by geology, equipment, reclamation, economics, and environmental control requirements is provided. Methods, techniques, and alternatives are recommended for selecting and designing mining systems that include soil handling and storage, overburden removal, and minimization of environmental effects of drilling, blasting, off-highway transportation, coal loading, and steep slopes.

The information in the book is from the following documents:

User's Manual for Premining Planning of Eastern Surface Coal Mining. Volume 1—Executive Summary (EPA 600/7-78-180), by R.V. Ramani and M.L. Clar of The Pennsylvania State University, prepared for the U.S. Environmental Protection Agency, August 1978.

User's Manual for Premining Planning of Eastern Surface Coal Mining. Volume 2—Surface Mine Engineering (EPA 600/7-80-175), by R.V. Ramani, C.J. Bise, C. Murray and L.W. Saperstein of The Pennsylvania State University, prepared for the U.S. Environmental Protection Agency, October 1980.

Surface Coal Mining Reclamation Equipment and Techniques. Proceedings: Bureau of Mines Technology Transfer Seminars, Evansville, IN, June 3, 1980, and Denver, CO, June 5, 1980. (IC 8823), compiled by the staff of the U.S. Bureau of Mines.

Proceedings of the Second U.S.-Polish Symposium—Coal Surface Mining and Power Production in the Face of Environmental Protection Requirements (EPA 600/7-79-159), prepared by J. Libicki, symposium director, of the Central Research and Design Institute for Openpit Mining, Wroclaw, Poland, for the U.S. Environmental Protection Agency, October 1979.

The table of contents is organized in such a way as to serve as a subject index and provide easy access to the information contained in the book.

Contents and Subject Index

PART I: INTRODUCTION AND TECHNOLOGY OVERVIEW

PART II: SURFACE MINE ENGINEERING TECHNOLOGY

PART III: RECLAMATION EQUIPMENT AND TECHNIQUES

Part I

Introduction and Technology Overview

The information in Part I is from *User's Manual for Premining Planning of Eastern Surface Coal Mining. Volume 1—Executive Summary* (EPA 600/7-78-180), by R.V. Ramani and M.L. Clar of The Pennsylvania State University, prepared for the U.S. Environmental Protection Agency, August 1978, and *Proceedings of the Second U.S.-Polish Symposium—Coal Surface Mining and Power Production in the Face of Environmental Protection Requirements* (EPA 600/7-79-159), prepared by J. Libicki, symposium director, of the Central Research and Design Institute for Openpit Mining, Wroclaw, Poland, for the U.S. Environmental Protection Agency, October 1979.

Introduction

The material in this chapter is based on *User's Manual for Premining Planning of Eastern Surface Coal Mining. Volume I—Executive Summary* (EPA 600/7-78-180), by R.V. Ramani and M.L. Clar of the Pennsylvania State University for the U.S. Environmental Protection Agency, August 1978.

With the possible exception of agriculture, mining is the most basic of all industries. Even in agriculture, much of the fertilizer consists of minerals initially mined and processed, and finally enriched to the chemical form. In fact, virtually all aspects of human life as it exists today require the products of mining. Exploitation of mineral resources is a must if better standards of living are to be attained, since mining affects practically every phase of industrialization and provides the basics for sustained economic growth. The fundamental differences between mineral and other resources are that the former are highly localized, limited, and nonrenewable.

In the most elementary form, mining is materials handling. It is a process of extracting in-situ materials (solids, liquids, or gases) from the crust of the earth. In this sense, both the extraction of mineral ores from the earth and the removal of gas and oil from underground chambers and the recovery of subterranean water can be considered mining. Both types of mining share many similar environmental considerations, but this project examines only the surface mining of coal in the eastern United States.

THE ROLE OF SURFACE MINING

The various mining methods can be classified as surface or underground. Surface mining accounts for over 90% of all mined products. And of all land affected by surface mining, more than 75% is the result of surface mining of coal, crushed stone, and gravel.

No major increase is anticipated for the share of coal in future energy demands. But the total national demand for energy is predicted to rise so rapidly that coal production itself is expected to be around a billion tons by 1985. If recent trends are any indication, surface-mined coal should continue to take a major share of the coal production, at least to the year 2000.

Decisions as to whether an ore body will be mined and what mining methods will be used depend on many technological, economic, ecological, and social factors. But the big question is: How can one mine these riches and still preserve the capability of the land for sustaining other uses during and after mining? Though this question is frequently viewed as a problem, it should also be recognized as an opportunity to restructure the land surface for a desirable end use.

ENVIRONMENTAL CONCERNS

Mineral extraction affects the environment in three basic ways: It influences air, land, and water quality, which in turn affect plant and animal life (Table 1). The environmental effects of mining are, of course, interdependent. For example, air quality may influence water quality, and both combined can affect the capability of the land for sustaining habitation.

LEGAL CONSIDERATIONS

Surface coal mining has come under increasing legislative regulation, particularly with regard to the environmental effects. The use of surface-mined lands for other productive purposes was demonstrated and pursued voluntarily by some coal companies as early as 1920, when surface-

Table 1. Major Environmental Effects of Various Surface Mining Unit Operations

Surface Mining Unit Operations	Water			Air			Land		
	Surface Water Contamination	Groundwater Contamination	Runoff and Infiltration Alteration	Exhaust Emissions	Dust Emissions	Other Emissions	Soil Erosion	Soil Stability	Soil Productivity
1. Exploration			X	X	X		X		
2. Area Dewatering and Diversion	X	X	X				X	X	X
3. Drilling			X	X	X				
4. Blasting			X		X	X		X	X
5. Overburden Removal	X	X	X	X	X	X		X	X
6. Haulage	X			X	X		X		
7. Soil Storage	X		X	X	X		X		X
8. Maintenance	X			X	X	X			
9. Reclamation	X	X	X	X	X		X	X	X

mined coal accounted for less than 7% of the total U.S. coal production. Nevertheless, before 1965, there were only seven states that had specific laws requiring land reclamation. West Virginia enacted the first such legislation in 1939. The period since 1965 has been one of rapid growth in laws and regulations regarding surface mining control at the Federal and State levels. Presently, new or amended regulations are in force in 38 states. The most recent Federal legislation to affect surface mining is the Surface Mining Control and Reclamation Act of 1977 (Public Law 95-87).

There have always been other laws and regulations under which control was exercised over surface mining by such agencies as the Bureau of Land Management, the U.S. Mine Safety and Health Administration, State highway departments, and local zoning authorities. But, under P.L. 95–87, the Office of Surface Mining Reclamation and Enforcement was created in the Department of the Interior. Surface Mining Reclamation and Enforcement Provisions (rules and regulations) are contained in Chapter VII, Title-30, Mineral Resources of the Code of Federal Regulations (CFR). Several major aspects of the Environmental Protection Performance Standards can be found in Section 515 of P.L. 95–87 and Parts 715–718 of the Regulations (Chapter VII, Title 30).

Title I of the Act (Sec. 101) recognizes the many adverse effects of surface mining operations and the need to minimize them insofar as practicable. It further declares that responsibility for developing, issuing, and enforcing surface mining regulations should rest with the States. The purposes of the Act are identified as follows (Sec. 102):

(a) To establish a nationwide program to combat the adverse effects of surface coal mining operations;

(b) To assure that the rights of surface landowners and others with legal interests in the land are fully protected;

(c) To assure that surface mining is not conducted where reclamation is not feasible;

(d) To assure that surface mining is conducted so as to protect the environment;

(e) To assure that surface-mined areas are reclaimed as contemporaneously as possible with the mining operations;

(f) To strike a balance between the Nation's need for coal and protection of the environment;

(g) To assist the States in implementation of programs to achieve these ends;

(h) To promote the reclamation of areas degraded before enactment of this legislation; and

(i) To assure public participation in reclamation plans and programs.

LAND-USE PLANNING

Mine planning must include a careful determination and demonstration of how the mined land is to be used afterward. Such determinations will dictate the most effective plans for mining and reclamation.

Premining definitions of long-range use plans make it possible to minimize any disturbance of the land and also to limit reclamation activities to those required by the future-use plans. The integrated mining, reclamation, and land-use planning diagram in Figure 1 outlines a procedure for the evaluation of mining and reclamation plans and shows the various interactions that may take place before a mining plan can be put into practice. The process through which both the mining and reclamation have to pass include not only regulatory agencies, but special interest groups and the general public as well.

Integrated surface mining, reclamation, and land-use planning has several advantages. In mining areas characterized by the absence of any formal zoning or land-use activity, the mining operations are subject to the State or Federal laws regarding mining, and air, water, and land resources. In such a situation, the frequently determined postmining land use is the premining land use, even though great potential may exist for creating an alternate land use.

Many potential land improvement projects never get further than the planning stage because of the relatively high cost of earth-moving and rehabilitation equipment and processes, and the high costs of building the transportation and communication networks. On the other hand, the ongoing mining operation must have these facilities, which can be retained and improved for later use by the community when the mining operation is completed. The infrastructure development so necessary for a mining operation (haulroads, other transportation networks, communications, power lines, office and recreational buildings, water facilities, etc.) can easily be designed for later use by surrounding communities or by new developments that can be planned around it. This concept is one of multiple and sequential land uses.

Figure 1. Integrated Mining, Reclamation, and Land-Use Planning Diagram

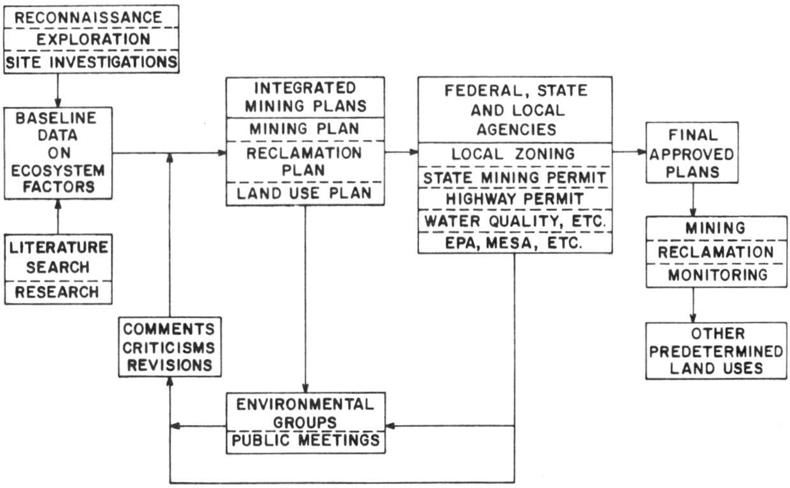

Land-use planning and control evoke rather difficult and emotional questions concerning public and private ownership, jobs, county development, taxes and earnings, local, State, and Federal economics, and authority over land-use decisions. Here the cooperation between mining companies and land-use planning authorities becomes very important. To make the reclamation planning process a success, each party must take into account the scope and limitations of the other. A selected list of these important factors is as follows:

(a) Local/regional planning, coordination, and development are usually the responsibilities of local, regional, State, or Federal agencies and not the mining companies.

(b) Government and other public land-use planning groups are concerned more with the surface utilization of land. Often their expertise may not extend to such matters as surface mine engineering and long-term impacts of mining.

(c) Though some large mining companies can perform the required premining and postmining land-use analysis, most mining companies, particularly the small operations, lack expertise in this area.

(d) Government planning for land use usually exists at a macro level dealing with large areas within local, State and Federal jurisdiction. On the other hand, company planning is usually at a micro level specific to a particular mining site and designed to operate at a profit, even though in compliance with the various legislative constraints.

(e) Land-use regulation is predominantly a function of local government. The domain of these local governments is often not based on physiographic considerations, but on political subdivisions. On the other hand, there is a growing recognition that the effects of mining and mining itself may transcend several political subdivisions, and that therefore reclamation activities must be bounded by physiographic considerations and may encompass several local governments.

In short, the planning of surface mines, reclamation, and postmining uses of land has become very complex.

PREMINING PLANNING

From the premining planning point of view, two important characteristics of mineral resources are that they are highly localized and nonrenewable. Choices are limited for siting a mine, and once it is in operation, it has a finite life. Thus surface mining, like any type of mining, is a temporary and nonrenewable use of land. But the land itself is a permanent resource with many concurrent and sequential uses. Only through proper premining planning can such a resource be preserved and its possible multiple uses enjoyed. Otherwise, some of its uses may not be possible.

In extreme cases, the land may be of no use and might even become a major source of environmental problems.

The ultimate objective of premining planning is to ensure a mining venture that has fully considered and provided for environmental capabilities and sensitivities on a national, regional, and site-specific basis. Environment, by definition, encompasses all aspects of life and living and includes both natural and cultural aspects. Environment is the net result of a complex, dynamic, and intimate interrelationship among the physical, biological, and social factors of its various systems (geologic, physiographic, hydrologic, demographic, etc.), components (rocks, soils, climate), and processes (erosion, productivity, behavior).

Current legislative, administrative, and public attitudes toward surface mining and premining planning can be interpreted as an attempt to evaluate the multiple resources of an area (minerals, water, range, recreation, forestry, etc.) and to determine the location, size, and timing of their exploitation so that resource extraction (and/or utilization) that has preemptive or conflicting effects on other resources and uses is not allowed to occur except as planned.

The search is under way, not only for an understanding of the total environment and the needs for preserving it, but for the application of existing knowledge and technical skills to the total resource management for optimum benefits to present and future generations. Increasing concern and attention to premining planning is but one of the results of this change in the values and attitudes towards the extraction, utilization, and conservation of all the material resources.

This volume presents a comprehensive summary of the analyses required to lay out a surface mine, taking into consideration the limitations imposed by geology, available equipment, and environmental control requirements. Methods, techniques, and alternatives are recommended for selecting and designing mining systems that include (1) soil handling and storage, (2) overburden handling, segregation and disposal, and (3) minimization of the adverse environmental effects of drilling, blasting, off-highway truck transportation, coal loading, etc. The major topics covered are: exploration, surface mining equipment, surface mining methods, fragmentation practices, surface mine haulroad design, slope stability, operational considerations, and surface mining costs. In addition sections are included on reclamation equipment and techniques and environmental aspects.

Technology Overview

The material in this chapter is based on "Present and Future Surface Coal Extraction Technologies in the United States," by L.D. Loy, Jr. of Skelly and Loy Consulting Engineers, in *Proceedings of the Second U.S.-Polish Symposium—Coal Surface Mining and Power Production in the Face of Environmental Protection Requirements* (EPA 600/7-79-159), prepared by J. Libick, Symposium Director, of the Central Research and Design Institute for Openpit Mining, Wroclaw, Poland, for the U.S. Environmental Protection Agency, October 1979.

INTRODUCTION

It is an honor for me to present a paper to this Second Polish/EPA Symposium. "Present and Future Coal Extraction Technologies in the United States" is a challenging topic because, as you know, there are many different types of terrain and geology in America; and there are also many different surface mining methods.

The United States may be divided into three mining regions: (1) Western United States—including Montana, North Dakota, Colorado, Wyoming, New Mexico, and Arizona; (2) Central United States—consisting of western Kentucky, Illinois, Indiana, Ohio, Missouri, Oklahoma, Kansas, Arkansas, and Iowa; and (3) Eastern United States (Appalachia)—covering eastern Kentucky, West Virginia, Tennessee, Virginia, Pennsylvania, Maryland, Alabama, and southeastern Ohio. Major mining technologies utilized and under development within each region vary due to the necessity of adapting to the predominant geology and topography of the area. For each mining method discussed, there is a separate captioned illustration. Most of the new mining methods now under development and described here are being accomplished by the United States Department of Energy.

WESTERN UNITED STATES

Present Mining Methods

Only two major types of surface mining methods are presently employed in western United States. These are: (1) Area Type Mining; and (2) Open Pit Mining.

Area Type Mining in the West is characterized by removal of one or two moderately thick coal seams (usually between 6 and 10 feet thick), with extraction activities ultimately progressing over a relatively large tract of land. The topography is usually flat or gently sloping, with an arid climate. As shown by the illustration, spoil is directly placed in adjacent previous cuts, using either shovels or draglines for excavation. Lack of substantial precipitation is the most difficult problem confronting mine operators during post-mining revegetation.

In contrast, two basic types of Open Pit Mining are employed in western United States, depending on the thickness and pitch of the coal seams to be mined. These are both shown on one illustration. Open Pit Mining of thick coal seams (up to 100 feet thick) involves removal of thin amounts of overburden in flat terrain. Shovels are the most common equipment used to remove overburden and coal into a series of benches 40 to 50 feet high. The low ratio of overburden to coal adds to reclamation difficulties associated with the arid climate.

With the second approach, Open Pit Mining removes steeply pitching seams, either single or multiple, which vary in thickness from as little as 4 feet to as much as 90 feet. This type of Open Pit Mining is performed by having the stripping operation follow the coal down-dip.

Figure 1: Area Type Mining—Western U.S.

Figure 2: Open Pit Mining

One particular Open Pit Mine in western United States is characterized by a whole series of steeply pitching seams which are mined by a series of pits on each specific coal seam; that is, each seam has its own pit, and overburden is placed both above and below the seam. Direct backfilling is not feasible and bulldozers must be employed to cover the exposed pit areas prior to revegetation.

Mining Methods Being Developed

Methods being developed with regard to western Area Type surface mining include application of bucket wheel excavators and conveyors. These pieces of equipment are used to extract coal and/or overburden, in contrast with central United States where they will more than likely be used to extract and move overburden only. In part this is due to the higher ratio of overburden to coal found in most central United States mines.

A series of four Cross-Pit Conveyor Systems have been conceptualized for western Area Type Mining, and the United States Department of Energy is working to select the best of these for development.

Future mining methods related to Open Pit Mining of steeply pitching seams include applications of more mobile equipment and use of an existing highway construction machine, the Gradall. Since it can operate on steeper grades than front-end loaders or shovels, the Gradall may be useful for extracting coal from steeply pitching Western coal seams. Under a new terrace pit project, research is being conducted to find a different and more efficient extraction technology for both types of Open Pit Mining in the West.

CENTRAL UNITED STATES

Present Mining Methods

In central United States, Area Mining is used almost exclusively. The terrain varies from relatively flat to slightly rolling, and the technique is similar to western Area Mining. The climate of central United States is wetter than western United States, permitting easier reclamation, and the overburden is thicker. In rolling terrain, cropline to cropline coal mining is permitted: this slight alteration is called Modified Area Mining. Stripping shovels are used where the overburden is approximately 40 to 50 feet deep and draglines where overburden ranges as deep as 150 feet.

Mining Methods Being Developed

Portable conveyor units have been used in Europe for a number of years, but so far they have not been put to use in central United States stripping operations. However, their potential for application in this region is excellent. Conveyor units are described later in this paper under the topic of eastern United States.

Another technique under development which shows promise for central United States is called Trench, Strip, and Auger Mining. This system disturbs much less land than other methods of mining. It is environmentally sound and allows recovery of deep resources with fairly thin coal seams. The method is very simple. A large trench or ditch is excavated down to the coal seam, from which the coal is then removed. Large augering machines bore into both sides of the trench previously excavated. With this method there is no problem with drainage from augered seams, because when reclamation is accomplished and the mined area backfilled, there is no place for water to go. When the trench has been excavated and augering completed on each side of the trench, the process is advanced by repeating the entire sequence over and over.

Now let's turn to eastern United States where much more complicated methods of mining exist.

EASTERN UNITED STATES (APPALACHIA)

Present Mining Methods

There are a number of different mining methods employed in Appalachia. This is principally because Appalachia has three distinct topographic areas: "flat or gently rolling terrain," "hilly or moderately rolling terrain," and "mountainous or very steep terrain."

Flat or Gently Rolling Terrain: There are two major mining methods utilized in flat or gently rolling terrain such as is common to Pennsylvania and Ohio.

Block Area Mining — The first is called Block Area Mining, used where overburden is usually less than 70 feet deep. This mining method employs bulldozers and front-end loaders almost exclusively. Successive 70 foot wide cuts are taken, with overburden from one cut placed into the mined out cut behind it. Very thin seams can be recovered with little equipment. Flexibility and ease of mining are advantages of this type of mining operation.

Modified Area Mining — The second method used in flat or gently rolling terrain is called Modified Area mining, which was already discussed as a major surface mining method in central United States. As the illustration shows, mining can take place from cropline to cropline, usually with 150 feet maximum overburden. One of the requirements of the new Surface Mine Law is that the land be regraded to approximate original contour. However, there are exceptions that can be made to this requirement. In this case, because Modified Area Mining is accomplished in rather flat or gently rolling terrain, return to approximate original contour is fairly easy to accomplish.

Hilly or Moderately Rolling Terrain: Where "hilly or moderately rolling terrain" is present, primarily in northern West Virginia, Pennsylvania and Western Kentucky, two major techniques are employed.

Modified Block Cut — One is called Modified Block Cut, which is a contour method of mining. All spoil is moved laterally along the bench, mainly by direct placement, with dozer/loaders. The Modified Block Cut Method is primarily applied in Pennsylvania. The advantage here is that there is no spoil placed downslope following the initial cut removal and, in fact, the new surface mine law prohibits placement of spoil downslope. This ensures that environmental disturbances are restricted to the actual area being mined. The mined area also must be returned to approximate original contour as required by the new law.

Haulback — Haulback is also a contour mining method, and is sometimes called Controlled Placement or Lateral Movement Mining. Trucks move the material along the bench, requiring longer open high-walls than with dozer/loader Modified Block Cut operations. A valley fill or head-of-hollow fill or old mine bench is required for placement of waste material from the first cut, because no material is allowed downslope. Overburden is transported by trucks after its excavation by bulldozers, front-end loaders, shovels, hydraulic shovels, or even an occasional dragline. As with Modified Block Cut Mining, this method is environmentally sound. Reclamation is fairly easy to accomplish, and the approximate original contour can be returned after mining is completed.

Mountainous or Very Steep Terrain: In "mountainous or very steep terrain" such as is common to eastern Kentucky, southern West Virginia, and Tennessee, there are several methods used:

Haulback — Use of Haulback in mountainous or very steep terrain is much more difficult. The new law requiring return to approximate original contour means that no highwall can be left exposed. However, when the terrain is steep, it is very difficult to completely backfill the highwall. Less than half of the steep terrain areas are now being mined with the Haulback technique; most are being mined with the mountaintop removal method.

Mountaintop Removal — Mountaintop Removal is a technique where, as the name implies, the entire top of the mountain is taken away, permitting total resource recovery of both single and multiple seams. This type of mining operation is allowed under the new strip mine law, and with good premine and future land use planning, it is a desirable and environmentally sound method. Previously, mine operators took contour cuts, ringing the mountain and leaving an "applecore," or unmined center section, because it was not economical to remove the overburden in the thicker center. Leaving these "applecores" is not permitted under the new law. Cross-ridge mining, discussed in the next section of this paper, is being developed in order to make removal of the entire top of the mountain economical. As with Haulback Mining, in Mountaintop Removal Operations, valley fills or old strip mining benches must be used to deposit the first cut of waste overburden removed from the mountain.

Mining Methods Being Developed

The most promising mining methods being developed for Appalachia involve use of Portable or Mobile Conveyor Systems. These Portable Conveyor Systems are being developed for all three Appalachian terrains: "flat or gently rolling;" "hilly or moderately rolling;" and "mountainous or very steep" terrain.

Portable Conveyor Systems — Taking them one at a time, we'll start with the "gently rolling terrain" where Modified Area Mining is used.

Modified Area Mining — The artist's sketch shows how conveyors can be used to help accomplish Modified Area Mining. Modified Area Mining has traditionally required double handling of overburden material where draglines have been working beyond their boom capacity. When draglines reach the limit of their boom length, they must work on an extended bench (create a bench for themselves), and then remove it, thus handling material twice.

Using conveyors, double handling can be eliminated and spoil placement controlled. Conveyors facilitate topsoil replacement and final vegetation, including seedling, liming, and mulching. All of these materials can be added to the conveyor at the appropriate time as the topsoil passes by. This type of mining, as mentioned earlier, is also being developed for application to central United States Area and Modified Area Mining Methods.

Haulback — Haulback is also being developed with Portable Conveyor Systems. This is primarily practiced in "moderately rolling to slightly steep" terrains. As shown here, the conveyor units are placed along the low wall, or outbound side, of the haulback operation. Excavated overburden is loaded on the conveyor units to be carried along the low wall and deposited at the reclamation end of the project. Since this restricts haulage of the coal in and out of the pit, an elevated type conveyor unit must be used to bring the coal up over the low wall conveyors and into the waiting haul trucks.

Use of conveyors provides many environmental benefits, including segregation of overburden, that is, the overburden can be placed back in the same sequence in which it was excavated. Since materials handling is continuous, reclamation can be accomplished quickly. Soil supplements, including lime, seed, fertilizer, and even mulch can be added to the topsoil as it moves along the belt. Use of conveyor units offers many safety benefits, since the truck congestion on short narrow benches associated with Haulback operations is eliminated.

Mountaintop Removal — The next category is Mountaintop Removal in very steep terrain. Continuous material handling afforded by mobile conveyors allows greater safety, facilitates valley fill construction, and enhances the environment if preplanning for future land use is done well. Aesthetic reclamation is a necessity in this type of operation; otherwise the final result is flat mountaintops which are unusable and undesirable from an aesthetic and environmental standpoint.

Cross-Ridge Mining

Because of the new law prohibiting exposed highwalls to be left after mining, a new technique of removing the entire top of a mountain is under development. The Cross-Ridge technique requires mining perpendicular to the long axis of the mountaintop. Since approximately two-thirds of the Appalachian landform consists of hilly and mountainous areas, this method has wide potential application in the eastern United States.

The equipment used is flexible and mobile. Most operations use dozers with front-end loaders, but hydraulic shovels and power shovels are becoming popular. By extracting the accesible outcrop coal and the thicker overburden at the center of the mountain simultaneously, the operator's cash flow is equalized over the life of the mine. Complexities of Cross-Ridge Mining require good planning of the entire operation from beginning to end.

One of the major advantages of Cross-Ridge from an environmental standpoint is that, by backstacking waste overburden on the area that has been previously mined, valley fills for overburden disposal disturb only one-fifth of the land that is normally disturbed by a Contour Mountaintop Removal operation. This is because Contour Mountaintop Removal requires placing nearly all of the waste overburden material in valley fills, whereas Cross-Ridge Mining requires use of a valley fill for only the first cut of waste overburden. This technology has been developed and is being demonstrated by the United States Department of Energy on several Appalachian sites with typical topography, facility typical geology, and conventional equipment usage.

Anthracite Open Pit Mining

There are large anthracite deposits in eastern Pennsylvania which have traditionally been mined by relatively small surface open pit operations. Some have been large enough to employ draglines. Because of the steeply pitching seams prevalent in this region, large scale open pits have excellent potential for application to anthracite coal mining. While they are used for iron ore and other hard rock mining in western United States, they have not yet been applied to anthracite mining.

Dimensions of anthracite open pits are expected to be 800 feet deep, one-half mile wide, and three-quarters mile long. As excavation takes place in the advancing face of the mine and reclamation takes place behind, the pit will "move" across the landscape. This is an environmentally sound method because reclamation follows mining right away. In eastern Pennsylvania, there are many old unreclaimed surface mine scars which would be removed and reclaimed as part of the mining operation. This method would also help eliminate deep mine subsidence problems as abandoned underground mines are removed by the large open pit.

Figure 3: Coal Removal Scheme Using BWE

Figure 4: Cross-Pit Conveyor System

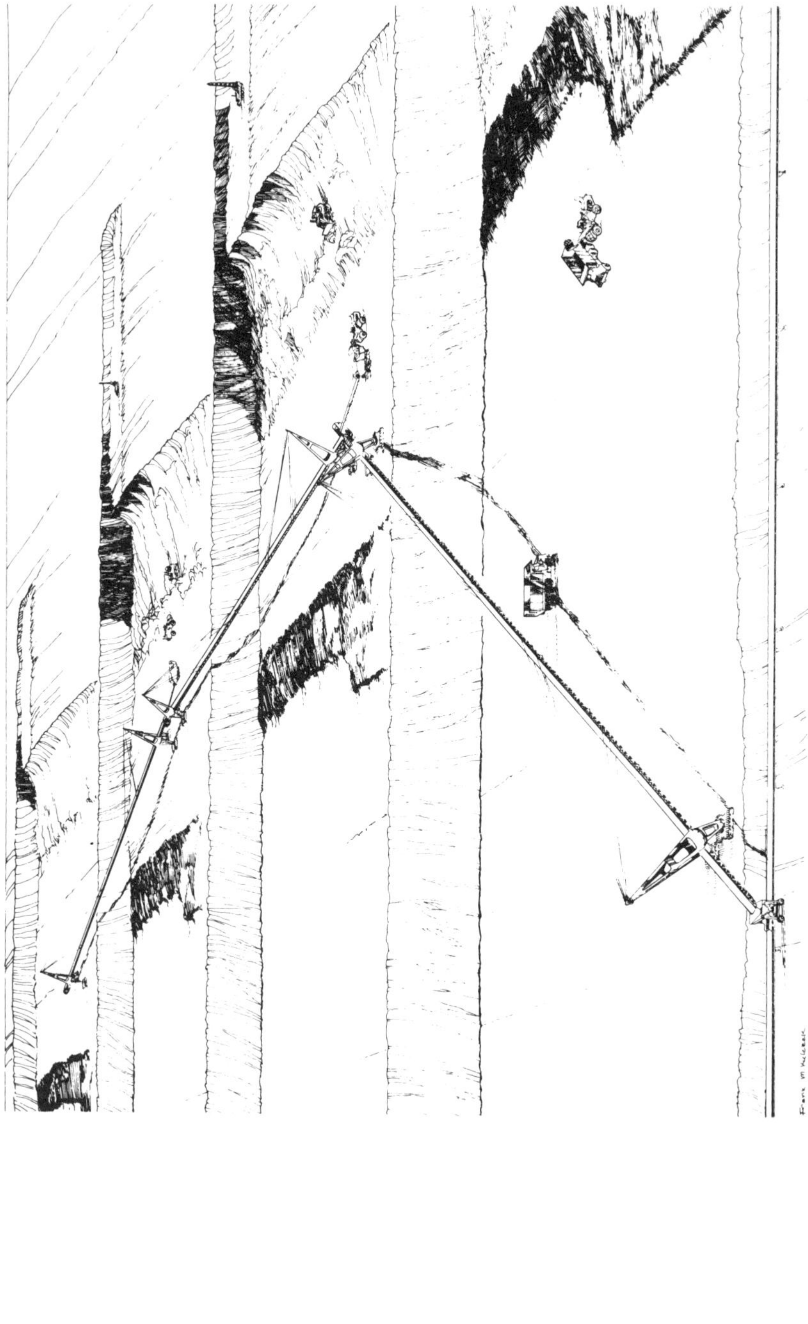

Figure 5: Gradall Coal Removal System

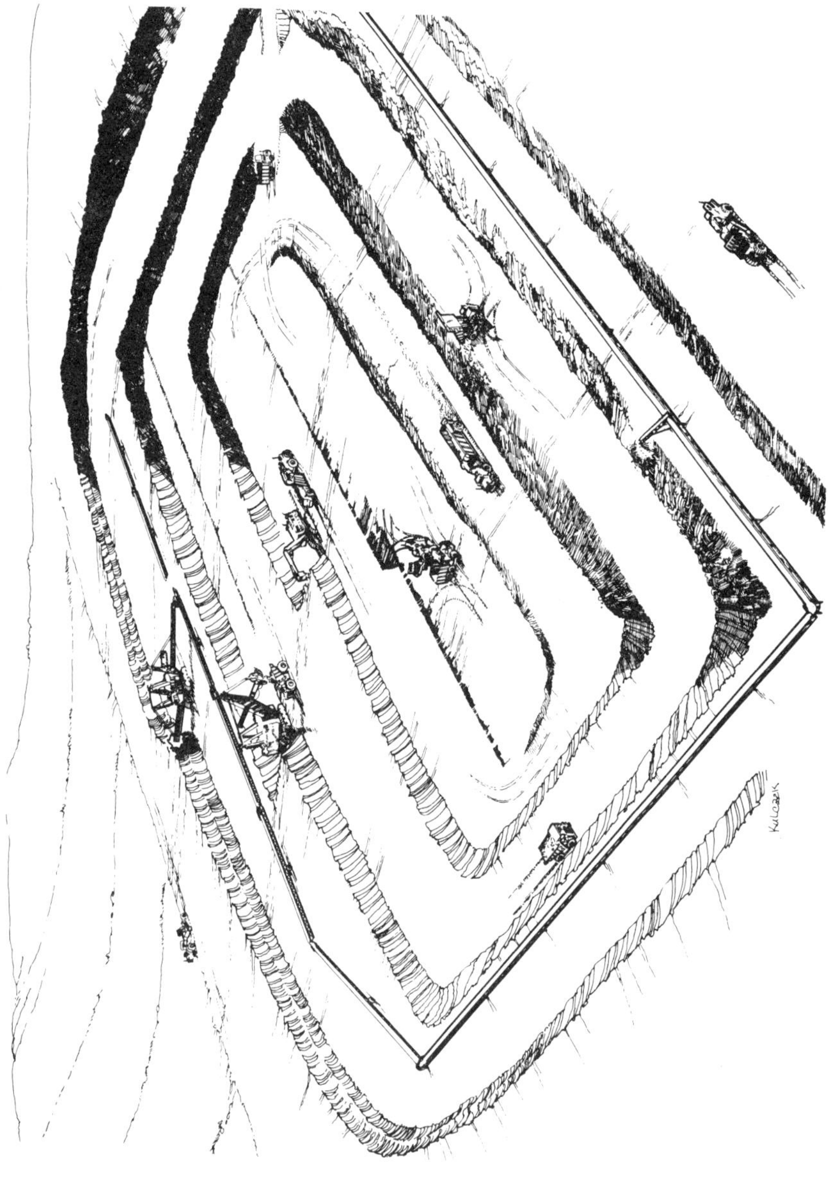

Figure 6: Terrace Pit Mining System

Figure 7: Modified Area Mining—Central U.S.

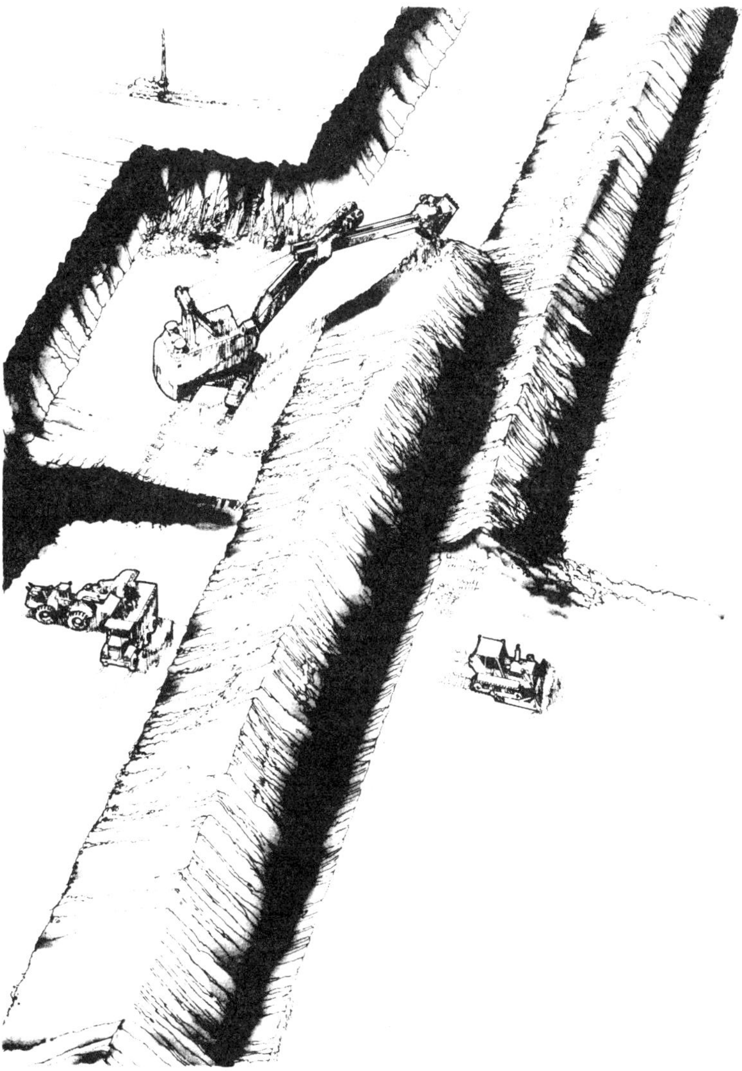

Figure 8: Trench Strip and Auger Mining

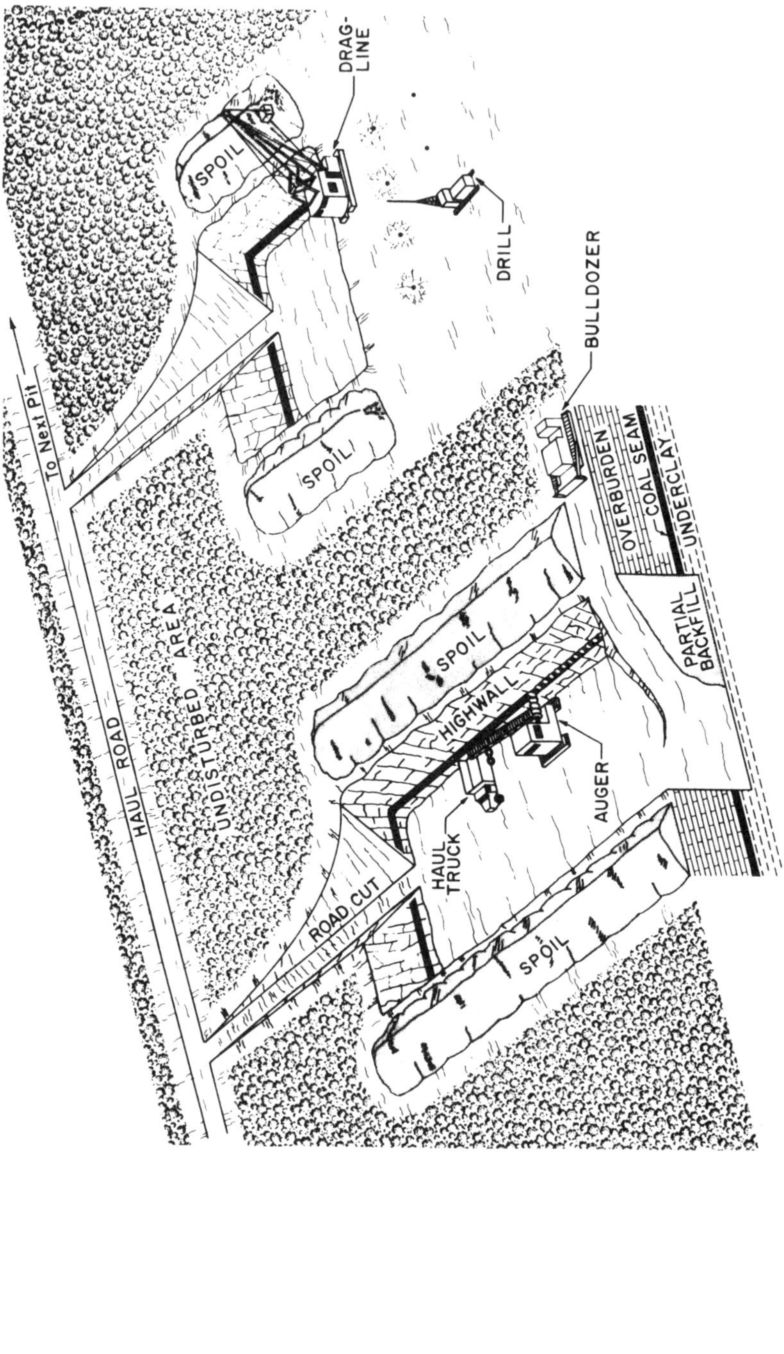

Figure 9: Block Area Mining

Figure 10: Modified Area Mining—Eastern U.S.

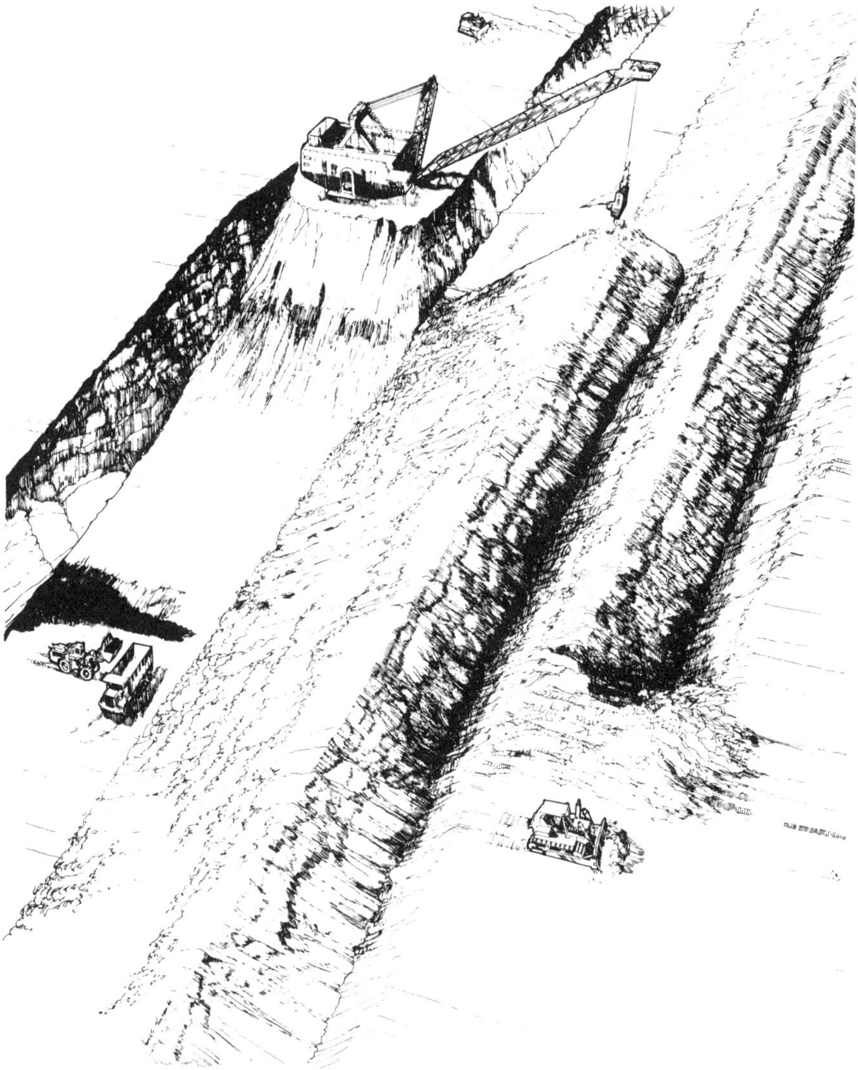

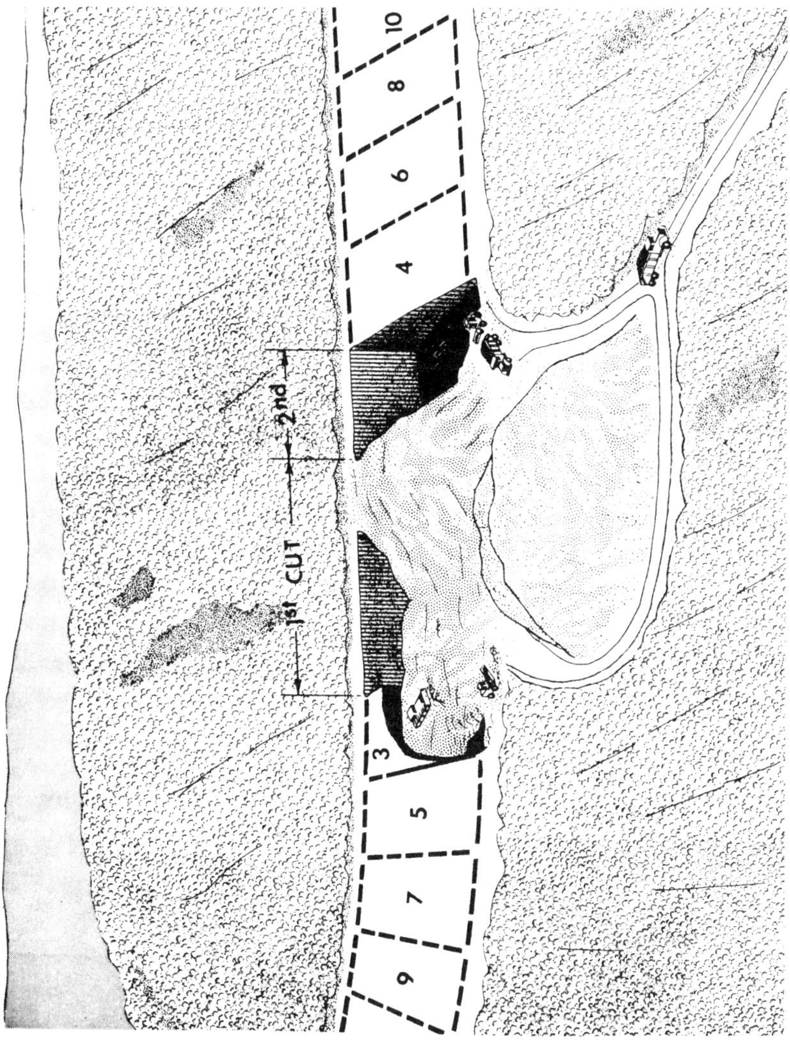

Figure 11: Modified Block Cut Mining

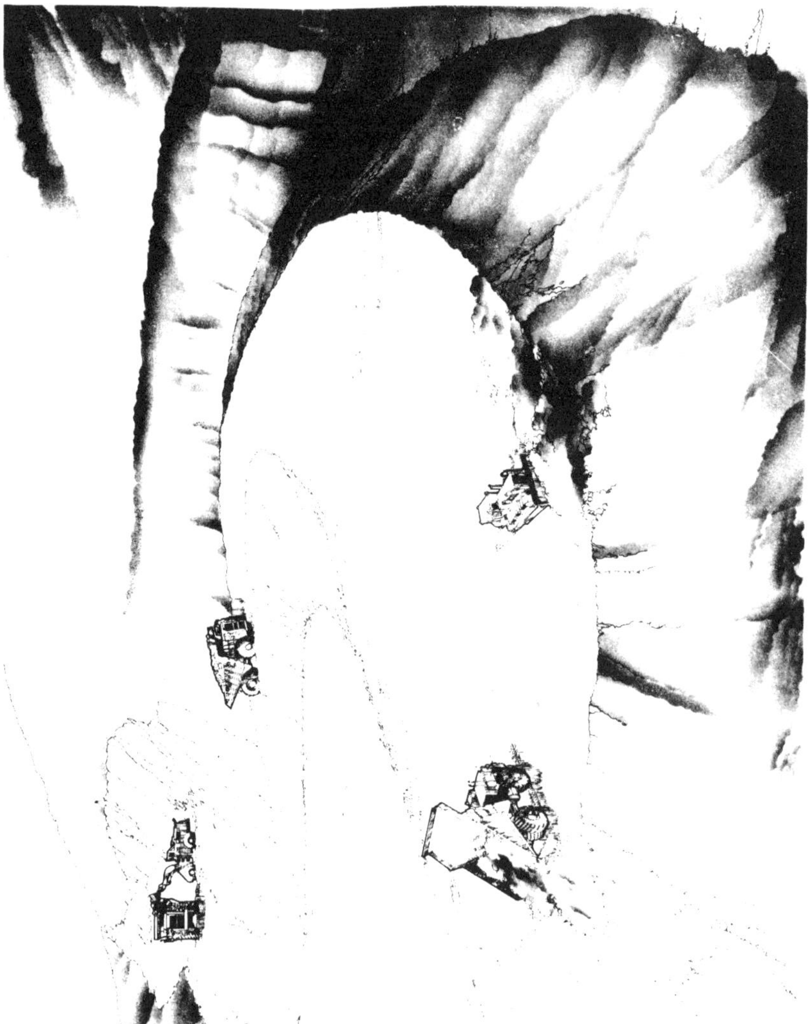

Figure 12: Contour Haulback Mining

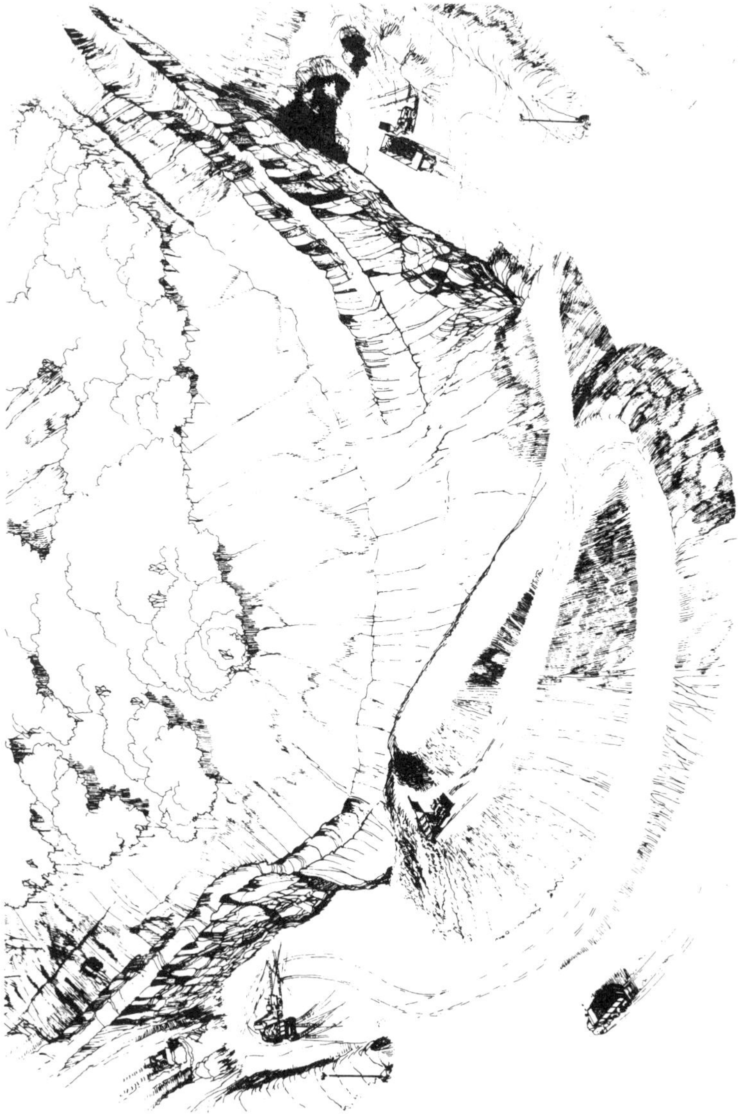

Figure 13: Contour Haulback Mining—Steep Terrain

Figure 14: Mountaintop Removal Scheme

Figure 15: Modified Area Mining with Conveyors

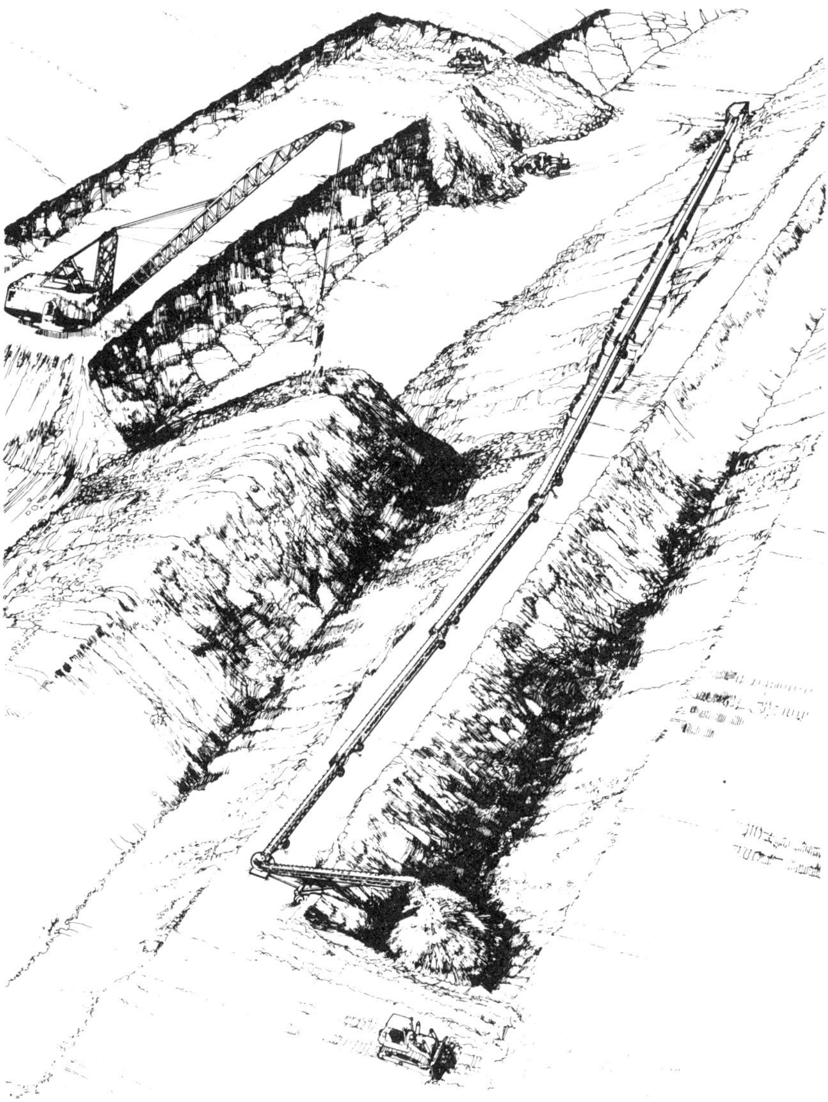

Figure 16: Conveyor Contour Haulback Mining

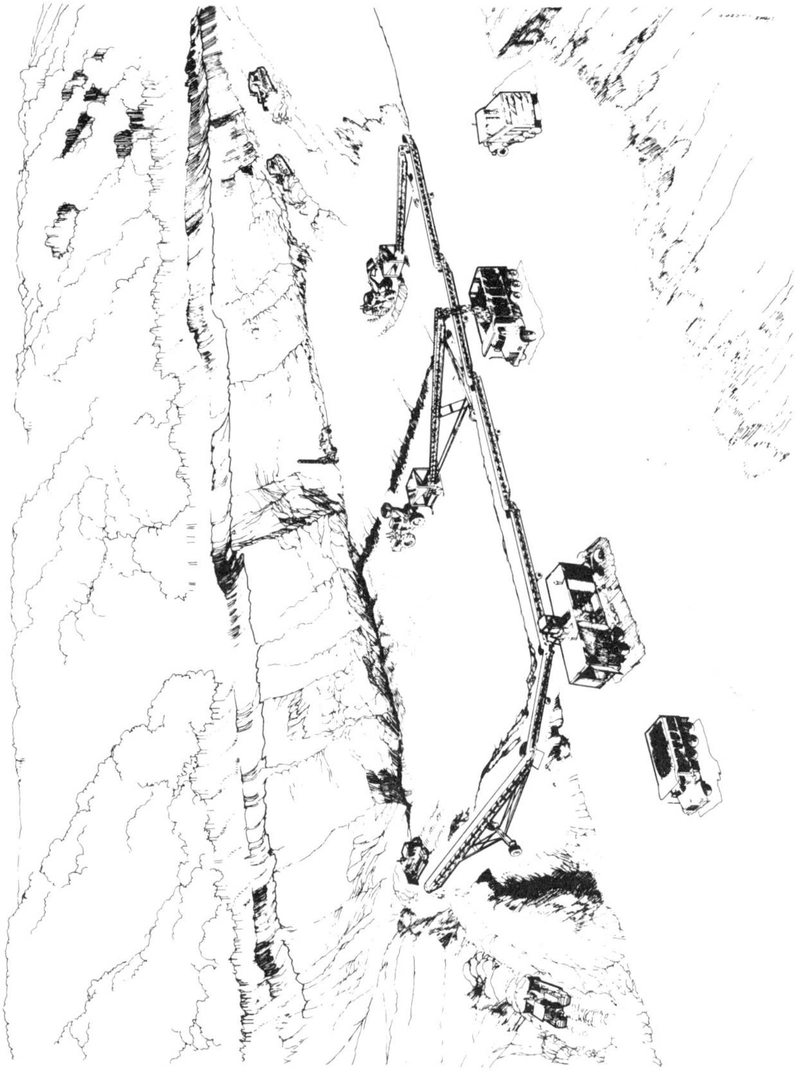

Figure 17: Conveyor Mountaintop Removal Scheme

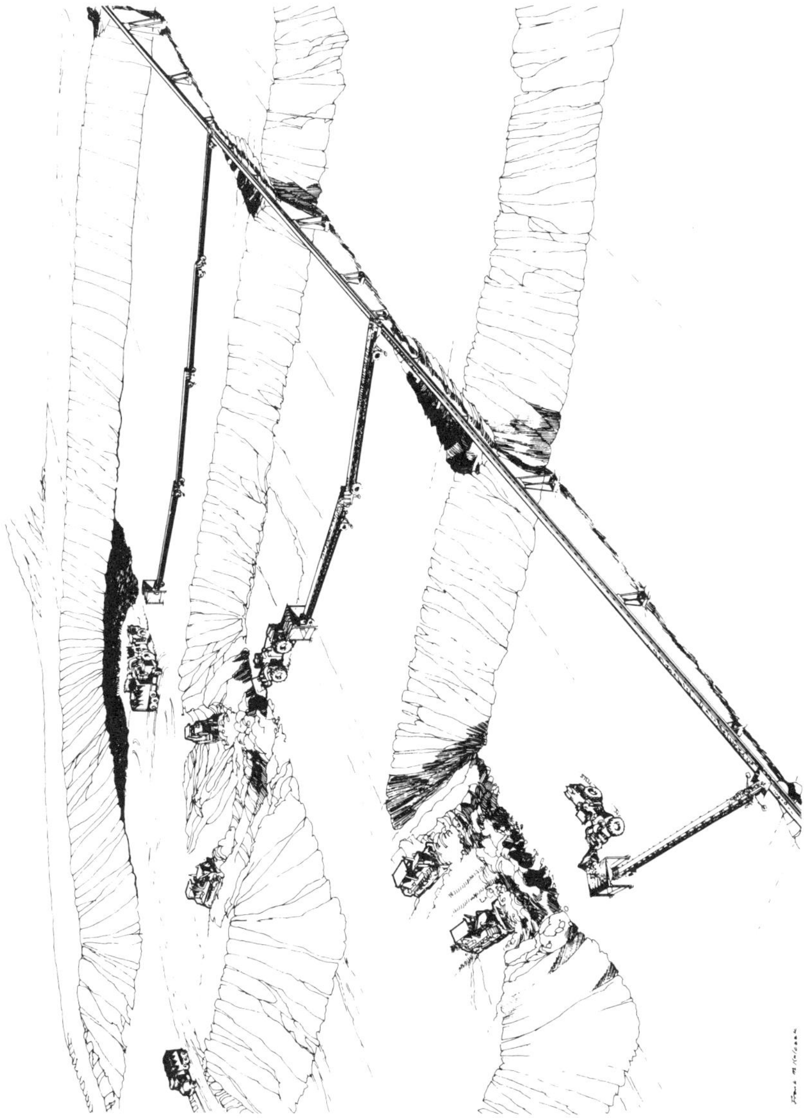

Figure 18: Cross-Ridge Mining

Figure 19: Anthracite Open Pit Mining

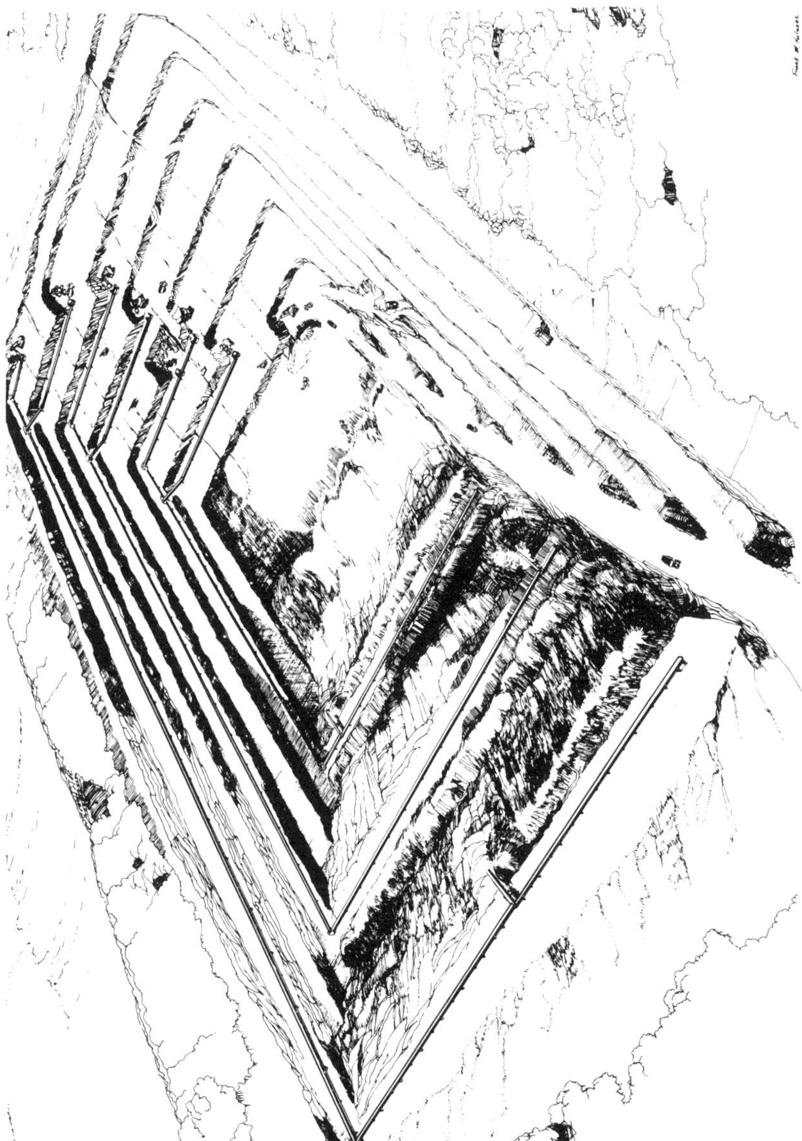

Part II

Surface Mine Engineering Technology

The information in Part II is from *User's Manual for Premining Planning of Eastern Surface Coal Mining. Volume 2—Surface Mine Engineering* (EPA 600/7-80-175), by R.V. Ramani, C.J. Bise, C. Murray and L.W. Saperstein of The Pennsylvania State University, prepared for the U.S. Environmental Protection Agency, October 1980.

Introduction

GENERAL

By the year 2000 A.D., the annual U.S. demand for energy is expected to double and that for the world, to triple (Starr, 1971). Gordon (1976) has summarized a sample of forecasts of U.S. energy consumption for the years 1985 and 2000 from several sources and task forces dealing with various aspects of the appraisal (Table 1). These projected increases will try man's ability to explore, discover, extract, and beneficiate the fuels in necessary volumes, to transport them safely, and to dispose of effluent wastes with minimum harmful environmental effects. Considering the difficulties experienced at present in mining without environmental damage, in finding acceptable locations for new power plants, and in controlling stack emissions from existing plants, the energy projections for the year 2000 indicate the need for a thorough evaluation of the available options and cautious planning.

ENERGY FORMS

In order to meet the growing demand for energy, all fuels will be required in increasing quantities. Economics and technology must also assure that each fuel is put to the use for which it is best suited. Coal has gradually overtaken natural gas and petroleum in the field of electricity generation, and the use of nuclear energy has increased. Over a longer period of time, nuclear energy may replace the fossil fuels (coal, oil shales, tar sands, natural gas, and petroleum) as the leading fuel in electricity generation, provided that technological advancements occur as predicted. The use of geothermal, solar radiation, and tidal energies is not yet widely practical and/or economical. However, coal, particularly surface-mined coal, may compete with nuclear energy for some time. Forecasts indicate that as technology in the field of coal gasification progresses, demand for coal may shift from electricity generation to synthetic gas and liquefied products. This would tend to remove some of the burden from natural gas and oil and extend the projected peaks of each fuel. If coal gasification becomes feasible by the early 1980's, the demand for coal may accelerate more rapidly than predicted. Although there is no exact accuracy in forecasting energy demand and fuel usage, coal should have an assured position throughout the future because of its abundant widespread distribution and chemical versatility. These forecasts are made in a complex ever-changing world where much depends upon international politics, technological innovation, and economic decisions. A projection of energy or fuel demands may be rapidly altered due to one or a combination of changes. Recent factors such as ecological conservation can also effectively block construction of new facilities and lead to a greater delay in supplying the required energy.

SURFACE MINING

Surface mining is a broad term which refers to the removal of the soil and strata over a mineral or fuel deposit and the removal of the deposit itself. The various unit operations in surface mining include:

1. preparing the surface
2. drilling
3. blasting
4. overburden removal
5. loading of the deposit
6. haulage of the mined deposit
7. reclamation

TABLE 1. SELECTED FORECASTS OF TOTAL U.S. ENERGY CONSUMPTION IN 1985 AND 2000 (QUADRILLION BTU) (GORDON, 1976)

	1985	2000
Dupree and West	116.63	191.90
National Petroleum Council -- Low	112.50	n.a.
High	124.90	n.a.
Intermediate	130.00	n.a.
Project Independence -- BAU* $ 7 Oil (no conservation)	109.06	n.a.
AS* $11 Oil (with conservation)	96.25	n.a.
Energy Policy Project -- Historic Growth	116.00	187.00
Technical Fix	92.00	124.00
Zero Growth	88.00	100.00
Organization for Economic Cooperation and Development --		
Base	114.32	n.a.
$ 6 Oil	105.83	n.a.
$11 Oil	100.09	n.a.
ERDA* -- No new initiatives	107.30	165.47
Conservation	96.97	122.48
Synthetic fuels	107.28	165.42
Electrification	106.77	161.16
Limited nuclear	107.05	158.01
Conservation, Synthetic fuels and Nuclear	98.14	137.03
EEI* -- High growth, Low energy prices	124.20	193.93
High growth, High energy prices	112.90	179.10
Medium growth, High energy prices	104.30	155.10
Low growth, High energy prices	98.90	104.50
Dupree and Corsentino	103.54	163.43
FEA* -- 1976 Reference $ 8 Oil	103.29	n.a.
1976 Conservation $16 Oil	92.51	n.a.

BAU: Business as Usual.
AS: Accelerated Growth Scenario.
ERDA: Energy Research and Development Administration.
EEI: Edison Electric Institute.
FEA: Federal Energy Administration.
1973 Energy consumption was about 76 Quadrillion Btus.
1 Btu = 0.2930 watts.

Other operations include the building of drainage systems, waste handling, and topsoil and subsoil storage. Surface mine planning and engineering, therefore, transcends the fields of engineering and physical sciences. Primary considerations include: geology, rock and soil mechanics, excavation and materials handling systems, and environmental aspects.

Surface mining of coal is an answer to many of the underground coal mining problems but in itself creates new problems. Recovery rates with surface mining are much higher than that from underground mining, generally 80 to 90 percent, and coal losses are limited to spillage and losses in transit. Safety problems are not severe in surface mining, but acid drainage from pit and spoil areas must be considered. Land reclamation is an expensive undertaking. A limiting factor with coal surface mining is the depth to which coal can be profitably mined. Currently in most areas the maximum depth of overburden for economic mining is slightly over 60.96 meter (200 ft). Surface mining at greater depths may be feasible in the future, depending upon technology, economics and environmental concerns. The rising trend in coal production in the United States should continue through the rest of the century. The contribution of surface mining to coal production has progressively increased. Today, surface-mined coal accounts for more than 50% of the total coal production (Figure 1). However, the average thickness of coal removed has remained constant, as opposed to an increasing average overburden thickness (Figure 2), a result of major advances in surface mining technology.

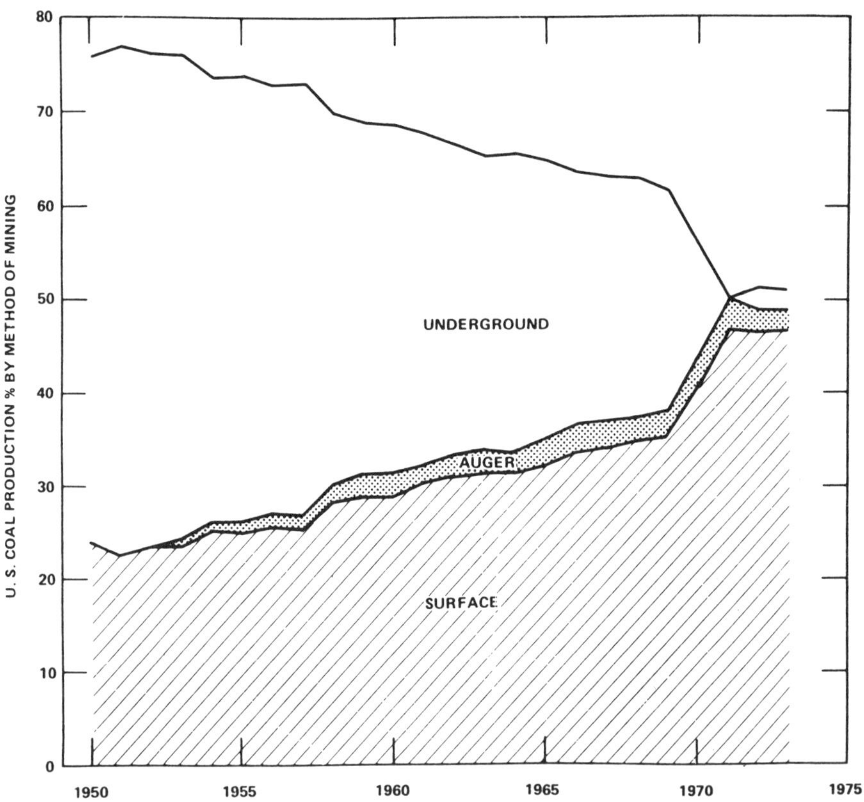

Figure 1. Historical trend of domestic coal production by method of mining (EPRI, 1977).

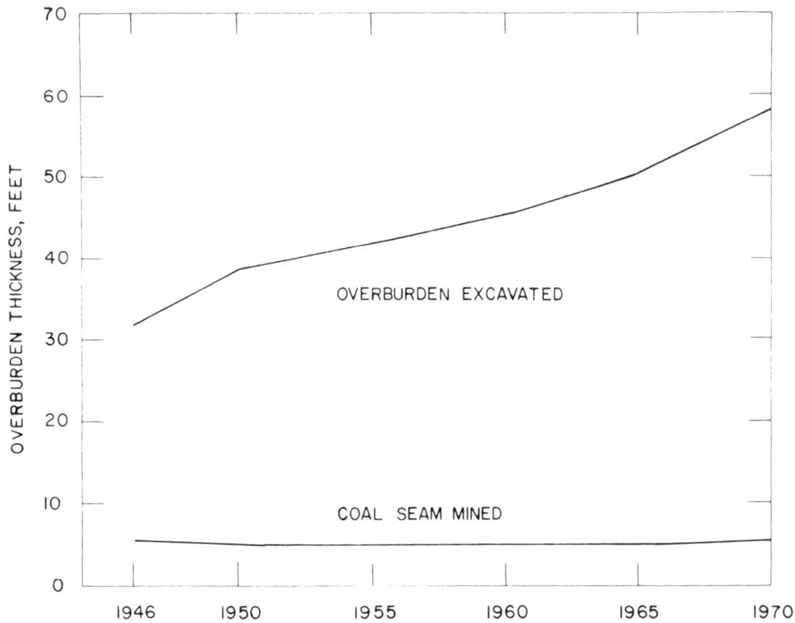

Figure 2. Average thickness of coal seam mined and overburden excavated in the U.S. (EPRI, 1977).

Considering the problems which confront the mining and electricity generation industries in their quest to meet the demands for energy, surface mining will be challenged to provide coal at a low cost with minimal disturbance to the environment. With respect to the regulation of surface coal mining operations and to the establishment of performance standards, the passage of the Surface Mining Control and Reclamation Act of 1977 represents a significant effort on the part of the federal government. The U.S. Congress, according to the Act, finds and declares that "surface mining and reclamation technology are now developed so that effective and reasonable regulation of surface coal mining operations . . . is an appropriate and necessary means to minimize as far as practicable, the adverse social, economic, and environmental effects..." Among the purposes of the Act, important ones are to assure that (1) surface mining operations are not conducted where reclamation required by the Act is not feasible, (2) surface coal mining operations are conducted to protect the environment, and (3) adequate procedures are undertaken to reclaim surface areas as contemporaneously as possible with the surface coal mining operations. While the increasing energy demand promises a good market for coal, increasing federal, state, and local statutes with regard to surface mining make the exploitation of coal deposits a difficult but challenging task. In order to meet this challenge, a need exists to critically study and evaluate surface mining of coal in detail and provide suggestions for solutions to these problems. Attempts to provide solutions to problems require a thorough knowledge of existing methods and criteria. It is the purpose of this report, therefore, to briefly discuss the various aspects of surface mining of coal and lay out important planning and design considerations.

Exploration

INTRODUCTION

Exploration includes activities and evaluations necessary to gather data for making decisions on such topics as the desirability for further exploration, feasibility of mining, size, initial flowsheet, and annual output of new extractive operations (Bailey, 1968). However, the main purpose of exploration activity is to facilitate the discovery and acquisition of coal deposits that may be extracted and processed economically, now or in the future. The extent of exploration programs will vary according to company practices and the area under investigation. The program should go to the point of detail where additional study would not provide greater reliability.

The bulk of exploration work is usually assigned to an exploration department staffed with geologists and engineers. Other staff departments that perform considerable work in exploration are legal, public relations, and accounting. The highly simplified, primary role of the exploration department is to determine the existence, location, quantity and quality of a coal deposit; assess the possible favorable and unfavorable factors of mining; and begin the acquisition program for mineral and surface rights and ownership of the required lands. A legal staff should determine prevailing federal and local laws and practices and provide backup for any required negotiations.

TYPES OF EXPLORATION

There are two general types of exploration approaches, regional and detailed. Regional exploration is a reconnaisance activity and is intended to determine if and where a deposit may exist. Detailed exploration is performed on a specific target area where a deposit is known to exist and delineation and verification are desired.

Areas subjected to exploration procedures may be broadly classified into lands where coal seams are known to exist and virgin areas. Coal exploration programs may be defined as follows:

1. Geologic prospecting with minimum drilling. This type of exploration is common in Appalachian regions where outcrops are abundant and geologic formations are well known.
2. Random drilling. This method is applied to previously explored regions with high geologic predictability.
3. Pattern drilling and statistical approaches. A program of this type is necessary in virgin areas or in areas of high geological variability.

The importance of a thorough exploration program to a coal mining company cannot be over-emphasized, particularly when viewed with regard to surface mine engineering and design. In the eastern coal mining areas with which this study is concerned, the presence or absence of coal is usually known. However, coal mining companies still require much information about overburden and coal characteristics, especially with respect to environmental considerations and present laws and regulations.

The initial step in all exploration programs is a literature search to obtain all public information and previous exploration data concerning a property. Publications and maps may be available from the U.S. Geological Survey, the U.S. Forest Service, state geological services, and other federal and state agencies, and county offices of clerk, assessor, and surveyor. Extensive drill hole information may be available in areas where oil and gas exploration has occurred. Records of mining companies that have previously operated in an area are helpful if they can be obtained.

Maps are important tools in all phases of a mining operation. Maps should at least show geology, political boundaries, and surface contours. Special maps may yield aerial views and paleotopographic and paleoenvironmental information. Good maps and mapping techniques will provide a means for planning and accomplishing exploration, development, reclamation, day-to-day operations, and equipment moves. Calculation of material volumes, location of physical elements, and determination of mining conditions are expedited by the use of maps. Maps provide a method for recording data so that they can be organized and analyzed for ready reference.

Maps suitable for use as base maps are available from a variety of sources. A wide range of map scales and types are available. Various laws require certain scales for reclamation and mining plans but this may vary from state to state. The most commonly used scales are 1" = 500', 1" = 400', and 1" = 100'. U.S. Geological Survey 7-1/2' or 15' topographical maps can be easily enlarged to 1" = 1000' and are the type most commonly used for exploration base maps. Other map scales are available but are suitable for more specific uses. These will be discussed in reference to the appropriate topic.

Aerial photography and mapping methods (photogrammetry) are increasing in usefulness in the surface coal mining industry. For some companies, they are indispensable. They serve as a base for mining plans, vegetation, and geologic and structural identification. Infrared aerial photography is useful for historical/archaeological investigations. Photogrammetric methods are relatively easy and cheap, can be adjusted to any scale, and are highly accurate in any terrain.

Aerial photography done at an altitude designed to produce maps of the scale 1" = 100' is quite popular. This method will provide for control that is accurate to one meter in 5000 meters horizontally and one-half of a contour interval vertically. The usual contour interval is 1.52 meter (5 ft) in normal terrain and 3.04 meter (10 ft) in rough terrain. An advantage of this type of mapping is that it may show drainage configuration, roads, buildings, lakes, streams, timber, power lines, railroads, and fences or other features that might be missed by a ground survey.

Table 2 provides a list of data that should be included or at least considered in a mapping program.

GEOLOGICAL ANALYSES

Geological analyses are employed by most companies and yield crucial information with regard to pre- and post-depositional forces and the effects of migrating chemical constituents from the overburden on coal. Physical features that can be identified include faults, folds, rolls, dikes, coal

TABLE 2. FEATURES TO BE IDENTIFIED ON AN EXPLORATION MAP

Topography

- Natural features
- Surface contours
- Drainage configurations
- Streams - continuous and intermittent
- Lakes
- Rock croppings
- Coal outcroppings
- Natural cavern openings

Man-made features

- Fences
- Wells
- Reservoirs
- Railroads
- Power lines
- Oil and gas lines
- Water and sewage lines
- Highways, roads and streets
- Buildings
- Old mining areas - surface and underground openings
- Location of drill holes - collar elevation, coal elevation, thickness of coal, depth of hole

Geographic features

- Private and political property boundaries
- Property ownership - surface and mineral
- Forests and vegetation
- General land use
- Annual precipitation
- Centers of population - labor supply and housing
- Unusual concentrations of plant or animal life
- Climate altitude

Geological features

- Washouts
- Burn lines
- Faults and fractures
- Major geological formation
- Groundwater - courses, water table classification, relation to coal
- Topsoil - thickness and condition
- Overburden - special characteristics, thickness, chemical features
- Coal - depth, thickness, physical and chemical properties
- Dip of beds
- Presence of methane concentrations
- Partings

Hydrology

- Properties distribution, and circulation of surface and groundwaters water courses, and the presence of concretions and sulfur. The occurrence of seam pinch-outs, local variations in properties and thickness of the coal, and the presence of erosion and channel sands can also be determined by a geologic analysis in conjunction with drilling information. The probability of soil siltation or dust problems can be assessed. Physical properties of coal and overburden that can be determined by physical inspection are specific gravity, hardness, color, luster, size stability, weathering, grindability, friability, fracture, cleat and cleavage. Geologic analysis will also identify chemical and geochemical factors such as toxic or neutralizing elements in overburden and coal.

EXPLORATION PROCEDURES

Good and rapid exploration programs will facilitate pit design and equipment selection whereby costly lead times for start-up can be reduced. Analyses for mining method, preparation processes, and market conditions will be completed sooner so that more profitable decisions can be made. General procedures for exploration are as follows:

1. Property for a new surface mine.

Step a. Acquiring and consolidating sufficient reserves to justify capital expenditures for a viable mine plan.

Step b. Gathering reliable information about coal beds and mining conditions. Exploration, especially drilling, must be performed with proper engineering control to ensure accuracy of the information required for evaluation. The correctness of reserve calculations, along with other factors such as size of property, topography, depth to coal, highwall stability, and thickness of the seam, will have a high influence on equipment selection. Occasionally, the target area may have surface features which preclude mining, thereby reducing available reserves. The scope of the study will vary from property to property and according to the apriori and actual information already available.

2. Property with an existing surface mine. This type of exploration is usually intended to expand reserves or further delineate known reserves. The analysis required may be different from the one above and involve merely the accumulation and organization of additional data.

Before actual field exploration can begin, several requirements must be satisfied. The two important items are the landowner's permission for entry and a prospecting permit.

Formal land acquisition procedures are usually initiated after exploration begins. Acquisition procedures, however, should be a continuing part of exploration procedures. Drilling programs are conducted under options to lease or purchase a particular property. Leases are negotiated with the federal government, state governments, and private landowners. Purchase agreements are negotiated with private landowners.

It is important that a mining company control the surface as well as the minerals during the period of mining activity. Complex land ownership patterns in eastern coal mining areas make this objective difficult to attain. Another problem resulting from the complex ownership patterns is the difficulty to put together a block of land containing reserves large enough to be mined economically. This problem is most common east of the Mississippi River.

After a thorough review of the available information from public and private sources, additional information may be obtained only from additional studies. Regional exploration may provide opportunities to use geophysical methods in conjunction with a drilling program. Gravity surveys and seismic reflection and refraction techniques may be helpful in determining the location and geometry of marker beds, faults, and basement rock. Large area exploration may also employ satellite or airborne photography, infrared imagery, and side-looking radar in connection with a drilling program. Side-looking radar is particularly helpful in the Appalachians where the surface growth is relatively dense.

The most common type of field exploration is a drilling program. The purpose of a drilling program is to provide a comprehensive data base covering the location of the coal and the overburden. Therefore, drilling must be planned to yield maximum information. Drill hole patterns will vary according to the terrain, geological formations, the presence of coal outcroppings, and the type of exploration program.

When very large areas are being studied, hole spacings will vary greatly and will probably not be in any set pattern. When the program is narrowed to a regional study, a grid pattern is most common. Detailed exploration is performed in areas where coal is known to exist, and therefore, closer spaced drill patterns are required.

One of the main advantages of a drilling program is that it can provide physical samples of the coal and the overburden for chemical and physical analyses. This is accomplished by coring. In most programs, a percentage of the holes are cored for overburden and coal. Usually, this ranges from 10 to 25 percent. Holes should be cored from the surface, through the coal, and then below it in order to maximize information about 'spoil' quality.

Air coring and wireline recovery allow for increased recovery up to 100 percent in favorable conditions. Photographing cores as they come out of the hole can provide greater reliability of data. Gamma ray and conductivity logs should be used. The coal core recovery should closely approach 100 percent. If it does not, the analytical information obtained should be considered suspect.

Prospecting along outcrops may also be accomplished with dozer cuts at regular intervals. This gives the attitude of the coal and will help to determine how much outcrop coal (degraded or oxidized coal) must be left as a barrier and the nature of the overburden with respect to machine operation.

Whatever method.of exploration is used, the data from drill holes or outcrop openings must be large enough to present an accurate picture of the deposit and conditions. All data should be plotted on maps as soon as it is obtained. Detailed exploration results should be plotted on large scale maps whose scales may vary from 1" = 100' to 1" = 2000'. The scale of 1" = 100' is the most popular. These maps indicate the depth to the coal and the thickness of the coal seam, isopachs (contour lines representing uniform thickness) for coal and overburden, and stratigraphic sections showing a representative profile of the overburden. There are too many instances where not enough information was gathered and investment decisions were made incorrectly.

SUMMARY

In summary, the details of exploration programs for locating and defining surface mineral coal deposits vary widely, depending on the prior availability of data and information about the area. The major steps that are involved, and some of the techniques that can be used are identified in Table 3. Basically, the major steps require a search of available material regarding the nature and occurrence of the coal, a regional exploration program using as many of the airborne geophysical methods as is practical or possible, and then site specific ground work using drilling and bulk sampling techniques.

TABLE 3. STAGES OF EXPLORATION FOR COAL (BAILEY, 1968)

Exploration Stage	Purpose	Method/Source of Data
Stage 1 Regional Appraisal	o-Availability for Mineral Extraction	Federal, State, Local, Zoning Laws, Land Use Plans
	o-Identification of known problem areas (i) water, (ii) historical/cultural features, (iii) endangered fauna/flora, etc.	Previous history, records, government and private studies
	o-Geologic compilation	U.S. Geological Survey State Geological Survey
	o-Market availability and needs	Intelligence data, growth potential, plans of utilities
Stage 2 Reconnaissance of the Region/ or Target Area	o-Infrastructure Development F-Field check on geologic compilation	Aerial photography Geophysical survey Infrared photography Pilot-hole drilling
	F-Identification of anomalies	
	F-Watershed identification	
	F-Regional land use plans	
	F-Detailed mapping of coal outcrops/coal quality, quantity	
	F-Overburden quality and thickness	
	F-Site location for plants and facilities	
	F-Waste disposal areas	
	F-Wildlife and vegetation	
	F&L-Geochemical analysis of streams/sediments	
	F-Historical and cultural features	
	o-Analysis to define more closely probable extraction problems	
	o-Analysis to identify major environmental problems	

(continued)

TABLE 3. (continued)

Exploration Stage	Purpose	Method/Source of Data
Stage 3 Detailed Target Area Investigation	F-Detailed drilling, core drilling of coal	Core drilling Geophysical surveys Áerial photo
	F-Logging (down the hole, geophysical)	
	L-Physical, Chemical and Washability of coal	
	F-Overburden depth, stratigraphic sequence	
	L-Physical, chemical and toxic characteristics of overburden	
	o-Ore reserve compilation	
	F-Groundwater table/ permeability/charge potential	
	F-Sites for box-cut	
	o-Preliminary economic evaluation	

o - Office study; F - Field study; L - Laboratory study.

While it may be possible to bypass some of the steps in some areas, it is becoming increasingly clear that exploration programs today must provide mine planners with information not only on the geology and hydrology of the coal deposits but also on factors such as land use, historical and archaeological features, and vegetation and wildlife. Since geological information is basic to many mine planning processes, and exploration is one of the first steps in acquiring this information, it is important to choose the right personnel for geotechnical investigations, identify the information needs and importance and their relationships to mine planning, and use suitable data acquisition systems or techniques. The use or the limitations of the collected data for mine planning purposes must also be known. Tables 4 through 8 are taken from a paper by Ellison and Thurman (1976) and can serve as a guide in this area. Finally, as shown in Figure 3, exploration is a sequential process where after each stage a decision must be made to either continue or abandon the exploration program.

TABLE 4. AREAS OF TECHNICAL EXPERTISE IN GEOTECHNICAL INVESTIGATION AND DESIGN (ELLISON AND THURMAN, 1976)

PLANNING PHASE	GEOTECHNICAL INPUT	TYPICAL PRINCIPAL[1,2] GEOTECHNICAL INVESTIGATORS	INVESTIGATORS PROVIDING[2] NECESSARY SUPPLEMENTAL DATA
Feasibility Studies Conceptual Planning	General Site Evaluation	Mining Engineer-Geologist Structural Geologist	Photogeologist Geotechnical Engineer Hydrogeologist Geochemist Petrologist Seismologist
Conceptual Planning Preliminary Design	Site-Specific Investigations	Mining Engineer-Geologist Geotechnical Engineer Hydrogeologist	Geophysicist Photogeologist Structural Geologist Geochemist Petrologist
Final Design	Geotechnical Designs	Geotechnical Engineer Hydrogeologist	Mining Engineer-Geologist Geophysicist Structural Geologist Engineering Geologist

(1) Principal Investigators may vary depending upon site conditions.

(2) An Investigator with proper experience and training may provide necessary expertise in more than one area.

TABLE 5. THE IMPORTANCE OF GEOTECHNICAL FACTORS TO SPECIFIC MINING CONSIDERATIONS (ELLISON AND THURMAN, 1976)

GEOTECHNICAL FACTORS / MINING CONSIDERATIONS		SURFACE MINING									SURFACE FACILITIES						
		CUT SLOPE STABILITY	FLOOR HEAVE	SPOILS STABILITY	HAUL ROAD DESIGN	HANDLING OF WATER	RIPPABILITY	REHABILITATION	MONITORING SYSTEM DESIGN	PONTANEOUS COMBUSTION	FACILITIES SITING	STRUCTURES	STORAGE PILES	CUT AND FILLS	TRANSPORT SYSTEMS	IMPOUNDMENTS AND SOLID WASTE DISPOSAL	SPONTANEOUS COMBUSTION
REGIONAL GEOLOGY	GEOMORPHOLOGY	2			2	2		2	2		2	2	2	2	2	2	
	STRUCTURAL GEOLOGY	2			2	2	2	2	2		2				2	2	
	STRATIGRAPHY AND LITHOLOGY	2			2	2	2	2	2		2				2	2	
	REGIONAL STRESS PATTERNS	2	2			2								3		3	
LOCAL GEOLOGY	STRUCTURE	1	1	1	2	1	1	1	1	1	1	1	1	1	1	1	
	STRATIGRAPHY AND LITHOLOGY	1	1	1	1	1	1	1	1	1	1	1	1	1	1	1	
	IN SITU STRESS	1	1	1		2	2		1	1	2	2		2	2	2	
	SURFICAL DEPOSITS	1		1	1	1	2	1	2		1	1	1	1	1	1	1
	NATURE OF EROSION	2	2	2	1	1	1	1	1		1	1	1	1	1	1	2
	MAN-MADE ALTERATIONS	1	1	2	1	1	2	1	1	2	1	1	1	1	1	1	2
GROUNDWATER CONDITION	UNCONFINED AQUIFERS	1	2	1	1	1	1	1	1	2	1	1	1	1	1	1	
	CONFINED AQUIFERS	1	1	2	2	1	2	1	1	2	1	2	2	2	1	1	2
	AQUICLUDES	1	1	2	2	1	2	2	1	2	1	2	2	2	1	1	
	EFFECTS OF FRACTURES	1	1	1	1	1	1	2	1	1	1	1	1	1	1	1	
SURFACE CHARACTERISTICS	TOPOGRAPHY	1		1	1	1	2	1	1	2	1	1	1	1	1	1	
	SURFACE HYDROLOGY	1		1	1	1	1	1	1	1	1	1	1	1	1	1	2
	EROSION	1		1	1	1	2	1	1		1	1	1	1	1	1	
	TRAFFICABILITY	1	2	1	1	1	1	1	1		1	1	1	1	1	1	
	MAN-MADE ALTERATIONS	1	1	1	1	1		1	1	1	1	1	1	1	1	1	1
	EXISTING AND PROBABLE DEVELOPMENT	1	2	1	1	1		1	1		1	1	1	1	1	1	
ROCK AND SOIL PROPERTIES	RESISTANCE TO WEATHERING	1	2	1	1	1	1	1	1	2	1	1	1	1	1	1	2
	LOAD-DEFORMATION BEHAVIOR	1	1	1	1			1	1		1	1	1	1	1	1	
	STRENGTH	1	2	1	1		1	1	1		1	1	1	1	1	1	
	TIME AND ENVIRONMENT DEPENDENCY	2	2	1	1	1	1	1	1	1	1	1	1	1	1	1	1
	EXCAVATION CHARACTERISTICS	1	2		1		1	1	1		1	1		1	1	1	
	TRAFFICABILITY	1	1	1	1	1	1	1	1		1			1	1		
	ERODIBILITY	1		1	1	1	2	1	1		1		1	1	1	1	2

LEGEND

PROBABILITY OF SUCCESS / APPLICABILITY	IN ALL AREAS	OFTEN	FREQUENTLY	OCCASIONALLY	VERY SELDOM
HIGH	1				1
DEPENDS ON SITE CONDITIONS	2	2	2	2	2
LOW			3	3	

TABLE 6. RELATIONSHIPS OF INVESTIGATION TECHNIQUES TO MINING CONSIDERATIONS (ELLISON AND THURMAN, 1976)

Investigation Techniques		Surface Mining: Cut Slope Stability	Floor Heave	Spoils Stability	Haul Road Design	Handling of Water	Rippability	Rehabilitation	Monitoring System Design	Spontaneous Combustion	Surface Facilities: Facilities Siting	Structures	Storage Piles	Cut and Fills	Transport Systems	Impoundments and Solid Waste Disposal	Spontaneous Combustion in Waste Piles
Background Studies	Literature Search	3	3	3	3	3	3	2	3	2	2	2	2	3	2	2	2
	Mining History	2	2	2	3	3	2	2	3	2	3	3	3	3	3	3	2
	Photogeology	2	2	2	2	3	3	2	3	2	2	2	2	2	2	2	3
Surface Mapping	Photogeology and Remote Sensing	2	2	2	1	2	3	2	3	2	1	1	1	2	1	1	3
	Detailed Geologic Mapping	2	2	3	2	3	2	2	2	2	1	1	1	2	2	2	3
Subsurface Investigation	Rotary Drilling	3	3	3	3	2	2	3	2	2	2	2	2	2	3	2	2
	Core Drilling	2	2	2	2	2	1	2	3	2	1	1	1	1	2	1	2
	Geophysical Logging	3	3	3	3	2	2	3	3	2	2	2	2	2	3	2	2
Geophysical Investigation	Seismic Reflection	3	3	3	3	3	3	3	3	3	3	3	3	1	2	2	3
	Seismic Refraction	2	3	3	3	3	2	3	3	3	2	2	3	3	3	3	3
	Seismic Cross Hole	2	3	3	3	3	2	3	2	3	3	2	3	3	3	3	3
	Electrical Methods	3	2	3	3	3	3	3	3	2	2	2	3	3	3	3	3
	Gravity Methods	3	3	3	3	3	3	3	3	3	2	3	3	2	3	3	3
	Magnetic Methods	3	3	3	3	3	3	3	3	3	2	3	3	2	3	3	3
Field Testing and Instrumentation	In-Situ Stress	2	2	2	3	3	2	2	3	3	2	2	2	2	2	2	2
	Modulus of Elasticity	1	2	2	3	3	2	3	3	3	2	2	3	2	3	2	3
	Groundwater and Gas Testing	1	1	1	2	1	2	1	1	3	1	1	2	1	2	2	3
	Vibration	2	-	1	2	-	2	-	1	-	1	1	1	2	1	1	-
Laboratory Testing	Stress-Strain	1	2	2	2	3	2	2	2	2	2	2	2	1	2	1	2
	Creep	2	3	3	2	2	2	3	2	2	3	2	3	2	3	2	3
	Strength	1	2	2	2	3	2	2	2	3	2	2	2	2	2	2	3
	Weathering After Excavation	1	1	1	1	2	3	1	2	2	2	1	1	1	1	2	2
	Rippability	1	1	3	1	3	1	2	3	3	2	2	3	2	1	2	3
	Geophysical	2	2	3	3	2	2	3	3	2	2	2	3	3	3	3	3
	Geochemical	2	1	2	2	2	2	2	3	2	2	2	2	2	2	2	2
	Soils Testing	1	3	1	1	2	-	1	-	-	1	1	1	1	1	1	2
	Water Quality	2	2	2	2	1	3	1	1	2	2	2	1	2	2	2	2

LEGEND

Probability of Success \ Applicability	In All Areas	Often	Frequently	Occasionally	Very Seldom
HIGH	1	1	1	1	1
DEPENDS ON SITE CONDITIONS	2	2	2	2	2
LOW		3	3	3	3

TABLE 7. ALTERNATE INVESTIGATION TECHNIQUES (ELLISON AND THURMAN, 1976)

Basic Investigation Technique \ Possible Substitute Technique		Surface Mapping		Subsurface Investigation			Geophysical Investigation						Field Testing and Instrumentation				Laboratory Testing							
		Photogeology and Remote Sensing	Detailed Geologic Mapping	Rotary Drilling	Core Drilling	Geophysical Logging	Reflection	Refraction	Seismic Cross Hole	Electrical Methods	Gravity Methods	Magnetic Methods	In-Situ Stress	Modulus of Elasticity	Groundwater Monitoring	Vibration	Stress-Strain	Creep	Strength	Weathering after Excavation	Rippability	Geophysical	Geochemical	Soils Testing
Surface Mapping	Photogeology and Remote Sensing		b/2																					
	Detailed Geologic Mapping	b/1																						
Subsurface Investigation	Rotary Drilling				a/2		b/1	b/1	b/1	b/1	b/1	b/1												
	Core Drilling			b/1		b/1																		
	Geophysical Logging				b/2		b/2	b/2	b/2	b/2	b/2	b/2												
Geophysical Investigation	Reflection			a/2	a/2	a/1		b/2	b/2	b/2	b/1	b/1												
	Refraction			a/2	a/2	a/1			b/2	b/2	b/1	b/1												
	Seismic Cross Hole													b/1			b/1							
	Electrical Methods			a/2	a/2	a/2	b/1	b/1			b/1	b/1												
	Gravity Methods			a/2	a/2	a/2	b/2	b/2	b/2	b/2		b/1												
	Magnetic Methods			a/2	a/2	a/2	b/2	b/2	b/2	b/2	b/2													
Field Testing and Instrumentation	In-Situ Stress																							
	Modulus of Elasticity								b/2								b/1							
	Groundwater Monitoring									b/1														
	Vibration						b/2	b/2	b/2															
Laboratory Testing	Stress-Strain				b/2				b/2				b/2	a/2										
	Creep																							
	Strength				b/2																			
	Weathering after Excavation																							
	Rippability								b/2															
	Geophysical					b/2	b/2	b/2	b/2	b/2	b/2	b/2												
	Geochemical									b/2	b/2	b/2												
	Soils Testing																							

Legend:

a/1 (cell key):

- a - indicates that the alternate technique can practically always be used in place of the basic technique.
- b - indicates that the alternate technique can sometimes be used in place of the basic technique.
- 1 - indicates that the alternate technique should cost less than the basic technique.
- 2 - indicates that the alternate technique usually costs more than the basic technique.

Note: The user of this table must recognize limitations in comparing techniques for generalized situations. This table is a planning guide only.

TABLE 8. QUALITATIVE EVALUATION OF PRESENT PREDICTABILITY OF MINING CONSIDERATIONS BEHAVIOR (ELLISON AND THRUMAN, 1976)

Basis For Predicting Behavior \ Mining Considerations	SURFACE MINING									SURFACE FACILITIES							
	Cut Slope Stability	Floor Heave	Spoils Stability	Haul Road Design	Handling of Water	Rippability	Rehabilitation	Monitoring System Design	Spontaneous Combustion	Facilities Siting	Structures	Storage Piles	Cuts and Fills	Transport Systems	Impoundments and Solid Waste Disposal	Spontaneous Combustion	
Transfer of Technology (1)	2	3	1	2	2	2	2	2	2	1	1	1	1	2	1	2	
Field Investigations (1)	3	3	4	3	3	2	2	3	3	2	2	2	2	2	1	4	
Analysis (1)	2	2	1	1	2	2	2	2	3	1	1	1	1	2	1	3	
Model Tests (1)	5	5	5					4	5	5	5					5	
Empirical Relations (1)	2	4	3		4	1	2	3	2	5	5	3	2	3	3	2	
Overall Predictability (2)	2	2	1	1	2	1	2	2	3	1	1	1	1	2	1	3	
Requirement For Monitoring (3)	2	3	5	5	2		1		1	5	5	5	4	5	1	1	

NOTES:

(1) Number designations for predicting behavior vary from:
  1 = Preditive technique is applicable in practically all cases
  5 = Techniques can be applied, but confidence level will be low

(2) Number designations for overall predictability vary from:
  1 = Very predictable in all cases
  5 = Technology and experience for prediction do not exist

(3) Number designations for requirement for monitoring vary from:
  1 = Formal monitoring program required in all cases because of importance, regardless of predictability
  5 = Monitoring can be limited to occasional observations

(4) Darker numbers indicate highest predictive basis for each mining consideration

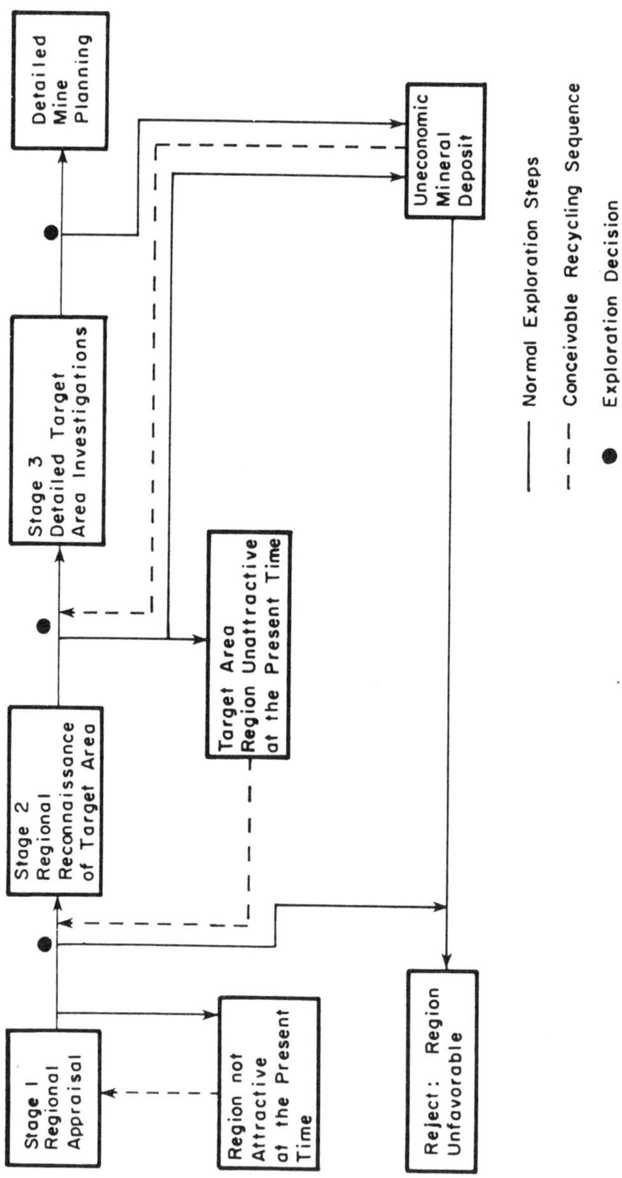

Figure 3. Exploration decision flow diagram (Bailey, 1968).

Surface Mining Equipment

Several types of equipment are available for surface mining, and because of this, a degree of selectivity can be exercised in choosing equipment for a particular job. A brief description of the equipment commonly used in surface mining and their performance characteristics is provided here. Extended descriptions of their physical construction and operational modes are available in many references (Peurifoy, 1970; Pfleider, 1968; Caterpillar Tractor Co., 1975, 1976).

STRIPPING SHOVEL

The stripping shovel (Figure 4) is a highly productive machine capable of handling all types of material, including large blocky material. However it is restricted to comparatively rigid operating conditions and has a limited ability to spoil. The shovel digs material from its operating base upwards. Factors affecting shovel productivity include low flexibility, low maneuverability, slow tramming rates, bank and bucket slides, and water seepage into the pit. Table 9 gives manufacturer specifications for various stripping shovels, and Figure 5 diagrammatically explains these relationships to the stripping shovel.

DRAGLINE

The dragline is an adaptation of the power crane for earthmoving purposes (Figures 6 and 7). These machines have the ability to dig above and below grade; can function under less rigid operating conditions than shovels; can perform optimally when digging an unconsolidated nonmuddy material; can handle blasted rock; and usually have greater reach than the shovel. Since the dragline operates on the ground surface, the problems of bank slides, water runoffs, and seepage are virtually eliminated. The dragline's ability to "chop down" some fraction of the overburden to dig its own bench (i.e., dig above its operating base) is an advantage for extending its operating range, though this method of operation should be avoided as far as practicable because of significant drops in the machine's production capacity. In any case, due to less precise motions, draglines are only 75% to 80 percent as productive as shovels of equal capacity. Table 10 gives manufacturer specifications for draglines, and Figure 8 shows the relationship between the dimensions.

BUCKET-WHEEL EXCAVATOR

The bucket-wheel excavator has its greatest application in unconsolidated overburden where little or no preparation is required. The excavator is adaptable to soft, thin or thick layers of overburden or broken ore, preferably consisting of earth or clay, or sand or soft shales with no hard rock formations present. The bucket-wheel is a continuous mining machine,

Figure 4. Stripping shovel.

TABLE 9. STRIPPING SHOVEL SPECIFICATIONS (STEFANKO, ET AL, 1973)

Manufacturer	Model	Boom Angle, Deg. (A)	Dumping Height, Maximum, Ft.-In. (B)	Dumping Radius @ Maximum Height Ft.-In. (C)	Dumping Height Over Spoil, Ft.-In. (D)	Dumping Radius Over Spoil, Ft.-In. (E)	Dumping Height @ Max. Dumping Radius, Ft.-In. (F)	Dumping Radius, Maximum, Ft.-In. (G)	Cutting Height, Maximum Ft.-In. (H)	Cutting Radius @ Maximum Height, Ft.-In. (J)	Cutting Height @ Max. Cutting Radius, Ft.-In. (K)	Cutting Radius, Max., Ft.-In. (L)	Radius of Clean-up, Ft.-In. (M)	Length of Crawlers, Ft.-In. (N)	Width of Crawlers, Ft.-In. (O)
BE	550B	45			68-0	100-3	43-0	106-6	92-0			114-0	66-0	17-3	9-0
BE	550B	45			64-0	96-0	43-0	101-6	88-0			110-0	66-0	17-3	9-0
BE	1050B	45			76-0	114-6	48-0	121-0	104-0			131-6	75-6	21-2 1/8	10-2
BE	1050B	45			71-0	109-6	47-0	115-0	99-6			126-6	76-0	21-2 1/8	10-2
BE	1850B	40	104-0	230-0			83-0	225-0				245-0	125-0	34-0	16-0
BE	1650B	47 1/2	106-6	140-0			69-6	147-6				158-0	97-0	22-7 1/2	13-5
M	5761	45	111-9	156-3	106-3	158-0	81-9	162-0	141-3	158-6	88-9	173-0	95-3	23-4	14-0
BE	1650B	47 1/2	96-0	127-6			64-0	135-0				145-0	84-0	22-7 1/2	13-5
M	5860	47	122-6	162-6	110-3	167-0	88-3	169-6	152-9	165-0	95-6	181-3	99-3	30-0	16-2
M	5861	46	133-9	180-0	123-6	184-0	92-6	189-0	163-6	182-0	99-6	200-9	1-5-6	30-0	18-0
BE	1950B	49	134-0	185-0			104-0	193-9				207-6	113-6	35-0	19-6
BE	1850B	47 1/2	101-0	150-6			76-4	157-0				170-0	109-0	34-0	16-0
M	5900	47	143-6	195-6	127-6	200-9	104-6	203-6	176-9	199-0	112-6	216-3	126-3	34-0	21-0
BE	3850B	50	150-0	210-0			108-0	218-0				232-0	135-0	41-0	25-8
BE	1950B	52	117-6	151-6			83-6	163-0				172-0	96-6	35-0	19-6
BE	3850B	50	135-0	192-0			102-0	200-0				215-0	129-0	41-0	25-8
M	5960	45	125-0	181-0	118-0	184-0	95-0	186-6	162-0	185-6	104-0	201-6	124-0	42-0	26-4
M	6360	45	153-0	211-6	133-6	217-9	113-6	219-9	189-6	218-6	122-0	236-9	156-9	45-0	30-0

(continued)

TABLE 9. (continued)

Manufacturer	Model	Width of Belt, Inches (P)	Length Over Both Crawlers, Ft.-In. (Q)	Width Over Both Crawlers, Ft.-In. (R)	Width and Length, Approximate Ft.-In. (S)	Rear End Radius Maximum, Ft.-In. (T)	Rear End Radius @ Lower Edge of Frame, Ft.-In.(U)	Clearance Under Frame, Ft.-In. (V)	Center of Rotation To Boom Foot, Ft.-In. (W)	Ground to Boom Foot, Ft.-In.(X)	Boom Length, Ft.	Dipper Handle Length Includ-ing Dipper, Ft.-In.	Dipper Capacity, Cubic Yards
BE	550B	30	47-0	38-9	22-9x48-1 1/2	33-6		19-0	15-7 1/2	21-7	95	61-6	25
BE	550B	30	47-0	38-9	22-9x48-1 1/2	33-6		19-0	15-7 1/2	21-7	95	56-6	28
BE	1050B	36	54-11 1/8	43-11	26-9x52-0	36-0		21-9	17-4	25-1	113	70-0	40
BE	1050B	36	54-11 1/8	43-11	26-9x52-0	36-0		21-9	17-4	25-1	11w	64-0	45
BE	1850B	66	75-6	57-0	58-0x79-6	57-0	53-0	29-0	26-1	35-0	235	146-0	65
BE	1650B	54	58-7 1/2	49-5	36-0x61-8	43-6		24-6	19-3	31-0	145	86-0	70
M	5761	60	59-4	50-0	45-0x65-0	45-0	39-0	23-6 3/4	20-0	30-6	170	102-6	75
BE	1650B	54	58-7 1/2	49-5	36-0x61-8	43-6		24-6	19-3	31-0	135	80-6	75
M	5860	66	72-0	58-2	50-0x72-0	50-0	44-0	27-9	22-0	35-2	180	109-0	80
M	5861	72	72-0	60-0	52-0x80-0	56-0	50-0	27-9	22-0	36-0	200	118-0	90
BE	1950B	78	80-0	68-6	64-0x85-2	62-0	58-0	33-1 1/2	29-0	40-0	200	122-0	100
BE	1850B	66	75-0	57-0	58-0x79-6	57-0	53-6	29-0	26-1	35-0	150	90-6	100
M	5900	82	78-0	65-0	60-0x86-8	62-0	56-6 1/2	33-2	26-0	44-2	200	126-0	105
BE	3850B	100	89-0	73-8	70-0x87-0	68-0	63-0	38-6	31-6	48-6	210	137-0	125
BE	1950B	78	79-0	63-6	64-0x85-2	62-0	58-0	33-1 1/2	29-0	40-0	170	102-6	130
BE	3850B	100	89-0	73-8	70-0x87-0	68-0	63-0	38-6	31-6	48-6	200	125-0	140
M	5960	104	92-0	76-4	74-0x95-0	67-0	59-0	35-6	28-0	46-6	190	117-0	150
M	6360	120	103-0	88-0	72-0x104-0	74-0	66-6	39-0	32-0	50-0	215	133-0	180

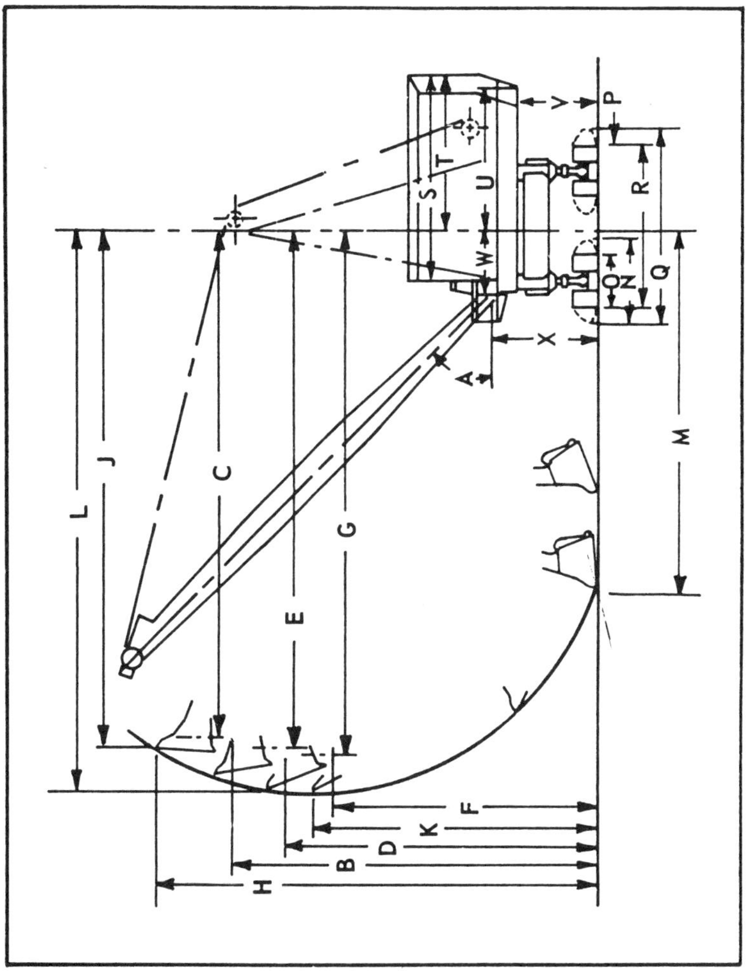

Figure 5. Stripping shovel working dimensions.

Figure 6. Dragline.

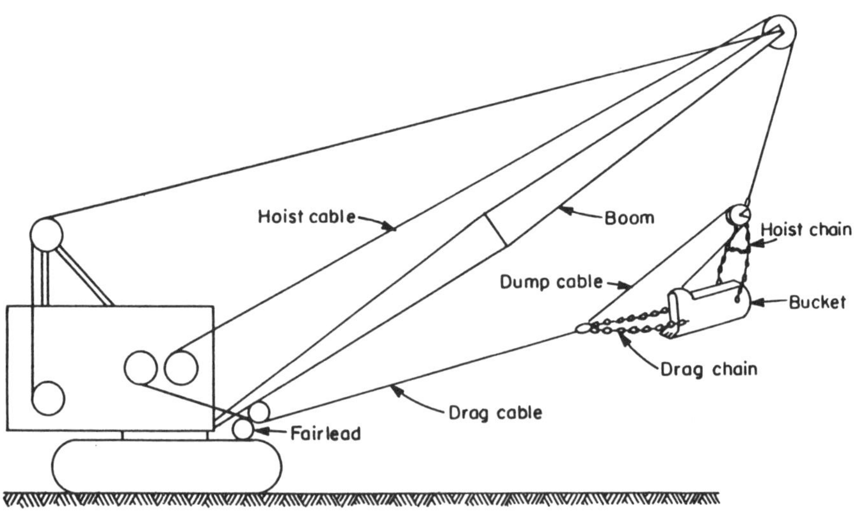

Figure 7. Dragline cable configuration.

TABLE 10. DRAGLINE SPECIFICATIONS (STEFANKO, ET AL, 1973)

Manufacturer	Model	Boom Angle, Degrees (A)	Boom Length, Feet	Dumping Radius, Feet (B)	Dumping Height, Ft.-In. (C)	Digging Depth, Ft.-In. (D)	Outside Diameter, Ft.-In. (E)	Width of Shoe, Ft.-In. (F)	Length of Shoe, Ft.-In. (G)	Width Over Both Shoes, Ft.-In. (H)	Clearance Radius, Ft.-In. (J)	Boom Foot Radius, Ft.-In. (K)	Clearance Height, Ft.-In. (L)	Boom Foot Height, Ft.-In. (M)	Dumping Clearance, Feet (N)	Boom Point Height, Feet (O)	Point Sheave Pitch Diameter Inches (P)	Bucket Capacity Range, Cubic Yards
M	151M*	46	80	65	38-0	40-0		3-6	20-9	17-6	18-0	7-5	6-3	8-1			38	4-8 1/2
M	183M*	42	120	100	60-6	40-6		5-6	30-0	22-6	23-0	8-3	5-0	10-1 1/2			38	5 1/2-10
M	D/E 7400	30	235	220	104-0	117-6	31-0	6-0.	36-6	44-6	38-0	16-2	4-4	7-3			54	9-14
P	732	30	175	173	64-0	115-0	37-0				30-10	21-0	5-4	10-0			80	13-19
M	7500	30	220	209	84-0	130-0	37-0	6-0	36-6	51-2	39-0	16-2	4-4	7-3			54	10-20
BE	480W	40	175	154		80-0	36-0	6-0	42-6	49-6	38-0	16-8 1/2	4-8 1/2	8-6	33	120	66	12-18
BE	Diesel 480W	40	175	154		80-0	36-0	6-0	42-6	49-6	38-0	16-8 1/2	4-8 1/2	8-6	33	120	66	12-18
M	195M*	43	120	98	63-0	75-0		4-7	31-3	22-7	26-6	9-5	5-9	10-2			50	16-18
P	736	30	170	171	59-0	115-0	41-0	7-6	35-0		41-0	22-6	6-9	10-6			80	16-24
P	740	30	195	197	70-0	120-0	45-0	8-0	40-0		48-8	26-0	7-7	11-8			80	21-32
BE	800W	35	195	179		135-0	45-0	7-6	50-0	62-6	50-0	15-3	8-3	14-9	35	127	90	16-26
BE	1260W	30	225	215		130-0	55-0	10-0	56-0	77-0	52-0	17-2 1/2	9-0	17-2	42	129	85 1/4	25-40
P	752	30	215	221	75-0	135-0	56-0	10-6	49-0		57-0	32-6	7-8	12-0			80	36-48
M	7820	30	275	262	112-0	150-0	50-0	9-0	48-0	71-0	50-0	20-3	7-4	13-6			82	35-45
BE	1300W	34	235	215		150-0	50-0	9-0	54-0	70-0	57-0	17-2 1/2	9-0	17-2	46	148	85 1/4	42
M	7920	30	275	263	104-0	135-0	52-0	10-0	55-0	75-0	66-0	21-6	8-7 1/2	12-3 1/2			92	40-60
BE	1350W	38	267	241		150-0	52-0	10-0	55-0	75-3	66-0	25-0	8-6	15-2	48	180	120	40-50
BE	1370W	38	267	241		130-0	58-0	10-0	62-0	81-3	66-0	25-0	8-6	15-2	50	180	120	52
M	8000	32 1/2	325	299	136-0	180-0	65-0	13-0	65-0	94-0	66-0	21-6	8-8	13-9			96	45-65
P	757	30	255	259	95-0	155-0	60-6	11-6	54-6		63-0	36-6	9-2 1/2	15-0			90	45-64
P	762	38	245	230	113-0	135-0	70-0	12-6	59-0		75-0	35-0	9-7	16-0			110	62-74
P	862		275	275	115-0	165-0	66-0	12-6	59-0		63-0	39-0	10-3	17-5 1/2			102	55-68
M	8050	33	315	290	135-0	200-0	65-0	13-0	65-0	94-0	66-0	21-6	8-8	13-9			96	50-70
BE	1500W	38	267	241		115-0	63-6	12-0	70-0	89-6	68-0	25-0	8-6	15-2	52	180	120	65
M	8200	37	320	282	137-0	150-0	58-0	11-0	55-0	83-0	68-0	21-6	11-0	14-10			120	60-75
BE	2450W	38	275	249		130-0	63-6	12-0	70-0	90-10	69-0	25-0	8-0	16-0	55	185	144	75
M	8400	30 1/2	305	290	120-0	200-0	65-0	12-0	55-2	92-0	66-0	23-3	13-0	19-2			126	60-80
M	8750	37 1/2	275	246	132-4	126-6	75-0	15-0	70-0	109-0	77-0	24-0	15-10	21-4			110	80-115
BE	2560W	30	275	274		175-0	65-0	12-0	65-0	93-0	80-0	30-0	14-0	16-0	65	153	144	90
BE	2570W	30	285	283		165-0	74-0	14-0	72-0	104-6	80-0	30-0	14-0	16-0	71	158	144	115
M	8850	30 1/2	360	340	146-0	180-0	80-0	15-0	70-0	114-0	84-0	24-0	13-0	21-4			132	110-140
M	8950	34	325	300	137-0	158-0	80-0	16-6	70-0	116-6	78-0	26-0	16-0	22-0			132	140-160
BE	4250W	38	310	302		185-0	105-0	20-0	130-0	151-0	105-0	50-0	16-7	24-0	129	214	144	220

* Crawler Mounted; BE - Bucyrus-Erie Co.; M - Marion Power Shovel Co.; P - Page Engineering Co.; D/E - Diesel Electric. All other electric unless specified.

NOTE: All dimensions based on largest bucket capacity available. For other bucket sizes consult manufacturers' specifications.

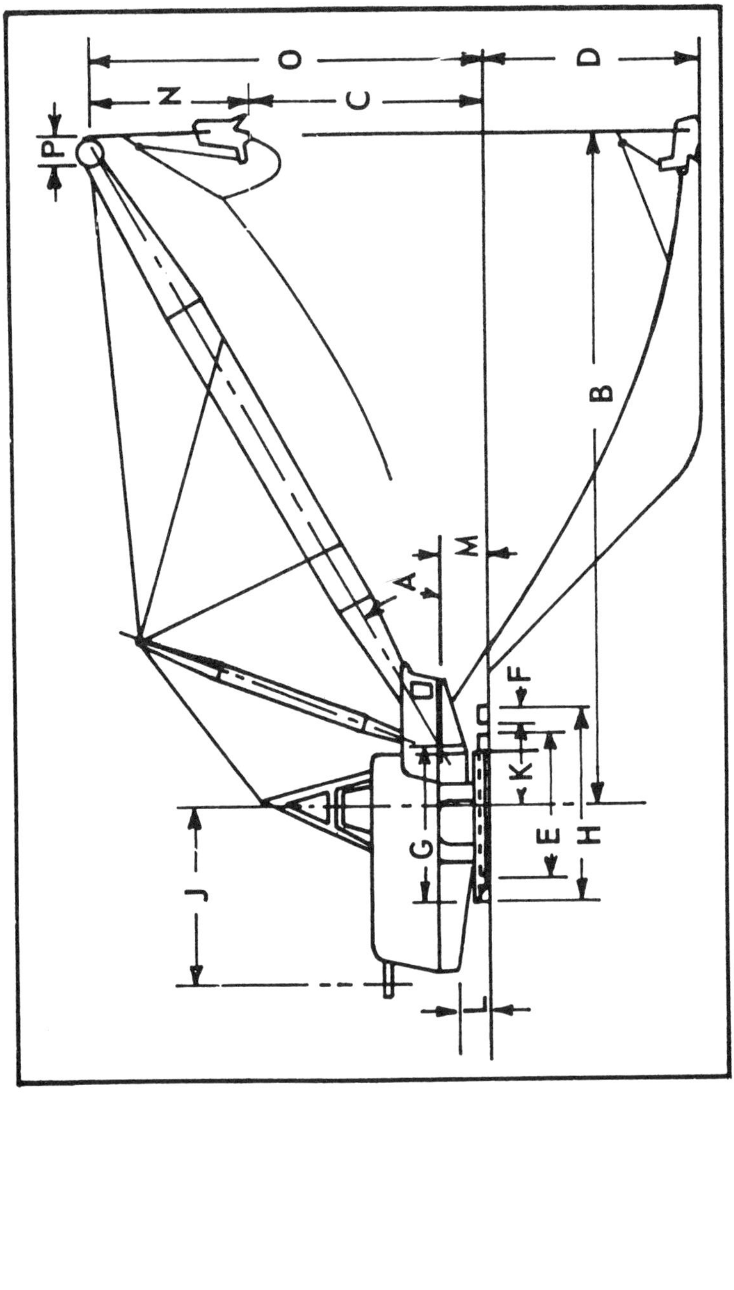

Figure 8. Dragline working dimensions.

independent of cyclic operation methods. Most wheel excavators are designed to allow the wheel to excavate below, above, or on the working level, and discharge into belt conveyors which are normally low-cost transportation devices. The long stacker allows a much greater discharge radius while consuming less power per cubic yard of material removed than other equipment (Price, Manula, and Ramani, 1973).

Bucket-wheel excavators for strip mining may be divided into two main types, German and American-Kolbe. The wheel, or digging component, of both types consists of a relatively large diameter wheel with buckets arranged around its circumference and an arrangement to transfer the mined material from these buckets to a conveyor system (Figure 9).

The German type was developed for use in their lignite fields. It is usually crawler mounted, with mountings arranged to give three-point support, and has a relatively low ground-bearing pressure, approximately 103.42 x 10^3 Pascals (15 psi). The ladder (loading) and stacker (discharge) booms swing independently of each other and therefore require counterbalances. Each boom is equipped with a conveyor belt and the transfer of material from the ladder boom to the stacker boom is accomplished through a somewhat complicated chute arrangement. These machines may excavate overburden, lignite, or both overburden and lignite.

Early model German-type bucket-wheel excavators were equipped with both independent swing of the ladder and stacker booms and with "crowd", a feature whereby the ladder assembly is mounted on rollers and a track so that it may be advanced or retracted. As the size of excavators increased, it became uneconomical to retain both independent swing and crowd, so most late model German-type excavators do not have the crowd feature.

Most American-Kolbe-type machines are crawler-mounted with four-point support and have hydraulic cylinders for leveling. Groundbearing pressures are high, approximately 314.76 x 10^3 Pascals (45 psi) because the machines are designed to operate from the coal surface. The ladder and stacker booms do not have independent swing and are not as heavily counterweighted as the German-type machines; since their horizontal relationship does not change, each tends to counterbalance the other. Transfer of materials from the loading to the stacker boom conveyor is simple, since the direction of material flow does not change. These machines usually excavate overburden only and may be used in conjunction with a stripping shovel. Overburden is usually spoiled but may be transferred to an external haulage system.

The Kolbe-type bucket-wheel excavators have crowd. Due to mechanical design restrictions, these excavators do not have independent swing of the stacker boom. Several other types of bucket-wheel excavators have been developed in the United States for special applications. These may be either crawler, rubber tire, or rail mounted and have a wide range of capacities.

In contrast to a shovel or dragline, the wheel excavator has a lower instantaneous power demand, no shock loadings, and less weight for greater outputs. Bucket wheels are also an asset for land reclamation, because spoil piles are more regular in size and contour and loose soil can be placed over the hard rock removed from above the coal seam in tandem operations.

Limiting factors for the wheel excavator include less flexibility due to the great bulk of the machine, high maintenance costs, and lower operating efficiency. Other limiting features in bucket-wheel utilization are the considerable working surface preparation required, and the difficulty encountered in handling hard materials or rocks sometimes present in unconsolidated overburden. The loading problems combined with a low break-out force also limit the application of wheel loaders. Safe operation on high benches can also become a problem.

Figure 9. Bucket-wheel excavator.

LOADING SHOVEL

The loading shovel in most cases is a scaled-down version of the stripping shovel. The shorter boom of the loading shovel is equipped with either a bottom-dump or a rigidly attached side-dump bucket. The loading shovel, widely used for coal loading in surface mines, is an efficient machine when great mobility is not required (Figure 10). However, it is being replaced at many mines with front-end loaders which can perform better than shovels, especially in contour mining. The loading shovel can load above grade and over the end of a track; can handle coarser materials than the front-end loader; has a longer productive life and better availability; and can load selectively under various mine conditions (Wendertz, 1970).

SCRAPER

Scrapers are of two distinct types, the wheel-tractor scraper (Figure 11) and the track-tractor scraper (Drevdahl, 1961). The main distinction is that the wheel-tractor scraper or rubber-tired scraper, as it is sometimes called, has a speed advantage over the track type, making it more suitable for increased haul distances. The track type, on the other hand, has superior traction in mud; can make its own roads in rough terrain; and can handle larger amounts of material. In surface coal mines the wheel-tractor scraper is more commonly used, although it often requires pushers to assist in loading. When haul distances exceed one mile, scrapers are uneconomical when compared to shovel-truck combinations.

BULLDOZER

A bulldozer is a tractor mounted with a blade which is secured by pivot arms. The blade comes in different types and capacities: U-shaped, "S" or straight-blade, and "A" or angling-blade. Most surface mine dozer applications utilize special "U" shaped blades for light noncohesive material (Figure 12). The bulldozer is a versatile machine, normally limited to a short operating radius of about 152.4 meter (500 ft). It is applicable for site clearing and preparation; construction and upkeep of access routes and haulroads; cleaning up behind shovels and draglines; boosting trucks from pit; towing disabled equipment; maintaining spoil piles, moving cable skids and stumps; and ripping and disposal. It is also used extensively in stripping and leveling for reclamation.

FRONT-END LOADER

A front-end loader is a wheel tractor with a loading bucket mounted on the front (Figure 13). These machines are efficient for loading coal from seams of small thicknesses 0.304 to 3.048 meter thick (1 ft to 10 ft) where it becomes necessary to gather coal before loading. They have high mobility and can operate under favorable conditions with high productivity. They also have found increased application in contour stripping for load and carry functions over short distances.

HYDRAULIC EXCAVATOR (OR HYDRAULIC SHOVEL)

For mining and other earthmoving operations, hydraulic excavators are adaptations of the traditional backhoe for excavation and loading purposes. Among the advantages claimed for this equipment are the greater break-out force and the variable pitch dipper. Figures 14 and 15 show the use of a hydraulic excavator to remove overburden and load it into a truck.

Figure 10. Loading shovel.

Figure 11. Scraper.

Figure 12. Bulldozer.

Figure 13. Front-end loader.

Figure 14. Hydraulic shovel loading into a truck.

Figure 15. Hydraulic shovel digging into a bench.

TRUCKS

In surface mining, off-highway trucks are used primarily for haulage because of their inbuilt ability for continuous uninterrupted operation, while regular highway trucks are used only as service vehicles. There are mainly two types of off-highway trucks, the bottom dump and the rear dump truck. The bottom dump will handle aggregate and unconsolidated material and dumps while in motion. They operate efficiently on grades of less than four percent, and are primarily used for coal haulage in surface coal mines. The rear dump truck comes in various forms: 1) small to medium two-axle trucks, ideal for small pits and development haulage; 2) large two-axle trucks which can handle large loads and negotiate steep grades; and 3) large size three-axle trucks which have identical attributes as the earlier two, but are uneconomical on haul roads due to exhorbitant tire costs (Burton, 1975, 1976).

Off-highway trucks are characterized by engine or drive train type, usually diesel mechanical, diesel electrical, and gasoline internal combustion engines. There are trucks available with either mechanical transmission drive or electric wheel drive. Electrical drive has the advantage of downhill braking the vehicle to eight kmh (five mph) before mechanical or hydraulic brakes are applied for final stop. Off-highway trucks are available in ranges from 26.41 x 10^3kg to 355.61 x 10^3kg (26 tons to 350 tons). Figures 16, 17, and 18 show truck and scraper application zones and the match requirements for trucks and shovels.

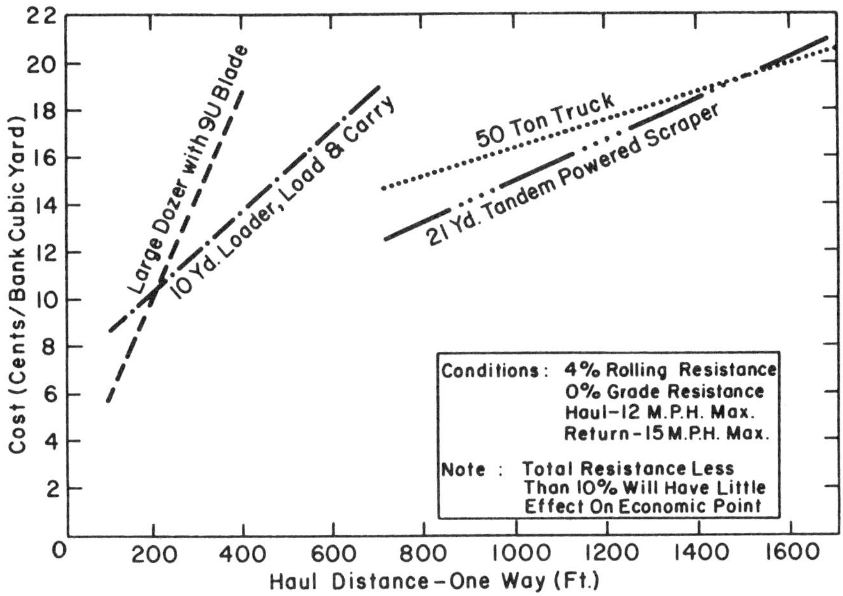

Figure 16. Equipment application zones (Haley, 1974).

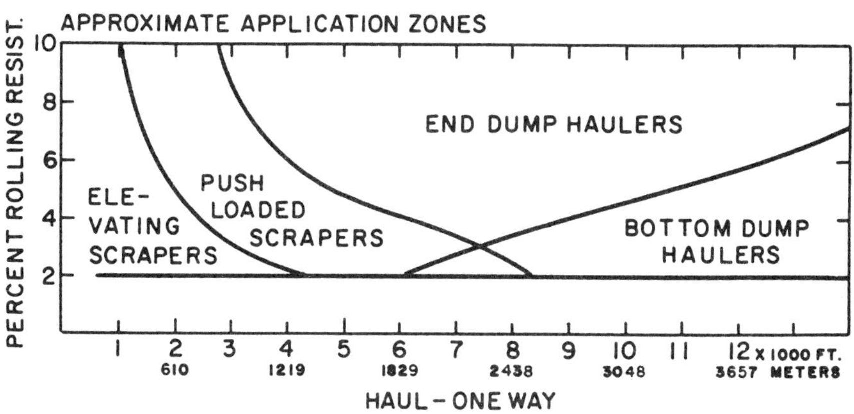

Figure 17. Approximate application zones (International Harvester, 1975).

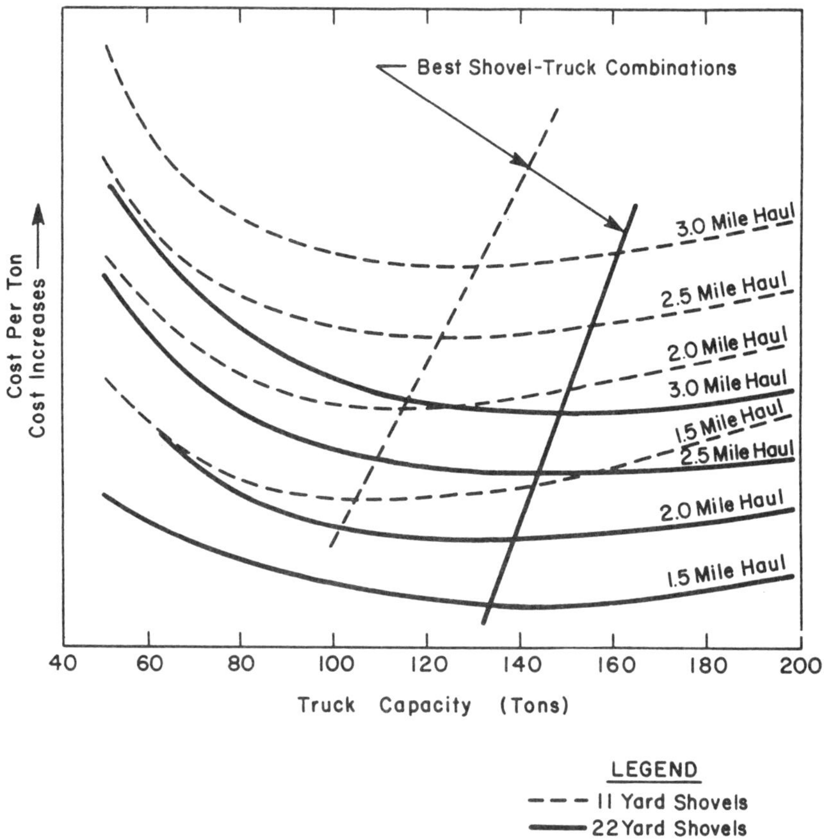

Figure 18. Optimum shovel-truck combinations (Pfleider, 1968).

EQUIPMENT TRENDS

It is difficult to generalize the equipment preference of operators and manufacturers. Depending on the type of equipment, the trend is toward the selection of medium and large capacities. For instance, the trend in draglines is towards medium capacity equipment with long booms. Long boom draglines have greater operating range, reduce the amount of rehandle, and have improved operating costs.

While no large stripping shovel has been selected in the recent past, a vastly increased use of quarry mine shovels is seen in surface coal operations for both overburden and coal handling. The trend in the loading shovels is towards large capacity machines, i.e., 15.29 cu m to 19.11 cu m (20- to 25-cu yd) shovels. This is a result of the shift to large truck capacities. While several truck manufacturers have 152.4 tonne (150-ton) units in service, in 1976, Wabco introduced its 3200B Haulpak, a 238.76 tonne (235-ton) truck (Felix, 1976). It is evident that the normally small secondary equipment are increasing in size for application with the already stabilizing medium to large size primary stripping equipment.

For secondary equipment such as dozers, front-end loaders, and scrapers, the trend is also toward larger sizes. In contour mining, there is an increase in the use of this equipment for methods which are referred to as "haulback" methods. In these methods, the overburden is transported by scrapers, and front-end loaders over short distances; when the haul is long, usually front-end loaders and trucks are employed. Haulback methods have become more common because of their flexibility in selective spoil placement and in reclamation. While small front-end loaders have not been displaced, large loaders are being employed in coal surface mines. The Michigan 6750, a 18.34 cu m (24-cu yd) loader, has proven useful in eastern surface mines.

Large diameter drill holes with longer drill rods are possible with modern rotary drills equipped with air filtering and ventilating devices. Additional efforts have been directed toward improving the productivity of the current models by increasing the rotary speed and adding automatic controls, thereby allowing operators to fully utilize the drill capabilities.

SUMMARY

This chapter has briefly reviewed the major equipment currently used in surface mining of coal. The need for overburden segregation, topsoil removal, storage and replacement, and selective placement and burial of toxic materials has dictated the use of several pieces of equipment that have loading and carrying capabilities. From a reliability point of view, the largest available machines have not been widely accepted. At the present time, there are several research projects funded by the Department of the Interior, Bureau of Mines, for equipment and methods development in surface coal mining. For the near future, draglines, trucks, and shovels are seen as the main overburden removal equipment.

Surface Mining Methods

OVERVIEW

Surface mining of coal is conducted in a relatively simple sequence of operations which includes: 1) preparing the surface, 2) drilling, 3) blasting, 4) overburden removal, 5) loading the deposit, 6) haulage of the mined deposit, and 7) reclamation. Mining techniques for a particular region are largely dictated by geologic and topographic conditions. Even where the techniques are generally comparable, economics of alternative equipment choices and utilization are not easy to generalize. The surface mining techniques can be broadly classified into: 1) contour mining, 2) area mining, 3) open-pit mining, 4) quarry mining, and 5) auger mining. Other methods include dredging and hydraulic mining. Auger mining is usually associated with contour mining where coal near the final highwall is recovered by boring into the seam to a depth of 60.96 meter (200 ft), with cutting heads up to 2.13 meter (seven ft) in diameter. Since quarrying, dredging, and hydraulic mining are not employed for surface mining of coal in the east, they will not be reviewed.

Contour mining is commonly practiced where the coal seams occur in rolling or mountainous terrain. Basically, this method consists of removing the overburden above the coal seam by starting at the outcrop and proceeding along the hillside. After the exposed seam is loaded out, additional cuts into the hillside are made until the economic limit for surface mining is reached.

Contour mining, as practiced in the past, created a bench on the side of the hill (Figure 19). On the inside, it was bordered by the highwall while a high ridge of spoil, the slope of which may cause severe erosion and landslides, was on the outside of the bench. It was not uncommon to punch into the coal seam from the highwall and extract the seam by underground methods (punch mining). It was also common, where there was not enough coal for punch mining, to recover the coal by auger mining (Figure 20). This method of contour mining had come under great criticism. Demands for grading the mined-land to original contour and for the burial of the exposed highwall has made it necessary that the spoil be hauled back into the mined-out pit. Therefore, coal companies are experimenting with several extraction and reclamation techniques that aim at controlling adverse environmental impacts and increasing land values after mining. Several new methods with innovative "spoiling" sequence have been proposed and are being practiced in the steep slopes of eastern United States. In many instances, haulback methods have replaced conventional overcasting methods. These methods provide the flexibility to separate top soil, segregate toxic materials, and generally facilitate placement of spoil in a desired sequence. Two such methods are the block-cut and mountain top removal methods.

Area mining, applied where the terrain is flat, commences with a trench or "box cut" made through the overburden, which exposes a portion of the

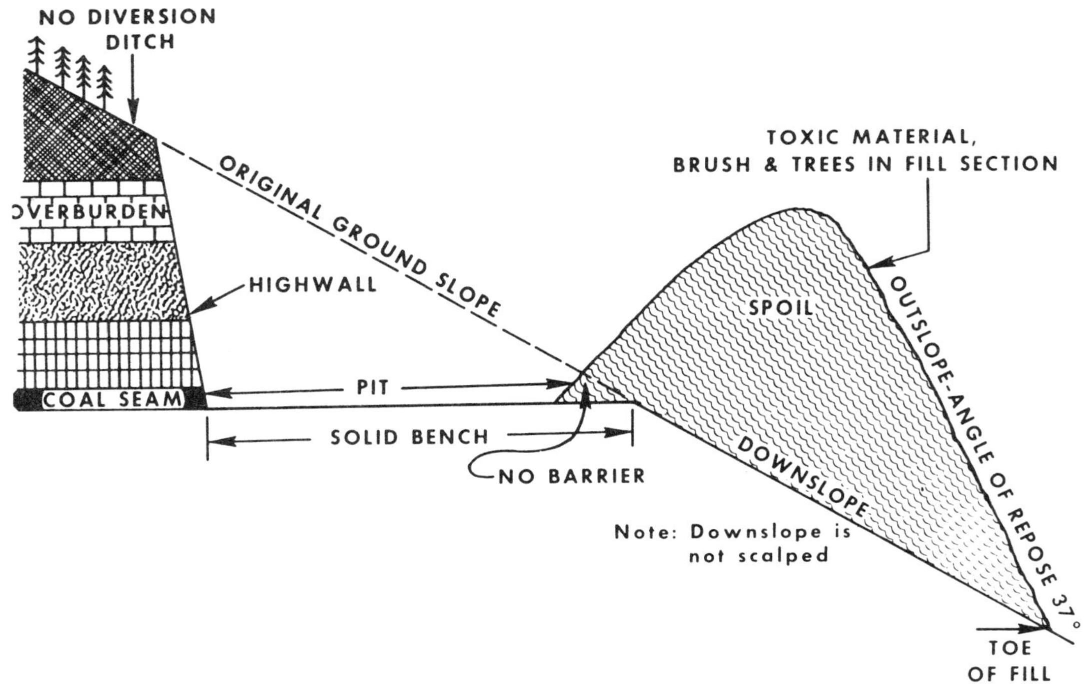

Figure 19. Past practice of conventional contour mining which is no longer an allowable method.

Figure 20. Auger used to drill laterally into a coal bed exposed in contour mining.

coal seam (Figure 21). This first cut may be extended to the limits of the property in the strike direction. As each succeeding parallel cut is made, the overburden is deposited in the cut previously excavated.

Simple overcasting, explained in Figure 21, is still the most common form of stripping. Shovels and draglines continue to be popular, with draglines increasingly favored over shovels. However, the requirements of topsoil removal, storage, and respreading on the top of graded spoil, and the requirements of textured material in the graded spoil for safe operation of farm machinery have dictated the choice of scrapers or other mobile machinery to handle the upper layers of the soil in a haulback mode of operation.

In open-pit mining of coal, the amount of overburden removed is small when compared with the amount of coal recovered (Figure 22). Thick western coals under shallow cover are often removed by open-pit mining. Open-pit mining may also be used for steeply dipping seams (Figure 23).

Geology of the deposit is a big factor in the selection of the method. All non-stratified deposits and steeply pitching and thick stratified deposits require an outside dump or around the pit transportation of waste. The general method can be classified as open-pit mining. Area mining is applicable to gently pitching, thin deposits. This influence of the geology and the topography are illustrated in Figure 24.

UNIT OPERATIONS

As previously mentioned, surface mining consists of a sequence of seven operations. A discussion of these unit operations follows. This section will also mention the types of equipment that can be employed.

Preparing the surface includes the removal of all vegetative cover in preparation for other mining operations. While this operation is not always necessary, it may be necessary to clear the land of trees or other obstructions. Sometimes, this operation may produce commercial quality timber or may be required for relocation purposes.

It is not easy to define what is topsoil though regulations in several states and federal statutes require that topsoil be removed and stored for later use. Topsoil is generally understood to be the soft layers of soil over which the current vegetation has established its roots. Since this material is soft, elevating scrapers, bulldozers, or front-end loaders are used for its removal. According to the Permanent Regulatory Program under the Surface Mining Control and Reclamation Act of 1977 (Title 30, Chapter VII, Subchapter A):

> Soil horizons are contrasting layers of soil lying one below the other, parallel or nearly parallel to the land surface. Soil horizons are differentiated on the basis of field characteristics and laboratory data. The three major soil horizons are:
>
> 1. A horizon. The uppermost layer in the soil profile often called the surface soil. It is the part of the soil in which organic matter is most abundant, and where leaching of soluble or suspended particles is the greatest.
> 2. B horizon. The layer immediately beneath the A horizon and often called the subsoil. This middle layer commonly contains more clay, iron, or aluminum than the A or C horizons.
> 3. C horizon. The deepest layer of the soil profile. It consists of loose material or weathered rock that is relatively unaffected by biologic activity.

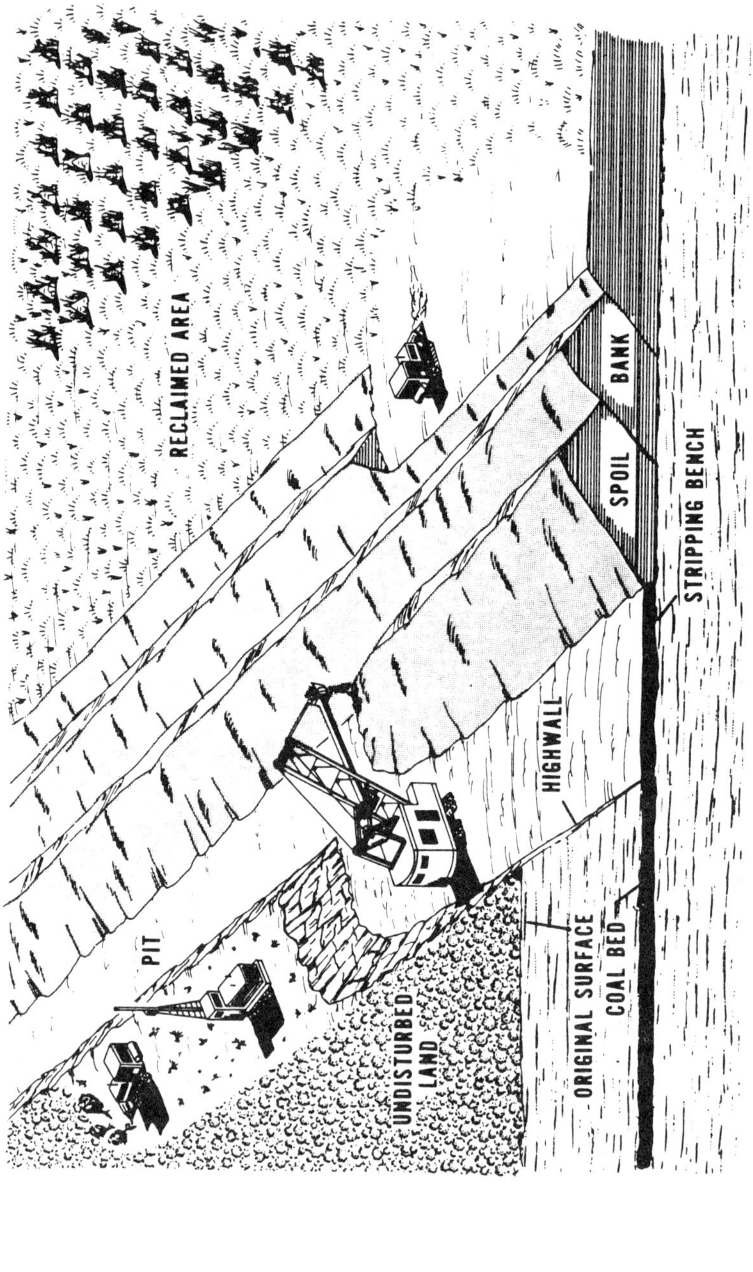

Figure 21. Typical area mining method with stripping shovel (Grim and Hill, 1974).

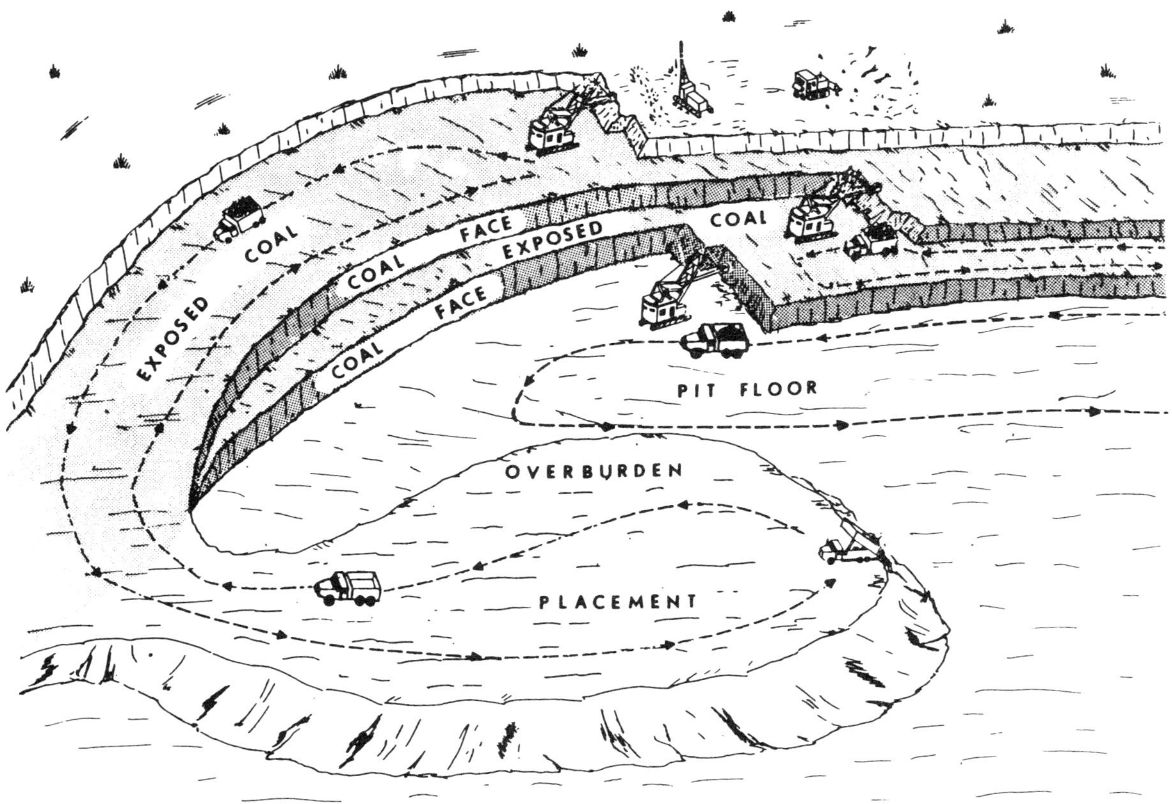

Figure 22. Typical open-pit mining method (thick seam) (Skelly and Loy, 1975).

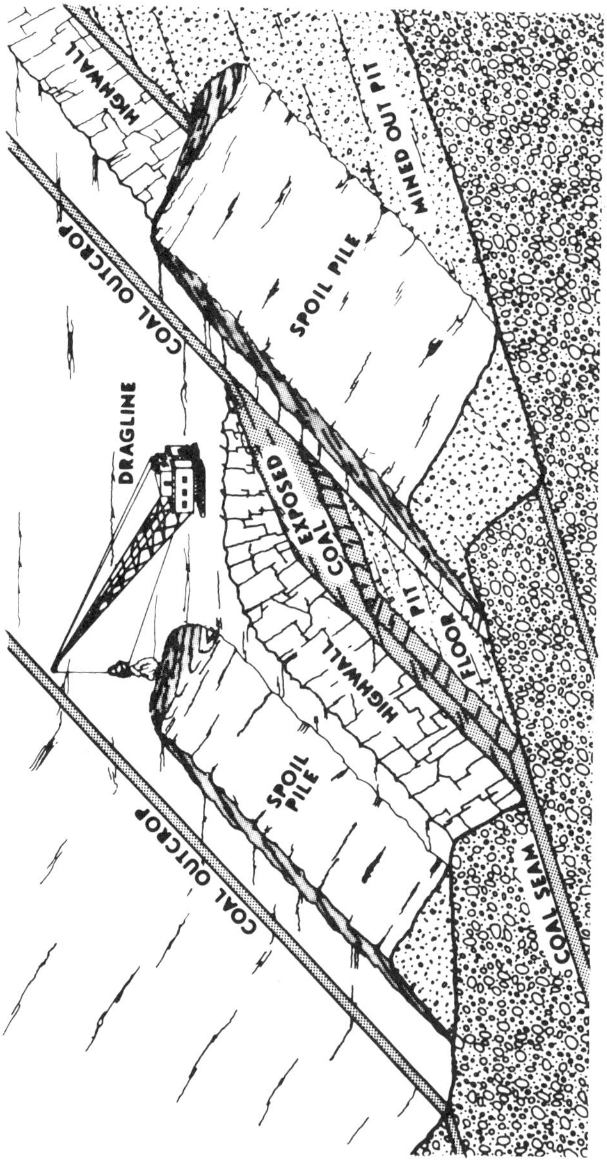

Figure 23. Typical open-pit mining method (dipping seams) (Skelly and Loy, 1975).

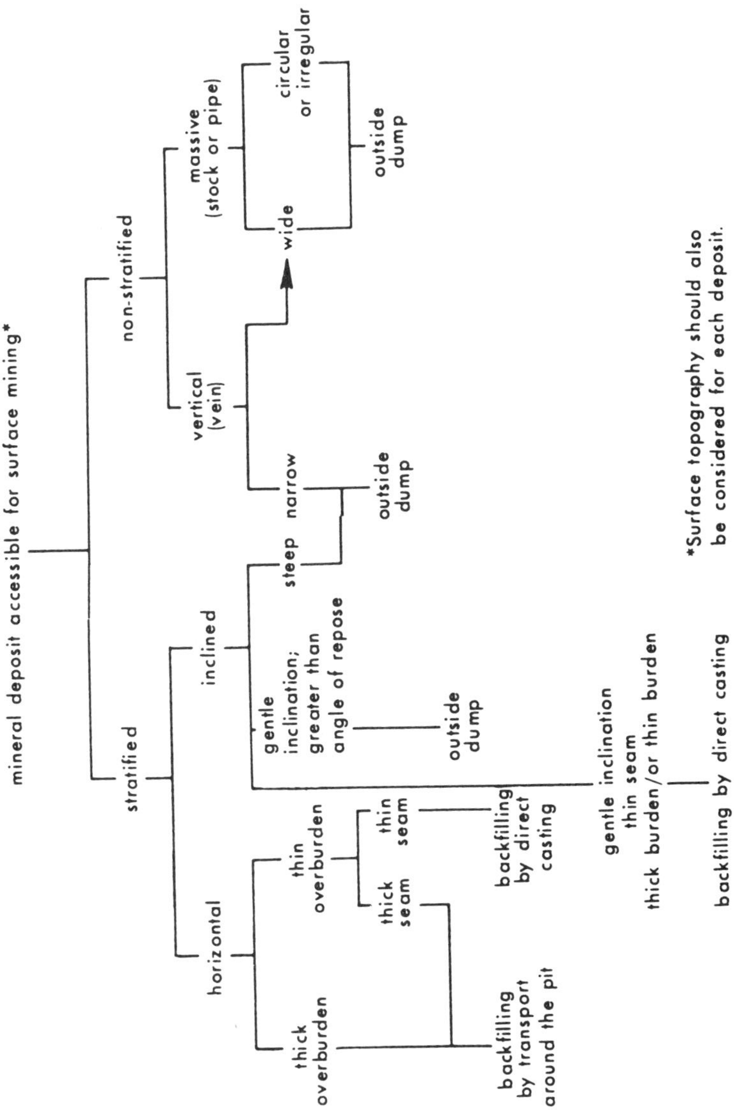

Figure 24. Influence of geological factors in surface mine design (Thomas, 1973).

To allow easier handling by stripping equipment later, the unit operations of drilling and blasting to fracture rock masses may or may not be required. Where the ground cover is hard, it is usually drilled and blasted. Soft strata may often be directly excavated. Some strata can also be sufficiently prepared through use of ripper-bulldozers.

Overburden removal is the most important aspect of the mining system. The equipment and methods used to remove overburden must be carefully chosen to provide the required production at the minimum cost. The stability of the highwall and the spoil is an important consideration in pit design. The equipment used varies from fleets of mobile equipment such as end-loaders to giant draglines, shovels, and bucket-wheel excavators. A factor that has become important is the ability to segregate the soil into stratigraphic layers at desired locations for reclamation and environmental control purposes.

Coal loading is usually done by a loading shovel. Other methods include use of end-loaders, hydraulic excavators, or fine graders. Coal transport is typically by bottom dump trucks.

Reclamation includes backfilling, regrading, surface stabilization, revegetation and restoration operations.

Backfilling is achieved through use of virtually any of the equipment used in overburden handling. However, approaches to segregation of material, specific burial, and layering and compacting can be distinguished as distinct backfilling steps.

Regrading is typically achieved by wheeled or tracked dozers. The amount of regrading depends upon the care with which backfill is done, and on the degree of topsoil restoration and grading, which is either desired or practicable.

Considerations of surface stabilization include provision for water quality maintenance, compaction and layering, and sealing the spoil layers. It also encompasses amendments and therapeutic actions which revitalize the surface or prepare it for revegetation, such as mulching, fertilizing, liming, stabilizing, disking, harrowing, etc. One phase is usually the spreading of topsoil on the regraded soil. Scrapers are often used to reclaim soil from the stockpile, transport it, and spread it on the spoil. End-loaders may be used with bottom-dump trucks which spread the topsoil "on the run." Seeding is actually the final stage of rehabilitation.

Revegetation and restoration takes place over a longer time period. The growth of plants and establishment of cover is an involved process. Careful attention over several growing seasons may be required. Of course, this step is not necessary where alternative land use, such as a shopping center or a school, is envisioned.

Tables 11 through 17 summarize the unit operations, types of applicable equipment, and ranges of operation, and Figure 25 provides some combinations of equipment that can be employed in surface mining systems for these unit operations.

AREA MINING

Area mining is practiced in the eastern and interior provinces as local conditions permit. It is commonly practiced in Ohio, western Kentucky, Illinois, Kansas, and Oklahoma. In this method, a trench, usually called box cut, is made through the overburden to expose the coal seam. Overburden material (or spoil) thus removed is placed on adjacent unmined land, stabilized and contoured. As the coal is removed from the first cut, a second cut

TABLE 11. EQUIPMENT RATING - TOPSOIL REMOVAL AND REPLACEMENT (SKELLY AND LOY - 1975)

LEGEND

1. Should be considered
2. May be considered
3. May be considered under certain conditions
4. May be considered special situation

A. High
B. Moderate
C. Low

		Dozers	Front-End Loaders	Scrapers: Elevating	Scrapers: Full-Power	Scrapers: With Push Tractor	Dragline	Shovel & Truck	Bucket Wheel Excavator	Front-End Loader & Truck
Topsoil Thickness	0 -2'	1	1	1	1	1	3	-	-	1
	2'-5'	1	1	1	1	1	3	4	3	1
Haul Distance	0 -300'	1	1	2	2	1	3	-	-	-
	300'-500'	2	2	1	1	1	4	-	3	-
	500'-1000'	-	-	1	1	1	-	4	-	3
	1000'-1500'	-	-	1	1	1	-	4	-	2
	1500'-5000'	-	-	2	2	2	-	4	-	1
Flexibility Under Varied Field Conditions	good	A	A	A	A	A	A	A	B	A
	fair	A	A	A	A	A	A	B	B	B
	poor	B	B	B	B	B	A	C	C	C

1 ft. = 0.3048 m

TABLE 12. EQUIPMENT RATING - OVERBURDEN PREPARATION (SKELLY AND LOY - 1975)

LEGEND

1. Should be considered
2. May be considered
3. May be considered under certain conditions
4. May be considered special situation

A. High
B. Moderate
C. Low

		Tractor-Ripper	Drills: Percussion	Drills: Rotary	Drills: Jet Piercing
Composition of Overburden	Very Consolidated	3	1	1	2
	Moderately Consolidated	1	1	1	1
	Slightly Consolidated	1	2	2	2
	Unconsolidated	1	4	4	4
Depth of Overburden	0 -30'	1	1	1	1
	30'-60'	2	1	1	1
	60'-90'	3	1	1	1
	>90'	3	1	1	2
Mobility	-	A	A	A	B
Flexibility Under Varied Field Conditions	Good	A	A	A	A
	Fair	A	A	A	A
	Poor	A	A	A	B
Hardness of Rock	Very Hard	3	1	1	2
	Moderately Hard	2	1	1	1
	Soft	1	1	1	1

1 ft. = 0.3048 meter

TABLE 13. EQUIPMENT RATING - OVERBURDEN REMOVAL (SKELLY AND LOY - 1975)

LEGEND
1. Should be considered
2. May be considered
3. May be considered under certain conditions
4. May be considered special situation

A. High
B. Moderate
C. Low

		Dragline	Shovel	Shovel & Truck Comb.	Front-End Loaders	Dozers	Front-End Loader & Truck Comb.	Bucket Wheel Excavator	Scrapers: Elevating	Scrapers: Full Power	Scrapers: With Push Tractor
Thickness	0-30'	2	1	1	1	1	1	1	1	1	1
	30'-60'	1	1	1	2	2	2	1	1	1	1
	60'-100'	1	1	2	3	3	3	2	2	2	2
	>100'	1	2	3	4	4	4	2	3	3	3
Characteristics	Poor Fragmentation (Blocky-Large Breakout Force)	3	1	1	3	1	3	-	-	-	-
	Moderately Blocky	2	1	1	2	1	2	-	2	2	2
	Good Fragmentation (Low Breakout Force)	1	1	1	1	1	1	3	1	1	1
	Unconsolidated	1	1	1	1	1	1	1	1	1	1
Transport Distance	50'-150'	1	1	-	1	1	-	2	-	-	-
	150'-300'	1	-	2	1	1	-	1	-	-	-
	300'-500'	2	-	1	2	2	3	1	3	3	3
	500'-1000'	-	-	1	-	-	1	-	1	1	1
	> 1000'	-	-	1	-	-	1	-	1	1	1
Coal Seam Support Characteristics	Good (Hard)	1	1	1	1	1	1	1	1	1	1
	Moderate	1	2	2	1	1	1	2	1	1	1
	Poor (Soft)	1	4	4	2	2	2	4	1	1	1
Segregation Capability	-	A	C	A	A	B	A	A	A	A	A
Production Capability	-	A	A	B	A	A	B	A	A	A	A
Flexibility Under Varied Field Conditions	Good	A	A	A	A	A	A	B	A	A	A
	Fair	A	B	B	B	A	A	B	A	A	A
	Poor	A	C	C	C	B	B	C	B	B	B
Mobility	-	B	B	B	A	A	A	C	A	A	A

1 ft = 0.3048 meter

TABLE 14. EQUIPMENT RATING - COAL LOADING (SKELLY AND LOY - 1975)

LEGEND
1. Should be considered
2. May be considered
3. May be considered under certain conditions
4. May be considered special situation

A. High
B. Moderate
C. Low

		Shovels	(Rubber Tired) Front-End Loaders	High Lift (Tracks)	Backhoe	Scrapers: Self-Loading Full-Powered	Scrapers: Elevating	Scrapers: Under-Powered With Tractor	Bucket Wheel Excavator
Coal Seam Thickness	1'-3'	2	1	1	2	1	1	1	4
	3'-5'	1	1	1	3	1	1	1	3
	5'-10'	1	1	1	-	2	2	2	2
	10'-25'	1	2	2	-	3	3	3	1
	>25'	1	3	3	-	4	4	4	1
Fragmentation or Ease of Digging	Soft or Highly Frag.	1	1	1	2	1	1	1	1
	Moderately Frag.	1	1	1	3	2	2	2	3
	Hard or Low Frag.	1	3	2	-	4	3	4	-
Floor Conditions	Very Soft	4	1	1	2	1	1	1	3
	Moderate	1	1	1	2	1	1	1	2
	Hard	1	1	1	2	1	1	1	1
Mobility	-	B	A	A	B	A	A	A	B
Flexibility Under Varied Field Conditions	Good	A	A	A	A	A	A	A	B
	Fair	B	B	B	A	A	A	A	B
	Poor	C	B	B	B	B	B	B	C
Production Requirements	High	1	1	3	3	1	1	1	2
	Medium	1	1	1	2	1	1	1	1
	Low	1	1	1	1	1	1	1	1

TABLE 15. EQUIPMENT RATING - REGRADING AND BACKFILLING (SKELLY AND LOY - 1975)

LEGEND

1. Should be considered
2. May be considered
3. May be considered under certain conditions
4. May be considered special situation

A. High
B. Moderate
C. Low

		Dozers	Graders	Scrapers	Draglines	Front-End Loader	Front-End Loader & Trucks
Spoil Configuration	High Peaks	1	-	2	2	3	3
	Moderate Peaks	1	3	1	2	1	2
	Low Peaks	1	1	1	2	1	1
% Spoil Rehandled	>75%	1	-	1	2	2	1
	75% - 50%	1	-	1	2	1	1
	50% - 25%	1	-	1	3	1	1
	<25%	1	2	1	-	1	1
Transport Distance	50'-150'	1	2	1	2	1	-
	150'-300'	1	-	1	2	1	-
	300'-500'	2	-	1	-	2	4
	500'-1000'	-	-	1	-	-	1
	>1000'	-	-	1	-	-	1
Final Surface Contour	Flat & Smooth	1	1	1	-	3	-
	Flat & Rough	1	2	1	2	2	-
	Steep & Smooth	1	-	-	-	3	-
	Steep & Rough	1	-	-	2	2	-

TABLE 16. EQUIPMENT RATING - REVEGETATION (SKELLY AND LOY - 1975)

LEGEND

1. Should be considered
2. May be considered
3. May be considered under certain conditions
4. May be considered special situation

A. High
B. Moderate
C. Low

		Airplane	Hydroseeder	Broadcast	Drill Seeding	Disc & Row Planting	Hand Planting or Spreading	Mulching Machine
Acreage	<3	-	1	1	1	1	3	1
	3-10	-	1	1	1	1	4	1
	10-25	-	1	2	1	1	-	1
	25-50	3	1	2	2	2	-	1
	>50	2	1	3	4	3	-	1
Seed Mixture	Grasses	1	1	1	1	3	3	-
	Crops	-	3	2	1	1	3	-
	Seedlings	-	-	-	-	1	2	-
Mulching	-	-	1	-	-	-	3	1
Fertilizer & Lime	-	1	1	3	1	1	3	-

TABLE 17. EQUIPMENT RATING - COAL HAULING
(SKELLY AND LOY - 1975)

LEGEND
1. Should be considered
2. May be considered
3. May be considered under certain conditions
4. May be considered special situation

A. High
B. Moderate
C. Low

		Trucks					Scrapers					
		Off Road			Highway							
		End Dump	Bottom Dump	Trailer End Dump	End Dump	Trailer End Dump	Full Power	Elevating	Under Powered w/Tractor	Conveyor	Train	Pipeline
Material	Rough-Blocky	1	-	1	1	1	-	-	-	-	1	-
	35 inch size	1	-	1	1	1	-	-	-	2	1	-
	24 inch size	1	2	1	1	1	1	1	1	1	1	-
	Fines	1	1	1	1	1	1	1	1	1	1	4
Length of Haul	300'-500'	1	2	2	2	2	2	2	2	4	-	4
	500'-1000'	1	1	1	1	1	1	1	1	4	-	4
	1000'-1500'	1	1	1	1	1	1	1	1	4	-	4
	1500'-5000'	1	1	1	1	1	1	1	1	3	-	4
	5000'-10,000'	1	1	1	1	1	3	3	3	2	2	3
	10,000'-15,000'	1	1	1	1	1	-	-	-	1	1	3
	>15,000'	1	1	1	1	1	-	-	-	1	1	2
Condition of Road	Good	1	1	1	1	1	1	1	1	-	-	-
	Wet, Soft	1	1	1	2	2	3	3	3	-	-	-
Type of Haulage Roads	Mine Roads Only	1	1	1	2	2	1	1	1	-	-	-
	Combination: Mine & Public	2	2	2	1	1	4	4	4	-	-	-
Daily Production Rate	High	1	1	1	1	1	1	1	1	1	1	4
	Medium	1	1	1	1	1	1	1	1	2	2	4
	Low	1	1	1	1	1	1	1	1	-	-	-
Flexibility Under Varied Conditions	Good	A	A	A	A	A	A	A	A	B	C	-
	Fair	A	A	A	A	A	A	A	A	B	C	-
	Poor	A	A	A	B	B	B	B	B	C	C	-
Maximum Adverse Grade	+3%	1	1	1	1	1	1	1	1	1	1	1
	+5%	1	2	2	1	2	1	2	2	1	2	1
	+10%	2	3	3	2	3	2	2	2	1	-	2
	+15%	3	-	-	3	-	2	3	3	1	-	2
	+20%	-	-	-	-	-	3	4	4	2	-	3
	>+20%	-	-	-	-	-	4	-	-	3	-	3

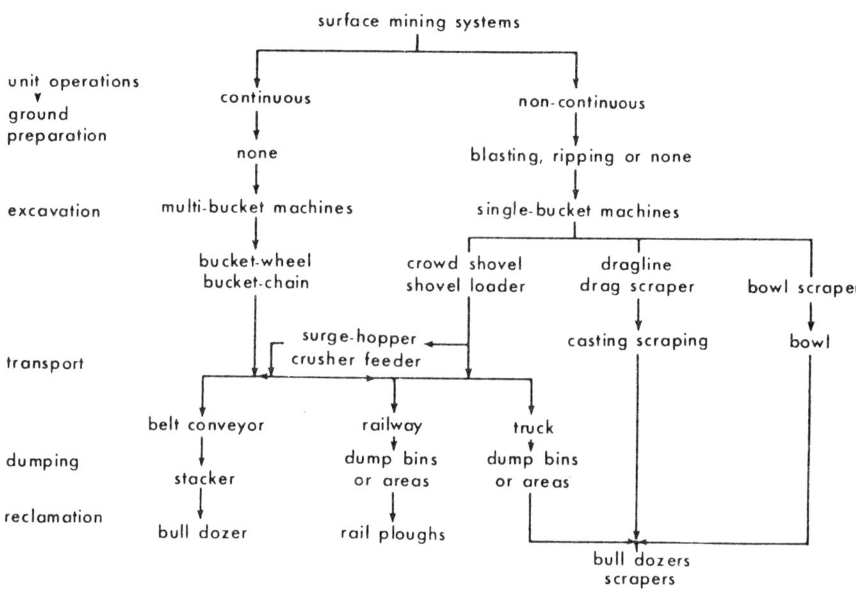

Figure 25. Equipment application for surface mining systems (Atkinson, 1971).

is started immediately adjacent and parallel to the first cut. Overburden removed from this cut is placed in that portion of the first cut from which the coal has been removed. Cut by cut, the mining process proceeds across the property. The final cut leaves an open trench, bordered on one side by the spoil and on the other side by the unbroken column of overburden or highwall. If restoration were not affected, the spoil banks from stripping would resemble a giant washboard or the ridges of a plowed field. The preferable practice in most cases, includes grading the spoil to a level terrain and planting grasses and/or trees.

The practice, depicted in Figure 21, is known as simple overcasting. Simple overcasting, so named because the overburden is cast over the pit to the spoil bank, is the most common form of overburden stripping. The suitability for reclamation, efficiency, and capability of equipment operation, and other considerations, have resulted in the development of alternatives to simple overcasting--passover, crossover, or haul-around techniques. Area mining incorporates a wide range of earth- and rock-moving equipment. The range of stripping applications is presented in Table 18. Table 19 presents some of the combinations which can be practiced in a two-part (multiple-seam) stripping operation. If a three-part stripping operation were to be undertaken, then the third lift would be taken by any method feasible under the second lift.

Dragline Overcasting

Figure 26 is a photograph of a 34 cu m (45 cu yd) dragline in western Pennsylvania. Up to 26 m (85 ft) of overburden are cast across an 18 m (60 ft) pit to exploit an 0.5 m (1.67 ft) coal seam. Figure 27 is a line diagram of the largest dragline operation. A 168 cu m (220 cu yd) machine removes 36.5 to 40 m (120 to 130 ft) of overburden to exploit a 1.2 m (4 ft) coal seam in a pit 76 m (250 ft) wide. Small dragline operations are common in the hilly terrain of Pennsylvania. Figure 28 shows a large dragline pit with other surface mining equipment.

Shovel Overcasting

Shovel overcasting is similar to dragline overcasting but incorporates a stripping shovel instead of a dragline. Figure 29 shows the second largest stripping shovel working in 45 m (150 ft) of overburden in Ohio. Figure 30 shows details of a shovel pit.

Bucket-Wheel Stripping

Stripping with a bucket-wheel excavator (BWE) alone is not practiced in the Eastern United States. The BWE can effectively handle soft material. Its performance is poor in hard material or strata that has to be fragmented. The applications common in the study area of this report incorporate a two-phase stripping system, in which the upper, unconsolidated material is removed by the BWE, and the lower, hard strata which must be fragmented is handled by either a stripping shovel or a dragline.

BWE-Dragline Stripping

Figure 31 is a line diagram, and Figure 32 is a photograph of an operation of this type in Illinois. Both machines work at the same bench elevation, 0 to 20 m (0 to 65 ft) below the surface. Up to 37.0 m (120 ft) of overburden are removed, 20.0 m (65 ft) by the BWE and 17.0 m (55 ft) by the dragline. The pit width is 28.0 m (90 ft), and the coal seam is 0.91 m (3 ft) thick.

BWE-Shovel Stripping

Figure 33 is a line diagram and Figure 34 is a photograph of an opera-

TABLE 18. SINGLE-PHASE STRIPPING POSSIBILITIES

Number	Approach Name	General Method	Equipment Used Excavation	Equipment Used Transport
(1)	Dragline Overcasting	Overcasting	Dragline	Dragline
(2)	Shovel Overcasting		Stripping Shovel	Stripping Shovel
(3)	Bucketwheel Stripping	Passover	Bucket-wheel Excavator	Belt Conveyor (Stacker)
(4)	Loader-Conveyor Stripping*		Front-end Loader	Belt Conveyor
(5)	Shovel-Conveyor Stripping*		Loading Shovel	Belt Conveyor
(6)	Bulldozer Stripping	Crossover	Bulldozer	Bulldozer
(7)	Scraper Stripping		Scraper	Scraper
(8)	Loader Stripping		Front-end Loader	
(9)	Dragline-truck Stripping	Haul-Around	Dragline	Truck
(10)	Shovel-truck Stripping		Loading Shovel	Truck
(11)	Loader-truck Stripping		Front-end Loader	Truck
(12)	Dredging*		Dredge	Dredge Pipeline

* These methods are not commonly practiced in coal in the Eastern United States.

TABLE 19. MULTIPLE-SEAM STRIPPING POSSIBILITIES* (RAMANI, 1968)

Upper Seam Method	Feasible Lower Seam Method
1	1, 2, 6, 7, 8, 9, 10, 11
2	1, 2, 6, 7, 8, 9, 10, 11
3	1, 2, 6, 7, 8, 9, 10, 11
4	1, 2, 6, 7, 8, 9, 10, 11
5	1, 2, 6, 7, 8, 9, 10, 11
6	6, 7, 8, 9, 10, 11
7	6, 7, 8, 9, 10, 11
8	6, 7, 8, 9, 10, 11
9	6, 7, 8, 9, 10, 11
10	6, 7, 8, 9, 10, 11
11	6, 7, 8, 9, 10, 11
12	1, 2, 6, 7, 8, 9, 10, 11

* Refer to Table 18 for the description of the numbered method.

Figure 26. A 45-cubic-yard dragline at work in a Pennsylvania coal operation.

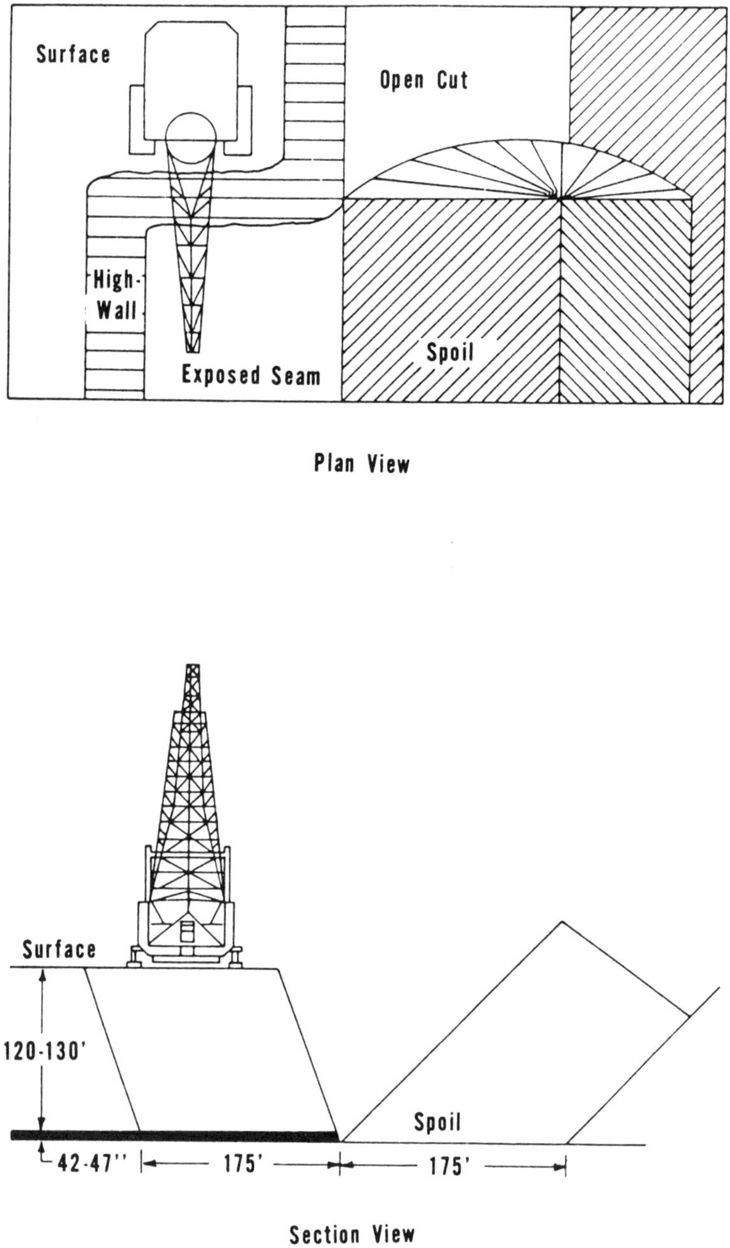

Figure 27. Pit layout at Bucyrus-Erie 4250-W operation (Stefanko, Ramani, and Ferko, 1973).

Figure 28. A 90-cubic-meter dragline with a 90-meter boom. Note the coal-loading shovel and truck in the pit (Ramani and Grim, 1978).

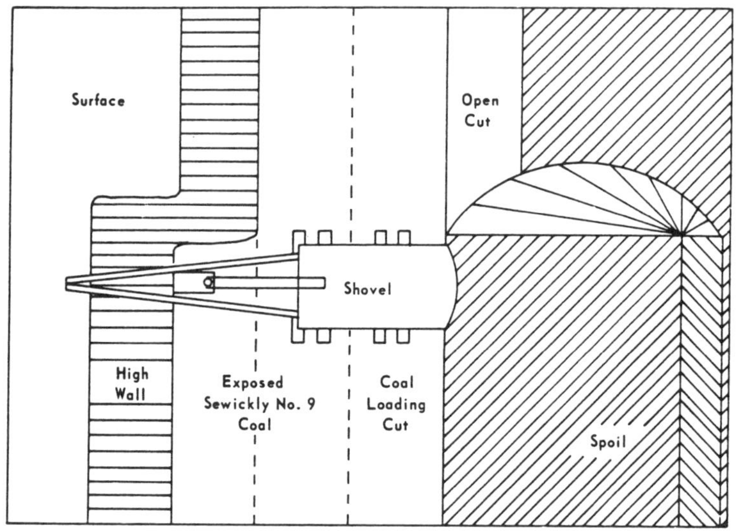

Plan View

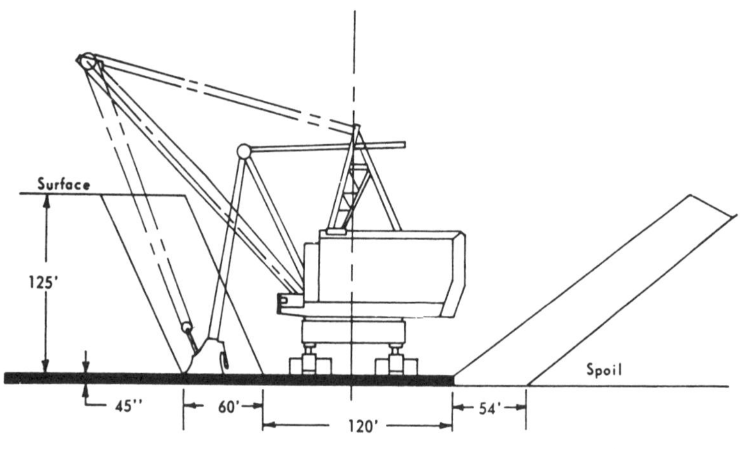

Section View

Figure 29. Plan and section views of a Bucyrus-Erie 1950-B pit (Stefanko, Ramani and Ferko, 1973).

Figure 30. Details of a shovel pit.

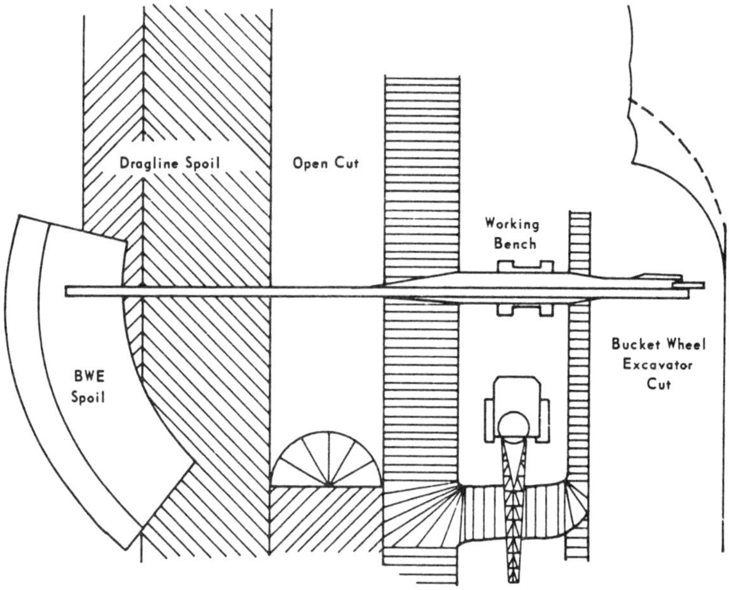

PLAN VIEW

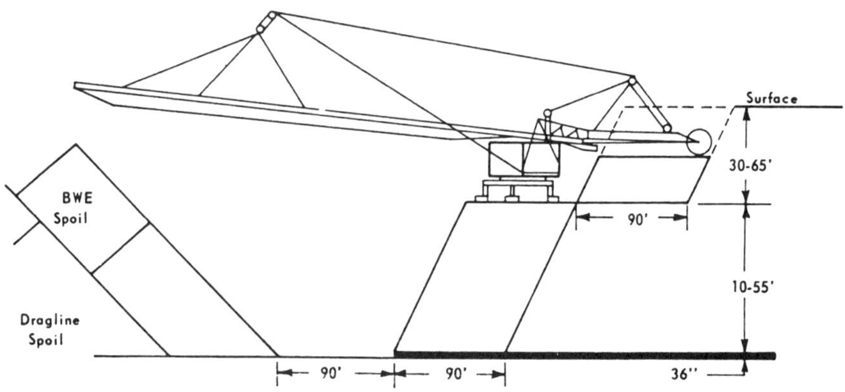

SECTION VIEW

Figure 31. Plan and section views of a dragline and bucket-wheel excavator in tandem operation (Stefanko, Ramani, and Ferko, 1973).

Figure 32. Bucket-wheel excavator, dragline, coal-loading shovel, trucks, and a dozer ripper are employed in this operation (Ramani and Grim, 1978).

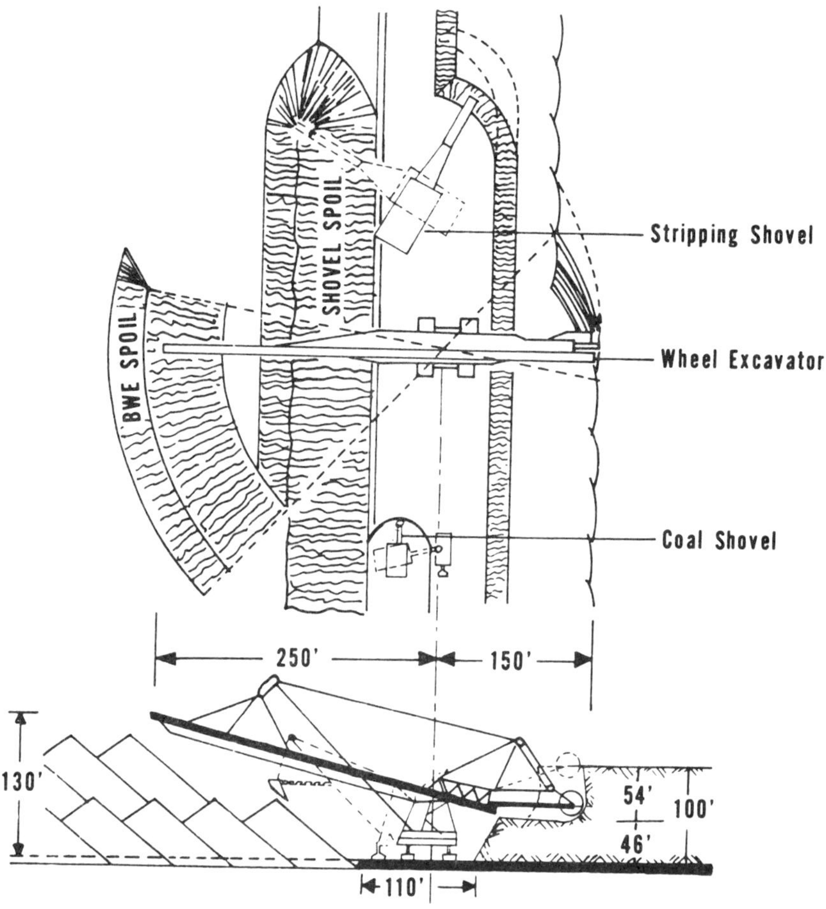

Figure 33. Pit layout at a shovel and bucket-wheel excavator tandem operation in Illinois (Ramani, 1968).

Figure 34. An operation in Illinois using bucket-wheel excavator, stripping shovel, loading shovel and trucks (Ramani and Grim, 1976).

tion of this type in southern Illinois. The BWE cuts its own bench from 0 to 16.8 m (0 to 54 ft) below the surface. The shovel operates in the pit at a depth of 14.0 m (46 ft) below the BWE bench. Pit width is 33.5 m (110 ft) and the coal seam is 1.37 m (4 to 5 ft) thick.

Smaller Scale Approaches

The preceeding discussions have centered on the largest scale mines, typical of Illinois, Indiana, and Ohio. In the east, area mines are usually of such limited dimensions that large-scale and fixed equipment are not used. A variety of approaches are taken in smaller-scale operations, approaches which incorporate virtually every combination of equipment and methods.

CONTOUR MINING

Contour mining is practiced where an outcrop of coal occurs on a hillside. The mining may progress into the hillside only to a certain point, after which there is too much overburden on top of the coal for economic removal. The mining may proceed around the hillside, however, at the given elevation. Successive cuts are removed along the contour line, hence the name contour mining. There are several methods of contour mining such as outslope disposal, outslope reduction, box cut, block cut, haulback or step blockcut, modified area, mountaintop removal, and head-of-hollow fill methods. For all practical purposes outslope disposal of spoil is no longer prevalent or allowed. The outslope disposal method is sometimes referred to as "shoot 'n shove" mining. As shown in Figure 19, the approach is to fragment the overburden by blasting from the outcrop into the stripping limit, and simply to shove the material over the edge. Outslope disposal contour mining (Figure 19) was the traditional approach but is legally limited today, due to its degrading effect on the environment of the mining area. Soil stability, erosion, sedimentation, water pollution, etc., are some of the problems that are enhanced by this method.

Outslope reduction methods are based on the theory that the stability of the outslope spoil bench could be enhanced if the slope of that bench were reduced and the spoil spread over a larger area. Grim and Hill (1974) discuss the various outslope reduction methods.

Outslope reduction was accepted in the recent past as an approach to mining in mountainous terrain and was often practiced as a means of reclaiming abandoned contour strip mines. It can still be used to reduce the slope of existing spoil pile or refuse bank areas. However, federal legislation and some state regulations have made clear the undesirability of any spoil disposal downslope from the outcrop or beyond the solid edge of the bench on slopes over 20 degrees. Therefore, these methods will not be discussed here. The interested reader is referred to Grim and Hill (1974) for detailed discussion of many of these methods. On the other hand, extended presentation of methods that have been developed to decrease or eliminate the outslope disposal of overburden are presented.

The Head-of-Hollow Fill Technique

Head-of-hollow and valley fills basically consist of storing the overburden material in narrow, steep-sided valleys according to a well engineered plan. Due to the swelling of material when broken, careful analysis of the cut-and-fill problem associated with the method is required. This method represents a controllable alternative to outslope disposal. Although the disposal is technically downslope and outside the solid bench, a specific, controlled situation is developed for such disposal.

While a wide range of valleys or hollows is acceptable for head-of-hollow fill purposes, v-shaped, narrow, steep-sided valleys located near the tops of ridges and free from connections to subsurface drainage (e.g., old underground mine openings, seeps, and springs) are preferred. The size of the valley selected for filling is a function of the amount of spoil which will be generated in the mining operation. The objective is to develop a valley fill which completely fills the prepared valley, all the way to the head.

While there are many references regarding the construction of a valley fill, they are either general in nature or specific in certain aspects relating explicitly to requirements of the state of operation. Therefore, dimensions and specifications for valley fill construction that are provided here should be interpreted only as generally indicative of current practices. Several important procedures are followed in adapting the head-of-hollow fill and mountaintop removal methods (to be discussed later in this section) to a particular area. Figure 35 presents a drawing of the head-of-hollow fill method. The head-of-hollow selected for filling is scalped of vegetative cover. Topsoil is removed and stored for later use. Check dams or silt control structures are built downstream from the hollow fill (Figure 36). The natural drainway in the hollow is deepened. The minimum width recommended for the drain is 4.56 meters. A French drain is built in this drainway with rocks no less than 0.36 meters in size in any one dimension (Figures 37 and 38). The fill is built in layers, 1.2 meters thick, beginning at the toe of the valley (Figure 39). The overburden material is deposited in uniform horizontal layers and then compacted with haulage equipment. Each layer of fill is built so that the face of the fill has an outer slope no steeper than two horizontal to one vertical, and usually has crowned terraces (Figure 40). Some operators have graded the face to approximately 22 degrees from the horizontal, eliminating crowned terraces. However, long slopes without interruptions, such as diversion ditches, are more prone to excessive erosion from surface runoff; therefore, where the face of the fill is not stepped, diversion ditches at a minimum of every 15 meters vertical height of the fill is recommended. Revegetation of the hollow-fill face, usually with hydroseeding, is done as soon as the fill height increases.

Block Cut

In the block-cut method, the general progress of mining is along a contour. However, mining is accomplished by taking successive blocks along the contour which may be from one to several cuts deep. Figure 41 shows how these successive block cuts are carried along the contour. The advantages of this method are that it allows overburden spoiling in the pit and on the bench and that each cut is worked to the ultimate limit of the stripping ratio once it is opened. In this second consideration, the small operator often finds the previously mentioned, against-the-contour advance of each block cut easier to handle on an economic basis.

Several extensions of block-cut mining are practiced. They have a number of different names, such as modified block cut, haulback, pit storage, and put-and-take. Figure 42 is a line drawing of the sequencing of cuts, and Figures 43 and 44 explain the stripping and backfilling phases.

Basic block-cut mining is frequently confused with these extensions. In this method, the width of each block is the same. The depth each block is carried into the hillside may vary, however, as each block is carried to its maximum allowable stripping ratio. One advantage is that the dragline can work, for example, cuts 3 and 4 from one location. Another is that the dragline can work uniformly along the highwall with short move time. Aside from the initial cut, the spoil is placed in a previous excavation, which limits rehandling and is environmentally preferable.

The modified block cut and haulback extensions capitalize on the pit-

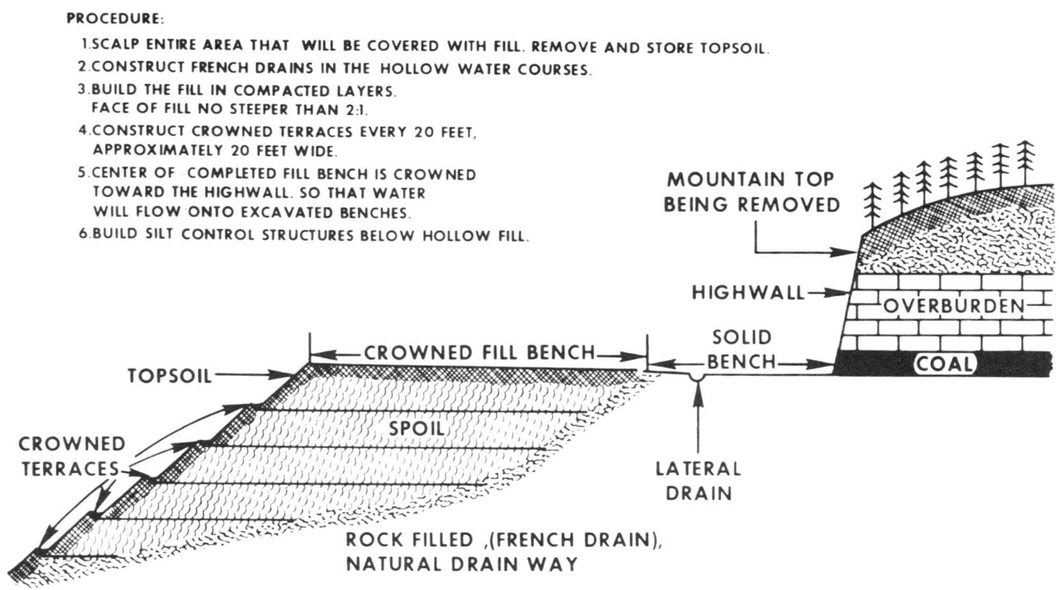

Figure 35. Head-of-hollow fill (Grim and Hill, 1974).

Figure 36. A check dam, downstream from a hollow fill.

Figure 37. A distant view of the French drain.

Figure 38. The size and gradation of the rock that form the core of the French drain is an important design consideration.

Figure 39. The valley fill is built from the bottom up in layers.

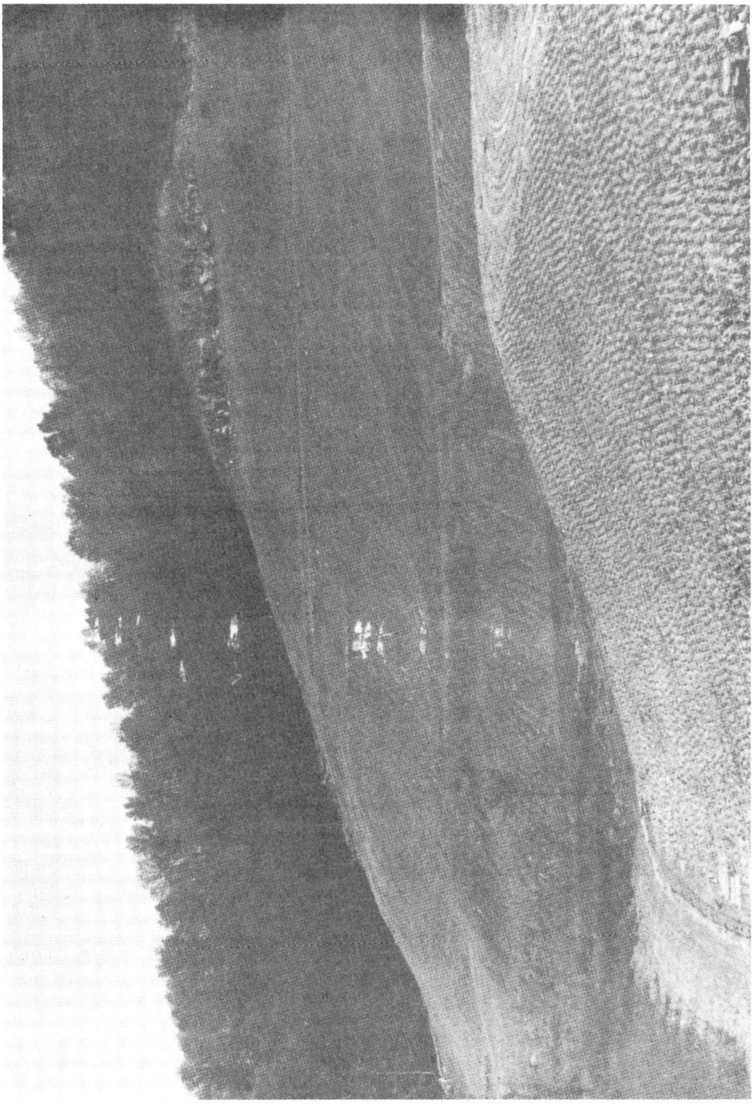

Figure 40. A completed head-of-hollow fill.

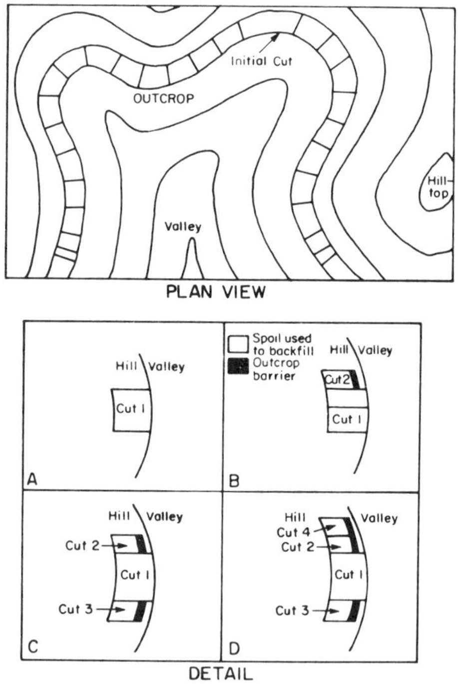

Figure 41. Contour mining around a hillside with the block system (Saperstein and Secor, 1973; Maneval, 1972).

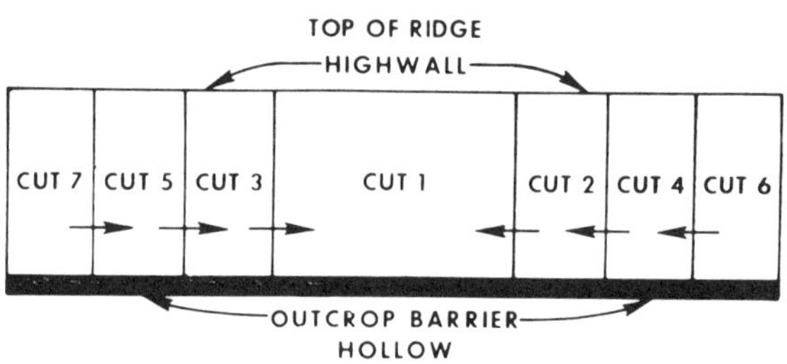

PROCEDURE:

1 SCALP FROM TOP OF HIGHWALL TO OUTCROP BARRIER, REMOVE AND STORE TOPSOIL.

2 REMOVE AND DISPOSE OF OVERBURDEN FROM CUT 1.

3 PICK UP COAL LEAVING AT LEAST A 15 FOOT UNDISTURBED OUTCROP BARRIER.

4 MAKE SUCCESIVE CUTS AS NUMBERED.

5 OVERBURDEN IS MOVED IN THE DIRECTION, AS SHOWN BY ARROWS, AND PLACED IN THE ADJACENT PIT.

6 COMPLETE BACKFILL AND GRADING TO THE APPROXIMATE ORIGINAL CONTOUR.

Figure 42. Block-cut method (Grim and Hill, 1974).

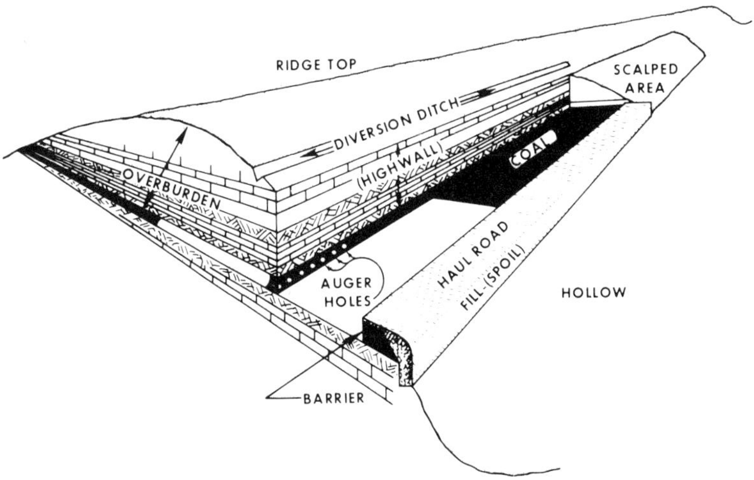

Figure 43. Block-cut method: stripping phase (Grim and Hill, 1974).

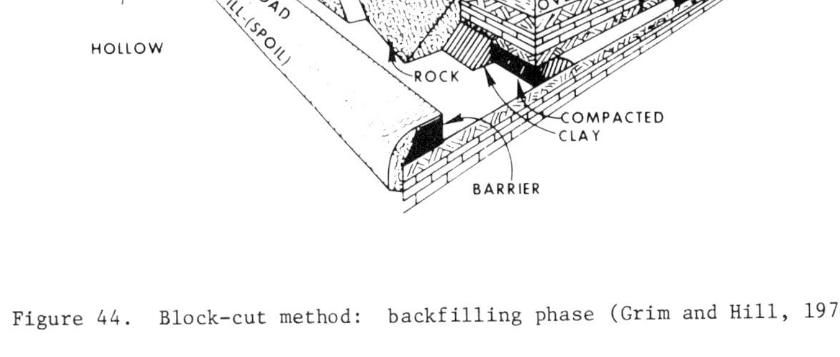

Figure 44. Block-cut method: backfilling phase (Grim and Hill, 1974).

spoiling aspects of block cut. In these, the first cut is sized larger than the others to assure that the volumes of broken and swollen overburden from the remaining cuts will be exactly equal to the mined-out volume. Therefore, once the spoil of the first cut is disposed of in a valley fill or other suitable manner, the remaining spoil can be stored in the pit. The difference in modified block cut and haulback is that modified block cut proceeds in one direction along the contour and may utilize any of several techniques for overburden handling, while haulback proceeds in both directions from the initial cut and utilizes load and carry or truck transport of overburden. The advantage of the haulback refinement of pivoting about a central location seems to be proven through actual practice in both methods. Haulback is unique, as it allows continuous overburden removal, coal removal, augering and reclamation in one pit.

Block-cut methods have been accepted by industry and government alike. Saperstein and Secor (1973, 1977) found that the block-cut methods can be cost-competitive with any other contour mining system.

One key to success in the block-cut methods is the determination of the ultimate stripping ratio. Given this ultimate ratio, the block depths must then be established. For modified block cut to haulback, extensive premining planning is necessary as additional considerations are brought in to assure that the proper allowances for swell have been developed and the first cut and valley fill have been properly sized. A rigorous analysis of the overburden handling unit operation is required. Many combinations of excavation and transport equipment are possible. The sizing and logistical configuration of many alternatives must be considered if an optimum is to be found. Grim and Hill (1974) list the following advantages and disadvantages of the block-cut method:

<u>Advantages</u>

1. Since the method is not limited by the slope of the area, currently available equipment can be used. Further, the method can be used in multiple-seam applications where unit operations must be closely concentrated.
2. Less coverage is disturbed than in other mining methods. The mined-out areas are completely backfilled, eliminating downslope spoil disposal and the final highwall.
3. Reclamation costs are lower, and revegetation costs are reduced. Also, overburden segregation is enhanced.
4. The problems associated with water quality control (acid mine drainage, siltation, and sedimentation), are reduced, since toxic or material which is difficult to handle can be selectively placed in isolation as the block cuts are filled. Fewer sediment control structures are required and the size of the disturbed-area drainage system is smaller.

<u>Disadvantages</u>

1. Extensive premining planning and scheduling is required. The location and size of the initial box cut must be precisely determined. Further, investment costs for overburden transport equipment are high, discouraging application in small operations.
2. The drilling and blasting processes are complicated and time-consuming even though they are standard approaches. Overburden must be controlled and fragmentation must be matched to the equipment used in excavation and transport.
3. The actual long-term environmental consequences are not known or documented.
4. Additional recovery techniques must be carried out concurrent with initial mining. Grim and Hill specifically mention augering.

Any future recovery of coal from the seam by a surface approach requires complete removal of the spoil that has been replaced.

The most attractive aspect of the block-cut technique is that the recovery of the mineable resource, and the restoration and rehabilitation of the surface are an integral part of the total mining process.

Mountaintop Mining

Mountaintop mining is an adaption of area mining methods to contour-mining situations. The two approaches used are modified area mining and mountaintop-removal mining. Where coal lies under ridges, mountaintops, and knobs, there is usually a stripping-ratio limit to the depth into the hill-side to which mining may proceed. Mountaintop methods are, however, the exception in which the overall stripping ratio is taken as the guide rather than local and ultimate maximum stripping ratios. It is based on the analysis that area mining methods, even at a higher average stripping ratio, allow the recovery of more coal (by removal of all the coal in a given tract) at a lower total cost per ton, due to economies of scale in operation, than is possible in contour mining. Figures 45 through 48 explain the sequence of operations.

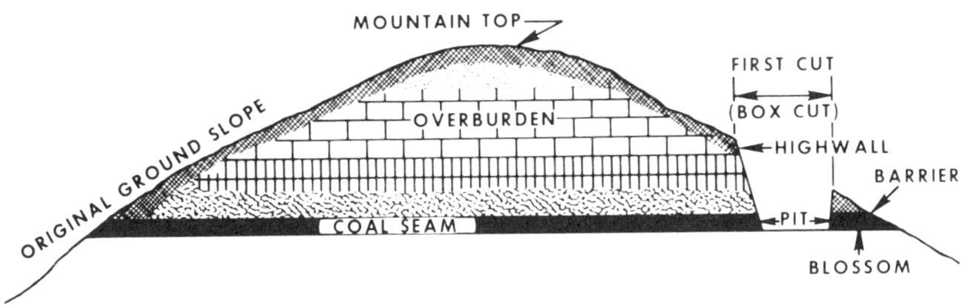

Figure 45. Mountaintop removal method: first cut (box cut) (Grim and Hill, 1974).

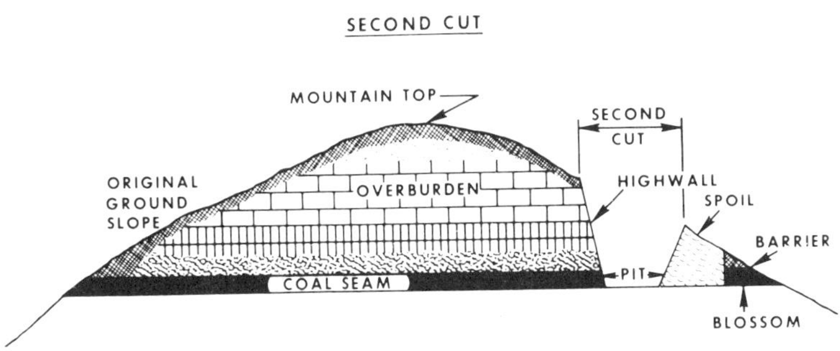

Figure 46. Mountaintop removal method: second cut (Grim and Hill, 1974).

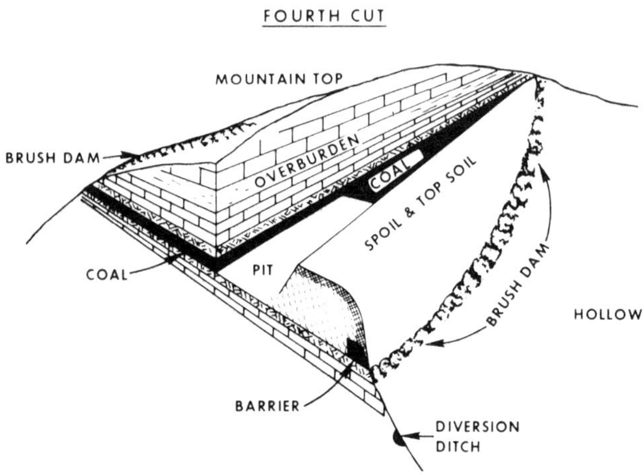

Figure 47. Mountaintop removal method: fourth cut (Grim and Hill, 1974).

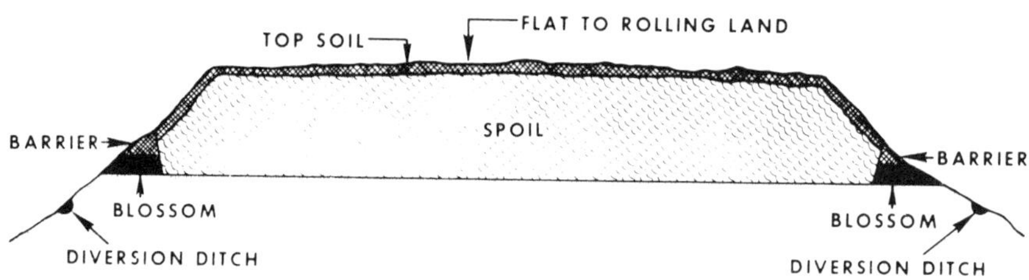

Figure 48. Mountaintop removal method: mountaintop after final grading and topsoiling (Grim and Hill, 1974).

These methods are applied where a coal seam is present on opposite sides of a hill or ridge to a great extent, and the overburden is sufficiently thin (less than 50 m [165 ft]) across the coal, to allow area mining. The method is applicable to previously contour-mined ridges and hills as well as to coal seams in virgin areas. When restoration is to approximate original contour, the best methods to follow are the dragline or shovel-casting approaches of area mining, and the technique is modified area. When the mountaintop is restored as essentially flat land, the best approaches incorporate load-and-carry, bulldozer, or loader-truck techniques, and the method is designated as mountaintop removal.

Modified area mining is practiced in areas such as western Pennsylvania and western Kentucky. The topography and the requirements of restoration to the original contour are two reasons for its practice. The spoil in modified areas is placed in the mine excavation. The value of the land is taken as better in original contour than in revised contour.

Mountaintop removal is practiced most extensively in West Virginia. Mountaintop removal incorporates the use of valley fills to which a significant portion of the spoil is committed. For example, by this method, it is possible to create from 50 hectares of coal land 75 to 100 hectares of "flat" restored land surrounded by stable and controlled outslopes. In most cases, such gently rolling land is worth much more in its state of revised contour than in its previous sloping state. Near Welch, West Virginia, 30.4 hectares (75 acres) of mountaintop removal land were donated as a high school building site, and the remaining land can be used for subdivision. In this area, residential land is worth as much as $98,684 per hectare ($40,000 per acre).

The steps in the mountaintop removal method are essentially the same as in valley fill and are as follows:

1. Determine the postmining volume of overburden. Given this volume, determine the amount that will be spoiled in the mined-out area and the amount that will be spoiled in valley fills.
2. Prepare the required valley fill areas (Figure 49).
3. Scalp (clear and grub) areas for temporary storage of topsoil.
4. Scalp the hilltop. Remove topsoil to temporary storage areas.
5. Initiate mining with a typical area-mining trench cut along one edge of the property. Usually a 5 m (16.5 ft) barrier of blossom coal is left along the outcrop edge.
6. Continue to mine relatively parallel to the first cut across the mountaintop. In the modified area, the dragline or shovel spoils most of the overburden in the previous pit to the approximate original contour. In mountaintop removal, much of the spoil in initial cuts is placed in valley fills by the mobile equipment fleets; when these valleys are completely filled, spoiling is carried out on the mountaintop.
7. Grade the mountaintop and fills to approximate original contour (modified area) or to essentially flat, revised contour (mountaintop removal).
8 Restore topsoil, amend and revegetate (Figure 50).

Mountaintop mining methods offer the same general advantages as the block cut and other practices that do not include outslope-disposal. The determination of the size and location of valley fills and the sequencing of cutting and filling operations are crucial for success. The increased value of the flat land created due to this method, may enable surface mining of coal which heretofore were recoverable by either underground or auger-mining techniques.

Open-Pit Mining

True application of open-pit mining techniques to the exploitation of coal resources is limited to two specific occurrences. In one case, the coal outcrops, but dips so steeply that simple overcasting cannot be easily practiced, due to limitations on equipment reach. The value of the coal is such that the overburden can be stripped in prismatic sections and the pit can be advanced along the strike of the vein. In the other case, the coal is very thick and is overlain by shallow or thick cover. Therefore, benches are required in the coal and the overburden (Figure 51).

While open-pit mining of coal is practiced in the United States, these are in places outside the geographic area of this report.

Figure 49. Mountaintop removal method. Note the head-of-hollow in the front being prepared for valley fill.

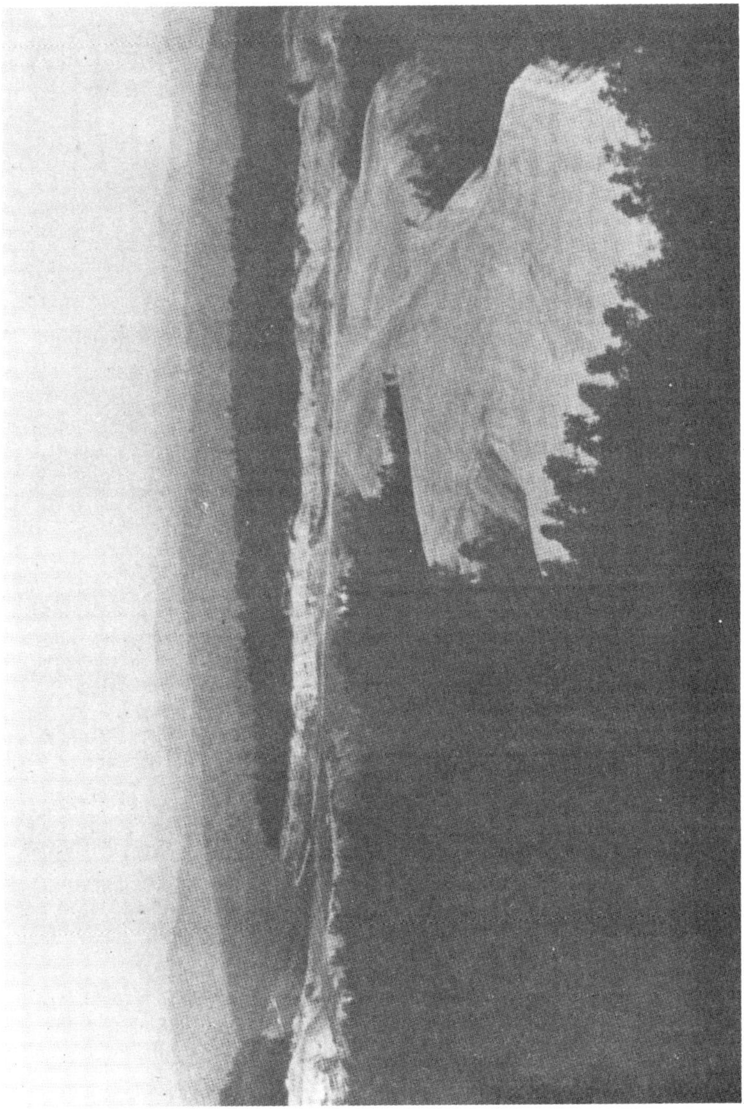

Figure 50. Large area of level ground created as a result of mountaintop removal mining in West Virginia.

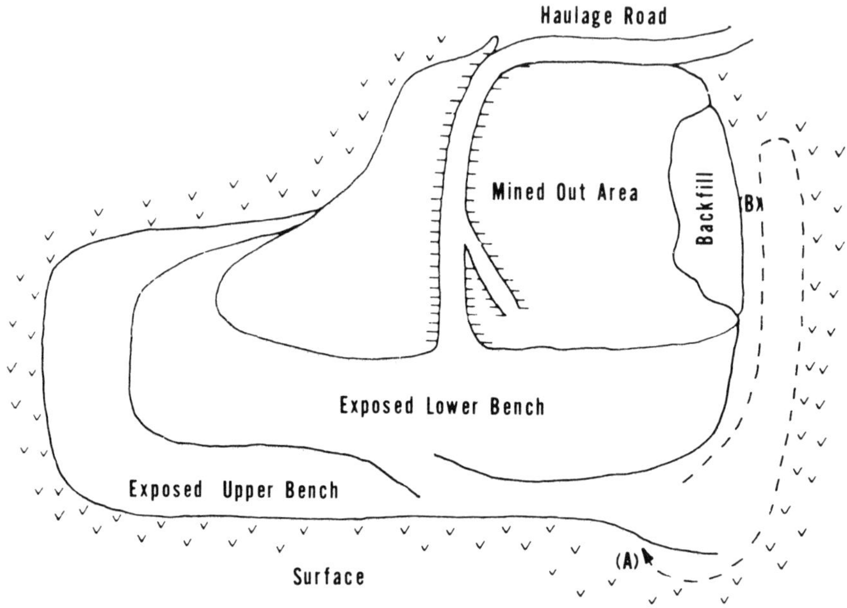

Figure 51. Plan view of an open-pit coal mine.

SPECIAL CONSIDERATIONS - MULTISEAM MINING

Multiple-seam surface mining is done with many types and combinations of equipment. The mining of a second seam of coal often reduces the stripping ratio and therefore provides an overall reduced stripping cost. More selective machines are constantly being designed for increased production and reduced operating costs in special applications, such as removal of both thin partings and coal seams. Flexibility and mobility of this new equipment, combined with lower capital investment, offer great inducement in multiple seam operations or multimine situations.

Mining a second seam of coal and handling the parting in accordance with economic and environmental controls presents a number of problems and poses a challenge to planners. In the context of total resource extraction, all available resources such as valuable claybeds, limestone strata, etc. are best extracted with the extraction of the coal seam under consideration. On the technical side, it is necessary to use a carefully planned sequence of mining and correct combination of available equipment. Loading and hauling units and overburden stripping must be matched.

Porter (1971) conducted a survey of multiple-seam strip mining operations and developed an economic feasibility model. All of Porter's case studies involve stripping shovels, draglines, or tandem operations from the East or Midwest. It is a comprehensive review of conventional multiple-seam (usually two-seam) operations in these regions. Another examination of multiple-seam operations is found in a report by Stefanko et al. (1973), (Figures 52 and 53). Although intended as a preview of all strip mining methods and equipment, the case studies presented cover some multiple-seam mines, particularly in the Midwest and Northern High Plains. Figure 54 shows a photograph of a multiple-seam operation with a 22.93 cubic meter (30 cu-yd) dragline in Pennsylvania.

The ratio of coal thickness to parting thickness is important in the consideration of haulback methods. When the thickness of the parting layer is large compared to the thickness of the coal, space for stoppage of spill is limited and the parting must be hauled upgrade for spoiling. Short, steep grades encountered with loaded haulage vehicles quickly nullify transfer and stowage benefits.

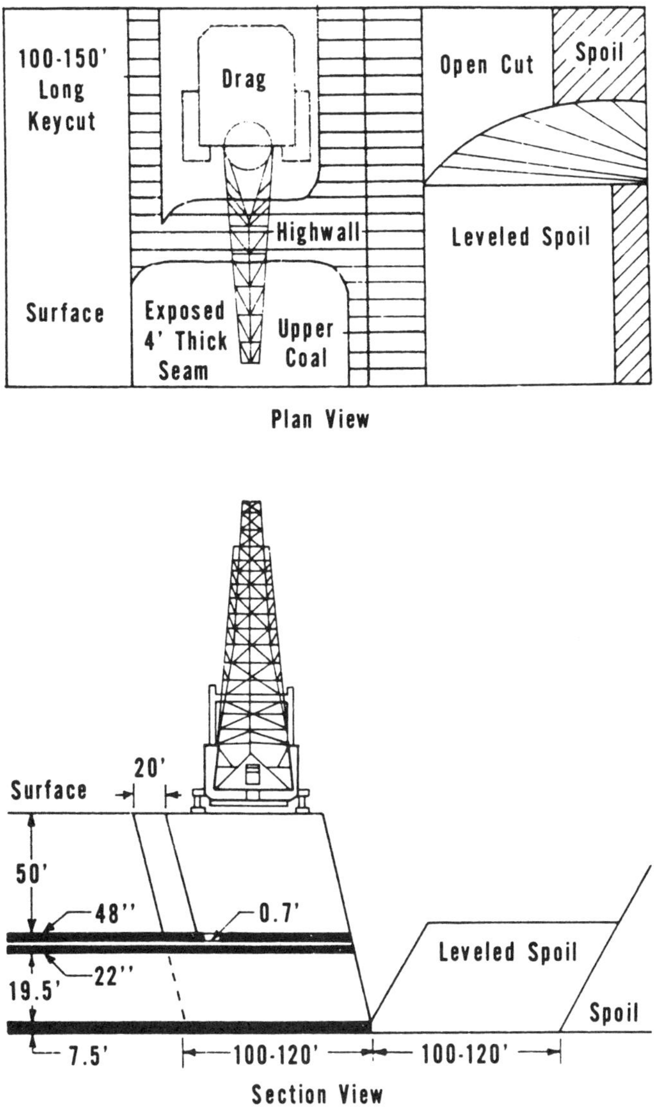

Figure 52. Multiseam stripping operation with a dragline (Stefanko, Ramani, and Ferko, 1973).

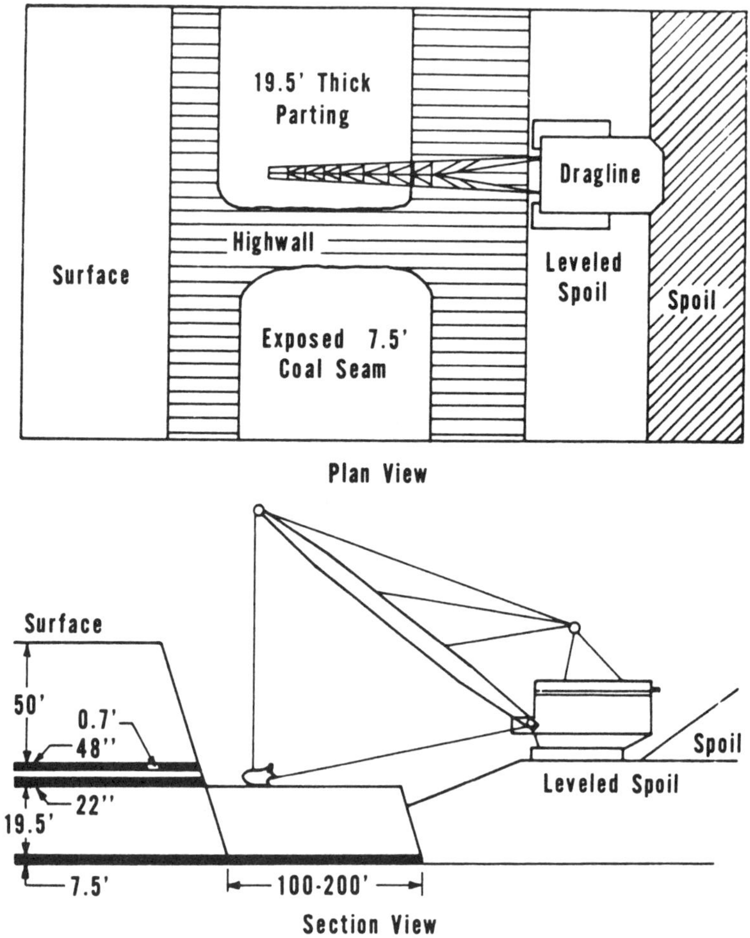

Figure 53. Dragline exposing lowest seam from leveled spoil (Stefanko, Ramani, and Ferko, 1973).

Porter (1971) details a tandem operation using a stripping shovel and dragline to mine three seams of coal. The operational configuration is shown in Figure 55. A 1050B shovel exposes the first two seams, and a Marion 8300 dragline exposes the bottom seam while simultaneously rehandling 36 percent of the shovel spoil. At intervals, the imbalance between the primary stripping machines necessitates the use of the dragline on the middle parting in order for the 1050B to maintain a proper lead. The rehandle is nonproductive work and the dragline output is about 30 percent less in the shown operating mode than would be expected in a conventional (highwall position) stripping situation. Some of this loss can be attributed to the 180 degree swing required while stripping from the spoil side. Another problem area noted, less related to overburden and parting handling but relevant to multiple-seam operations, is haulage from three levels. Haulage roads are brought into the pit about midway between the upper two seams. Access to the lower seam is gained by an incline dug in the face of the spoil parallel with the pit. The entire operation requires a longer than normal pit to allow maneuvering room. Although the pit is a good operation

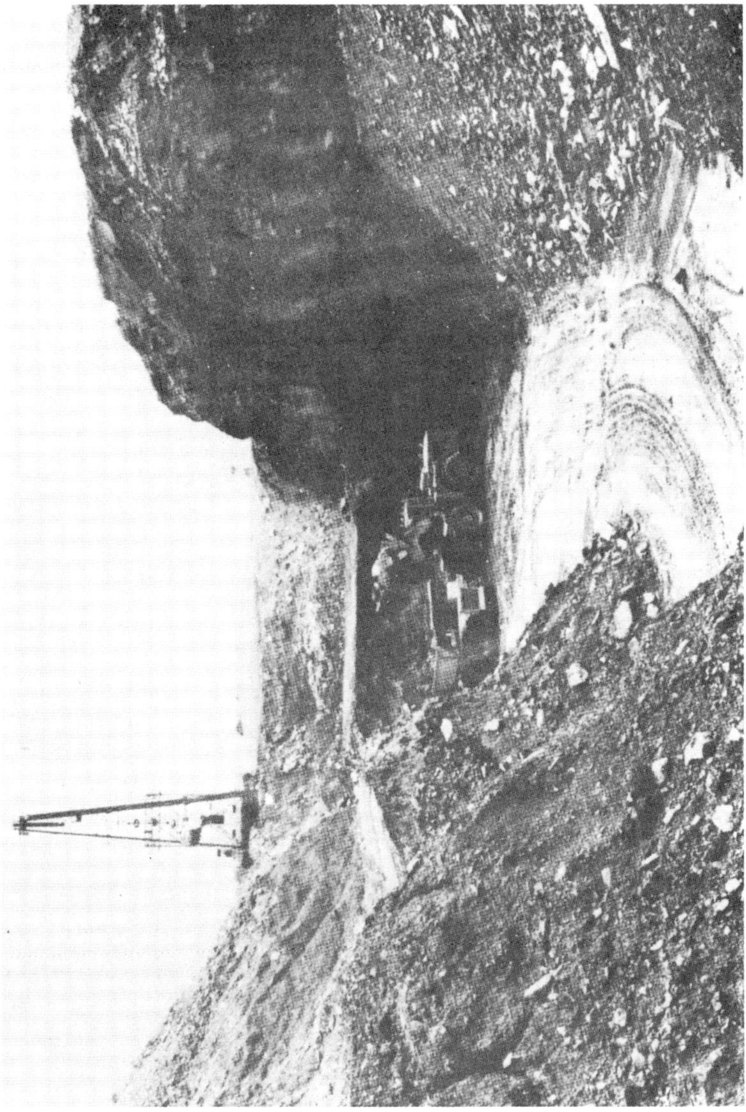

Figure 54. A two-seam operation with a 30-cubic-yard dragline in Pennsylvania.

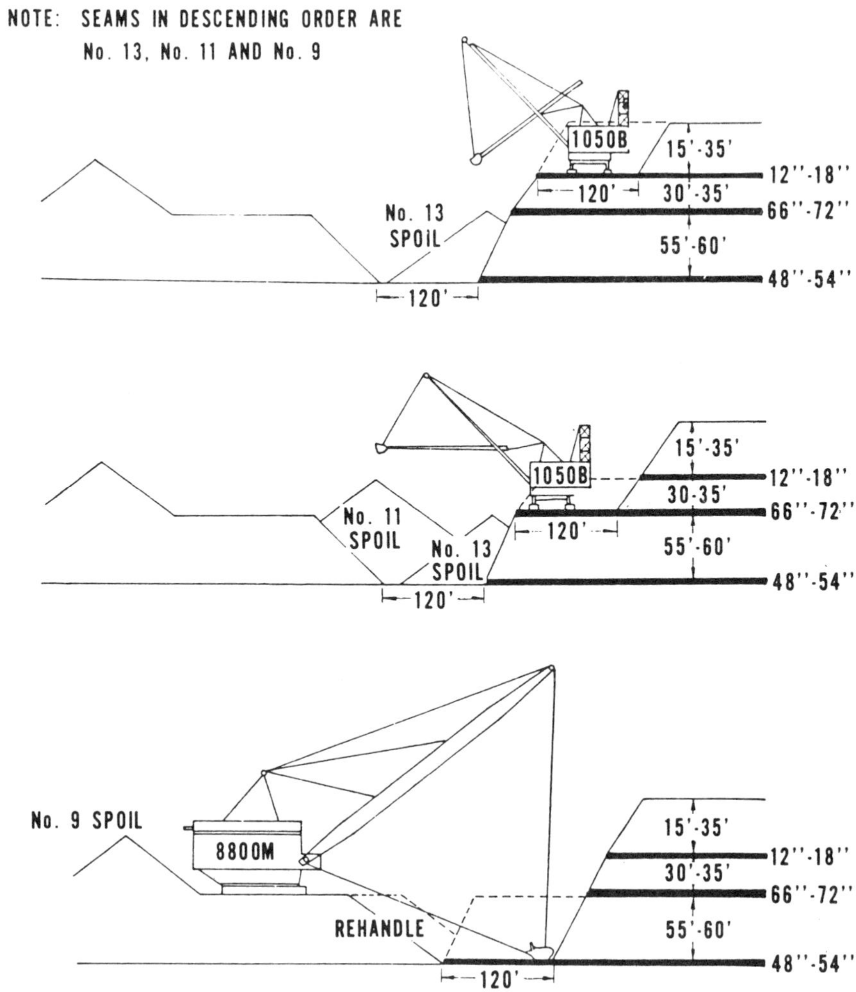

Figure 55. Shovel-dragline tandem operation for multiseam mining (Porter, 1971).

from a conventional view, the thicker seams of the West offer possibilities for stowing. Stowing would eliminate rehandle, ineffective dragline application, and complex timing sequences as machines pass one another. Effective stowing could shorten the required open-pit length. Although the thinner coal seams will not allow efficient backstowing, the problems encountered in conventional multiple-seam mining are evident.

One multiple-seam operation in Illinois utilizes one shovel to uncover two coal seams. A Marion 5761 shovel strips both seams as two active, parallel cuts. An average of 12.19 meters (40 feet) of top overburden is removed with a 180 degree swing and the seven-foot parting is stripped from directly in front of the shovel with a 90 degree swing. The 1.67 meter (5.5 ft) thick top seam and associated thin parting create this unique situation. The shovel can selectively remove overburden as the top of the parting is only 3.81 meters (12.5 ft) above the top of the second seam on which the shovel sits. The volume of material in the innerburden is minimal and leaves sufficient spoiling room for the top overburden. Although the seams are exposed simultaneously, they are effectively separate slices and the overburdens are shot separately. Utilization of the single unit is good in this particular case. The reach available permits easy handling in a pit 45.72 meter (150 ft) wide. One desirable result is burial of the parting material. Exposure of both seams is another benefit as this allows blending of the coals from the two seams.

Another example of the use of one shovel to mine two seams of coal is near Central City, Kentucky (Figure 56). Although the mining method is not identical to the plan proposed here, it is similar in that one shovel is used to excavate two separate strata of overburden material simultaneously.

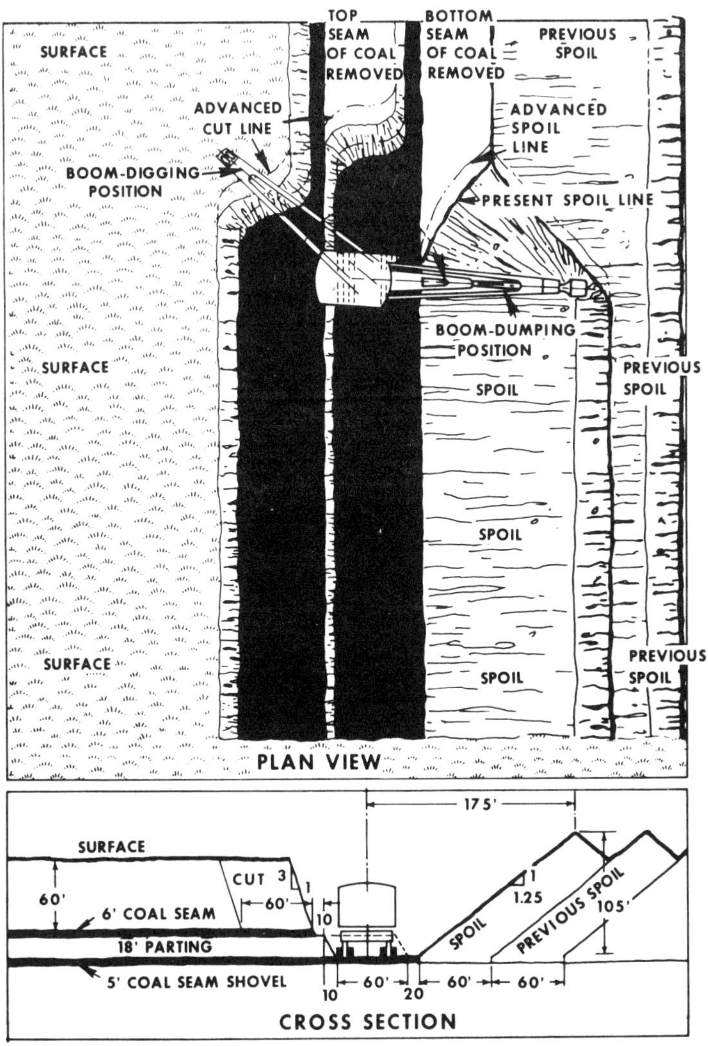

Figure 56. Multiseam stripping operation with a shovel.

The stripping shovel, a Marion 5761 with a 49.69 cu m (65 cu yd) dipper, operates on top of the lower coal seam to remove the overburden and the parting layer. Overburden thicknesses average 12.19 meters (40 ft) and the parting layer averages 2.13 meters (7 ft) in thickness. Two working faces in a double-wide cut are advanced at the same time. The parting layer is excavated from in front of the shovel and dumped into the empty cut. The 5761 with a 57.91 meter (190 ft) boom then digs to the highwall side to remove the overburden which is also spoiled into the empty cut and on top of the parting. This type of operation requires a large shovel with good reach, but the stowing and selective mining concept indicates that such a shovel application is feasible. A variable pitch dipper would increase the efficiency of such a mining method.

Grim and Hill (1974) outline three methods for multiple-seam mining which are applicable in steep terrains:

Method No. 1--If the overburden from the upper seam will not reach the bench of the lower seam, treat each seam as a separate mining operation, mining the lower seam first. This bench may be used to store spoil produced during stripping of the upper seam.

Method No. 2--If the overburden from the upper seam will reach the bench of the lower seam, mine the lower seam in advance of the seam above. Grading should be delayed on the lower bench in order to catch big rocks from the upper seam and bury them in the pit. In no instance can spoil from the upper seam extend more than one-half the distance from the highwall to the edge of the solid bench of the lower seam.

Method No. 3--If both seams appear in the same highwall, separated by more than 7.6 meters (25 feet), and two or more cuts are planned, the coal should be recovered from the bottom seam first. If the seams are separated by less than 7.6 meters (25 feet), mine from the upper seam down, recovering both seams in one systematic operation (Figure 57, Steps 1 and 2). Lateral movement of the spoil is recommended.

SUMMARY

This chapter has briefly reviewed the more commonly used surface coal mining methods. Several examples were presented on specific equipment applications, including multiseam mining operations. While haulback methods are becoming popular in the hilly terrains of the east, area mining with simple overcasting is most popular in the interior region. Tandem operation of large strip equipment (bucket-wheel excavator, shovels. draglines) is quite common in Illinois. The concern for concurrent land reclamation, and topsoil replacement has seen the introduction of a fleet of wheeled vehicles for topsoil and overburden removal.

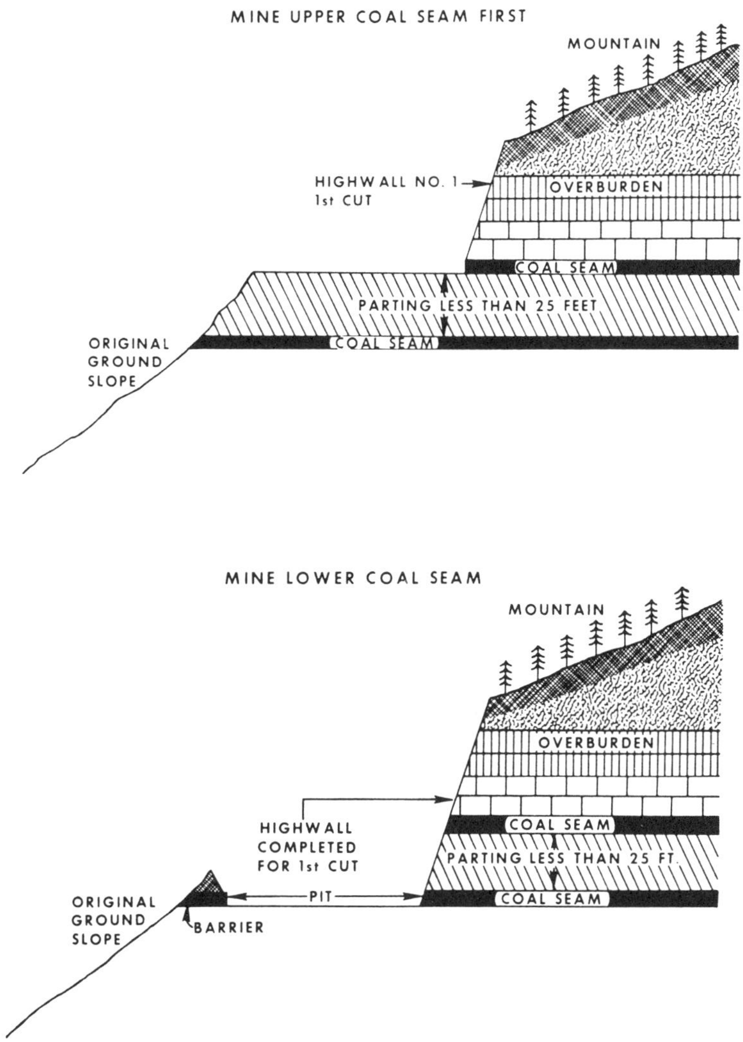

Figure 57. Multiple-seam mining method: two seams less than 25 feet apart in the same highwall.

Fragmentation Practices

INTRODUCTION

Fragmentation encompasses several different unit operations in the surface mining of coal. These operations combine to provide for the fracturing of both the overburden and coal, thereby simplifying the subsequent materials handling procedures. In some favorable conditions, neither the overburden nor the coal may exhibit such strength as to require fragmentation prior to actual excavation. In the most restrictive case, the overburden and the coal may both be so strong as to require noncurrent blasting. As is typical in the mining industry, composite situations exist whereby portions of the overburden or coal can be directly excavated while other portions must be prepared prior to the digging operation.

As far as overburden preparation is concerned, a range of possibilities is presented in the fragmentation unit operations. Overburden fragmentation and preparation may be accomplished in some situations by ripping with bulldozers. As an intermediate case, widely spaced drilling and blasting may be sufficient. In the limiting case, a regular and intensive drilling and blasting approach may be required. Where blasting is required, various considerations can affect the drilling layout. The size and nature of the drillholes may vary, thus affecting spacing. Further, the proper type of explosive, either high or low strength, must be carefully examined.

Although the stated purpose for rock fragmentation is to fracture the overburden and coal for loading, there are other secondary, but vital, purposes of a properly designed fragmentation process. These purposes include product size, and rotation and translation.

The size of the broken materials should be small enough to fit into the bucket of the excavating machine and to eliminate secondary blasting, yet small particle generation should be kept to a minimum for dust control. While the largest size is clearly dictated by the dimensions of the machinery, the lower limit is rather difficult to determine. Suffice it to say that fragmenting to a size smaller than that required is wasteful in explosives and cost.

Rotation and translation (throw) refers to the manner in which the blasted material comes to rest. If the explosive force achieved is too great, then the material may move beyond the affected area, thus causing difficulty in loading. Further, a properly designed blasting layout can rotate the fragments to improve their loadability characteristics.

Three additional factors in fragmentation relate to the precautions needed to minimize the environmental effects of blasting. The blast design must minimize the environmental effects of blasting. The blast design must minimize fines, noise and vibration. Critical to the environmental effects is fines production (dust). Over-designing the blasting rounds can, in addition to fines production, cause conditions of excessive noise and vibrations. If not properly controlled, the above-mentioned three problems can have as great an impact on fragmentation practices as the production requirements.

The parameters on which the design of fragmentation processes are based relate to two general areas. One is the previously mentioned total operation design, i.e., the dimensions of the area to be prepared, the desired fragmentation, and the economic considerations. The second area concerns the properties of the material to be broken, i.e., primarily the strength of the material and its response when subjected to high energy impulses.

A solid mass exhibits different responses to the application of energy. Basically, the response differs with the absolute quantity of energy introduced and the time over which the energy is applied. Rocks can absorb small impulses of energy, or even large quantities of energy applied evenly over a long period of time, and can deform without collapse. Therefore, the goal in breaking rock is to apply a large enough force in a short enough time to create fractures in the mass and to rotate and translate the resultant fractured mass.

The simplest case of fragmentation is that in which no explicit fragmentation process is required, or the case where the excavator (bucketwheel, dragline, shovel, loader, scraper, or bulldozer) can dig the material without prior treatment. In this case, the digging machine is capable of delivering an excess of the total amount of energy that the mass can withstand in the static case. The next extension of the simple case is that in which the excavator is just barely incapable of digging the material. In this case, it is possible to concentrate energy through use of a ripper. A ripper concentrates the tractive effort of the dozer (which can be considerable) at one point, the ripper tooth. The applicability of direct ripping or direct excavation without fragmentation is determined in the specifics of equipment selection for either coal or overburden handling. In many cases, small operators may choose to use such techniques at marginal conditions in an effort to save cost in drilling equipment and personnel. Large operators may choose to use direct excavation where mobility is a key to successful operation. In the Eastern United States, drilling and blasting are necessary prior to excavation in most operations.

DRILLING

Where the strength of overburden material is such that blasting is required prior to excavation, drilling precedes the use of explosives. Quite simply, drilling is the unit operation which produces the hole for the placement of explosives. Drilling, for the mining of coal by surface techniques, is categorized as rock penetration by mechanical attack. Mechanical attack is further subdivided into two methods - percussive and rotary action.

Beyond the mechanics of attack, drilling is also categorized by the manner of application. Specifically, the drilling of overburden for the surface mining of coal can be approached in three manners:

1. Vertical - conducted from the top of the highwall down to the coal perpendicular to the strata bedding planes.
2. Horizontal - conducted from the pit floor parallel to the bedding planes of the strata.
3. Inclined - since the shortest drilling distance to a pitching seam is perpendicular to strata bedding planes, inclined drilling is conducted from the top of the highwall normal to the coal seam's angle of pitch.

Since most coal seams in the East are tabular, the third option is rarely encountered.

PERCUSSION DRILLING

Penetration of overburden by percussion drilling is affected by successive impacts of the bit into the rock. The bit is rotated between blows to aid in the crushing and chipping process. Percussion drilling is used when small diameter holes must be driven through hard rock where rotary units are difficult to employ.

There are a number of advantages of percussion drilling. First, small diameter holes can be drilled through hard rock. The rigs themselves are light, less expensive, and are quite maneuverable. Finally, one man can operate the rig through a wide range of drilling angles (Morrell and Unger, 1973).

Percussive drills are subdivided into two categories, surface-mounted and down-the-hole. The percussive action of surface-mounted units is transmitted to the bit by a prime mover which remains outside the hole. In comparison, the prime mover for a down-the-hole drill actually strikes the bit and follows it into the drilled holes.

The most widely employed percussive drills for surface mining are surface mounted. Pneumatically powered, these drills also use the air for flushing the cuttings from the hole. Since the addition of drill steel can decrease the available drilling energy nearly 50 percent, surface-mounted percussive drills are used primarily where the drilling depth is less than 30.48 meter (100 ft) (Morrell and Unger, 1973).

Since the bit must be rotated between blows to effectively chisel the rock, surface-mounted percussive drills are either rifle-bar or independently rotated (Table 20). The drill steel in a rifle-bar rotation unit is automatically rotated by a fluted rifle bar and nut on the block stroke. Since rotation is automatic, the ability to vary the rotational speed is minimal. Independently rotated drills are, as the name implies, rotated by an independent motor. This adds flexibility to the system when rock of varying strength is encountered. For example, harder rock requires a decrease in rotational speed and an increase in the impacting force. Rotational speeds of 200 to 400 rpm are common for 50.8 mm (2 inch) bits, while 152.4 mm (6 inch) bits are usually operated at less than 100 rpm.

Surface-mounted drills offer the following advantages (Morrell and Unger, 1973):

1. Surface-mounted drills can be more powerful and can penetrate faster in shallow holes than down-the-hole drills because, unlike the latter, the drill-piston diameter can be greater than the hole diameter.
2. The minimum hole diameter for a down-the-hole drill is 101.60 mm (4 in), whereas a surface-mounted drill is capable of even smaller diameters.
3. Unlike a down-the-hole drill, the loss of the drill unit due to hole caving is not a problem for surface-mounted units.

However, surface-mounted drills have the following disadvantages:

1. Total penetration depth is limited to 30.48 meters (100 ft) because the loss of drilling energy, due to the addition of five joints of drill steel, can approach 50 percent.
2. Holes greater than 152.4 mm (6 in) in diameter are not common.

TABLE 20.· SPECIFICATIONS OF SOME RIFLE-BAR ROTATION AND INDEPENDENT ROTATION PERCUSSIVE DRILLS (AFTER MORRELL AND UNGER, 1973)

Specifications	Rifle-Bar Rotation			Independent Rotation		
Hole size, in.	$1\frac{1}{2} - 2\frac{1}{2}$	$2 - 3\frac{1}{2}$	$3\frac{1}{2} - 5\frac{1}{2}$	$1\frac{1}{2} - 2\frac{1}{2}$	$2\frac{1}{2} - 4$	$3\frac{1}{2} - 5\frac{1}{2}$
Piston diameter, in.	$3\frac{1}{2}$	$4\frac{3}{16}$	$5\frac{1}{2}$	$3\frac{1}{2}$	$4\frac{1}{2}$	$5\frac{1}{2}$
Drill weight, lb.	112	155	385	---	290	690
Drill length, in.	30	29	40	---	39	55
Blow frequency,[1] bpm.	1,760	1,975	1,616	1,970	1,640	1,250
Energy per blow,[1] ft-lb.	135	160	275	117	228	375
Energy output,[1] ft-lb per min.	238,000	316,000	444,000	230,000	374,000	469,000
Air consumption,[1] cfm.	195	340	900[3]	325[2]	560[3]	1,050[3]

[1] At 100-psi operating pressure.

[2] Includes air motor.

[3] Includes blow air for flushing.

3. Unlike down-the-hole drills, noise is not facilitated by hole penetration.
4. Drill rod life is reduced due to the continued transmission of impaction energy.

Down-the-hole percussive drills are employed where drillholes greater than 30.48 m (100 ft) are required and hole diameters of up to 304.8 mm (12 in) are preferred (Table 21). Since the percussive unit follows the bit into the hole, a clearance of approximately 12.7 mm (0.5 in) must be provided by the hole diameter for the passage of cuttings around the drill unit. The rotational motor is the only part of the operation that takes place out of the hole. A typical rotation range is 10 to 100 rpm (Morrell and Unger, 1973).

TABLE 21. DOWN-THE-HOLE DRILL SPECIFICATIONS (AFTER MORRELL AND UNGER, 1973)

Drill Specifications	Hole Diameter 4 In.	6 In.	8 In.
Overall diameter, in.	$3\frac{1}{4}$	$5\frac{1}{4}$	$6\frac{1}{8}$
Tool weight, lb	90-100	180-200	300-350
Tool length, in.	40-45	40-60	50-65
Piston diameter, in.	$2\frac{1}{2}$	$4\frac{1}{2}$	5
Operating pressure, psi	100-250	100-250	100-250
Air consumption, cfm	100-500	300-600	500-1,600

Down-the-hole drills have the following advantages:

1. Penetration rates are maintained because the drill-to-bit distance is constant, unlike that of a surface-mounted drill, and percussive energy is not lost.
2. The drill noise is dampened as soon as the unit enters the hole.
3. Since they do not transmit impact energy, drill rods have a longer life than those employed with surface-mounted drills.

Down-the-hole drilling has the following disadvantages:

1. Since the minimum hole diameter is dictated by the drill diameter, the minimum hole size is four inches.
2. Hole caving can make drill recovery difficult, if not impossible.
3. Since larger bit diameters are required for each drill size, the amount of energy that can be applied to a bit is limited.

Percussive drill bits are divided into two categories for surface mining applications, cross bits and button bits. The tungsten carbide cross bits are either of an "X" (at an angle nearly 90 degrees) or a cross (90°) configuration. Button bits, in sizes from 50.8 to 165.1 mm (2 to 6.5 in), are flat-bottomed with cylindrical tungsten-carbide inserts. Cross bits can usually be sharpened 20 times, while button bits do not require sharpening.

ROTARY DRILLING

Rotary drilling, the most versatile form of overburden penetration for surface mining of coal, is conducted by rotating a rigid, tubular string of rods to which is attached a rock-cutting bit. Thus, the drilling energy is supplied by rotation as well as thrust. The cutting action consists of either abrasion, scraping, spalling, or chipping, as well as a combination of the above. The size range for rotary-drilled blasthole diameters is from 101.6 mm (4 in) to 381 mm (15 in) with 228.6 mm (9 in) and 304.8 mm (12 in) being common sizes.

For the surface mining of coal, most rotary drilling rigs are truck-mounted, although crawler-mounted units are also used. All drilling rigs are equipped with jacks for leveling on uneven ground. The power source for the rotary drill rigs is either electric (for economy) or diesel (for flexibility in tramming).

Drill steel, of up to 15.24 m (50 ft) in length, is stored on a circular rod carrier which rotates in a mast for the automatic changing of drill steel. Although most drilling is vertical, the mast can be tilted up to 30 degrees off vertical for angled drilling.

There are two types of bits that are used for rotary drilling, drag or rolling cutter. Drag bits are applicable in soft to medium-soft shales or their equivalents. Since their effectiveness is dependent upon a successful scraping action, drilling thrust is lower for drag bits than for rolling cutter bits, but the opposite is true for torque requirements. The bits are usually fitted with replaceable tungsten-carbide cutters.

Rolling cutter bits, though available in the two-cone configuration, are usually of the three-cone style (Table 22). The rolling cones have either large, steel teeth (for drilling through soft formations) or small, tungsten-carbide inserts (for drilling through hard formations). Table 23 relates bit size and weight to the strength of the rock.

The major advantage of rotary drilling is its rapid drilling rate. This is due to the fact that, unlike percussion drilling, rock penetration is uniform with depth (Morrell and Unger, 1973).

VERTICAL AND HORIZONTAL DRILLING

There are basically two approaches to drilling overburden for the surface mining of coal, vertical and horizontal drilling. Although vertical drilling appears to be the overwhelming choice for surface mining applications, the horizontal method can be used effectively under certain conditions.

Vertical drilling consists of drilling from the top of the highwall perpendicular to the coal seam. This makes the use of bulk explosives possible. It is also the preferred approach where overburden, which consists of competent strata, exceeds 15.24m (50 ft) in depth or massive stone occurs at an elevation higher than that which can be reached by horizontal drilling. Because common practice in vertical drilling is to penetrate to the coal and then backfill approximately four feet, there is less shattering of coal with this approach.

Horizontal drilling, a relatively new development for the surface mining of coal, is usually conducted by a twin-mast rotary drill. Both holes, which can be drilled either simultaneously or independently, can be varied from 8.53 m (28 ft) to 11.58 m (38 ft) centers and can be drilled from 10 degree above to 12 degree below the horizontal. Typical hole diameters vary from 171.45 mm (6-3/4 inches) to 269.87 mm (10-5/8 inches).

TABLE 22. THREE-CONE ROLLING CUTTER BIT DIMENSIONS AND HOLE VOLUME PRODUCED (AFTER MORRELL AND UNGER, 1973)

Bit Diameter, In.	Pin Size API* Reg., In.	Bit Weight, Lb	Volume of Hole Drilled, Cu Ft per Ft
5 5/8	3 1/2	26	0.173
6 3/4	3 1/2	35	0.249
7 3/8	3 1/2	49	0.279
7 7/8	3 1/2	60	0.338
8 1/2	4 1/2	73	0.394
8 3/4	4 1/2	75	0.418
9..........	4 1/2	77	0.442
9 7/8	6 5/8	122	0.532
10 5/8	6 5/8	141	0.616
12 1/4	6 5/8	200	0.818
15	6 5/8	306	1.227

* American Petroleum Institute.

TABLE 23. BIT WEIGHT AS A FUNCTION OF ROCK STRENGTH AND BIT SIZE (AFTER MORRELL AND UNGER, 1973)

Rock Strength	Weight per In. of Diameter Lb	Recommended Bit Loading for Bit Size Indicated, Lb		
		4-6 In.	6-9 In.	9-12 In.
Very soft formations: overburden, soft shales, limestone and evaporites	1,000-4,000	4,000-24,000	6,000-36,000	9,000-48,000
Medium formations: limestone, dolomites and sandstones	3,000-5,000	12,000-30,000	18,000-45,000	27,000-60,000
Hard and very hard formations: basalt, granit, quartzite, taconite	4,000-8,000	16,000-40,000*	24,000-60,000*	36,000-100,000

* Maximum weight for largest bit size indicated.

Where the only portion of the overburden that must be blasted is located directly above or near the seam, horizontal drilling is applicable since less footage is usually required for the same pit width. Horizontal drills tram on the coal floor which, by eliminating the surface preparation required for vertical drills, allows for faster rates. Although applications are usually limited to a maximum of 30.48 m (100 ft) of overburden, 45.72 m (150 ft) deep holes (approximately two pit widths) can be conveniently drilled at one time (Nordness, 1976).

BLASTING

Effective bank preparation not only deals with the fragmentation of overburden, but also with the minimization of incurred costs. As such, the target for proper overburden preparation is efficiency with regard to the number of holes drilled and the amount of explosives utilized.

It has already been stated that operators must continually evaluate their fragmentation procedures. The amount of explosives employed must be adequate for fracturing the rock to a size which is acceptable for equipment removal. Since the character of the overburden in a large surface mine can change over a wide area, economics alone is a sufficient reason for the emphasis on studying and selecting good fragmentation practices.

SYSTEM COMPONENTS

Blasting systems are constructed of the following components: 1) detonators, 2) primers and boosters, and 3) explosives.

Detonators are devices for producing an explosive decomposition in a high-explosive charge. Initiation of the detonator is by either a safety fuse or by electricity. Detonators consist of two styles, blasting caps or detonating cord.

The most commonly used detonator is the electric blasting cap. It consists of a copper shell, closed at one end, which contains a detonating compound. Because it is placed in the explosive charge, the ignition of the cap, by a current or spark, sets off the entire reaction. The advantages of electric blasting caps are as follows (Dick, 1973): 1) can be safely handled, 2) accuracy of delay periods, and 3) controlled firing time. The following are hazardous conditions which dictate caution when using electric blasting caps (Dick, 1973): 1) lightning, 2) radio frequency energy, and 3) mixing of different brands of blasting caps.

Blasting caps can also be activated with a safety fuse. A safety fuse consists of a potassium-nitrate black powder core enclosed in a covering of textile and waterproofing compound. The fuse is inserted into the open end of the cap, which is then attached to the charge. Burn rates for fuses vary between 98.42 and 131.23 second per meter. Care must be taken when using a safety fuse because of the following conditions (Dick, 1973):

1. Since the fuse cap is open at an end, exposing the explosives, care must be taken in handling.
2. Altitude changes affect burning rates.

However, the following are advantages for using safety fuses:

1. Elimination of the hazard of stray electricity.
2. No need for a power source or associated lead wires.

The detonating cord is a fuse, consisting of a high-explosive core of pentaerythritol tetranitrate (PETN), enclosed in tape, and wrapped with textile yarns, that fires the charge without the assistance of any other detonator between it and the charge. However, an electric blasting cap is needed to start the reaction in the detonating cord. Although blasting caps with fuses can be used, electric blasting caps are preferred because of their ability to control the instant of firing. In fact, the use of two caps at each initiation point is common to guarantee proper firing. Since a detonating cord is insensitive to shock, friction, and lightning, it is quite applicable as a safe and convenient trunkline for firing blasts.

Initially, applications of detonating cord were instantaneously fired by means of a trunkline. To achieve a delay, electric caps were attached in each hole. Millisecond (M.S.) connectors were then developed to provide delay firing by surface, rather than downline, initiation.

With instantaneous firing, the layout must be designed with attention to the direction of the blast movement. For example, the charges nearest the initiation point should not "cut off" the firing in later holes due to ground movement. Because of this, the trunkline should run parallel to a free face and should also have an open end (Figure 58). With this configuration, detonation should be from the open end. By looping the trunkline ("double trunkline"), a second means of detonating the holes is available if a break occurs in the main trunk. In multiple row blasts (Figures 59 and 60), availability of secondary detonation is provided by cross ties, spaced on intervals of less than 60.96 m (200 ft).

Delay firing, resulting in reduced vibration, improved fragmentation, and less backbreak, is considered to be more desirable than instantaneous firing. Delay intervals where detonating cord is used vary between 5 and 35 milliseconds. For surface mining applications where large-diameter boreholes are spaced on approximately 9.14 m (30-foot) centers, 25-millisecond intervals are used to avoid cut-offs. In fact, to initiate detonating cord downline with millisecond delays attached at the surface, the minimum available interval is 25 milliseconds.

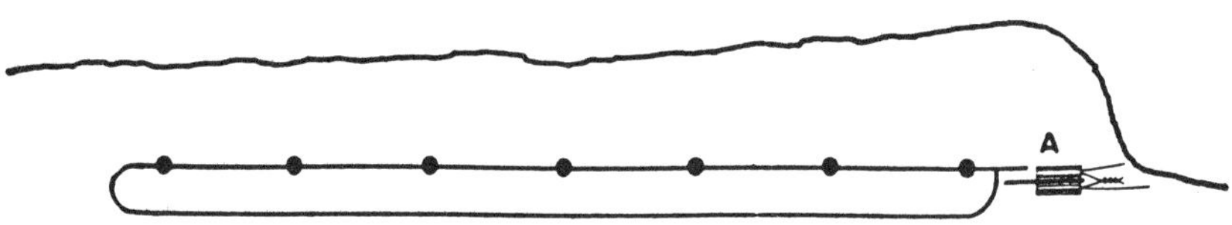

Figure 58. Single row of holes connected by "primacord" and detonated by electric blasting caps or caps and fuse attached to "primacord" at A. Also, note that the "primacord" loop is tied to the trunkline with a right angle connection (After DuPont Co., 1967).

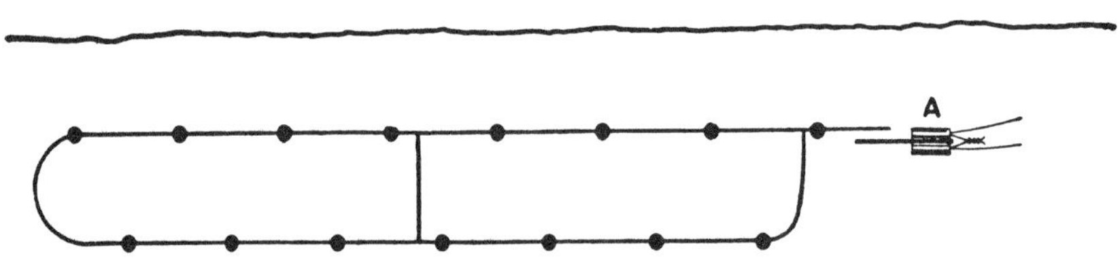

Figure 59. Double row of holes connected by "primacord" (After DuPont Co., 1967).

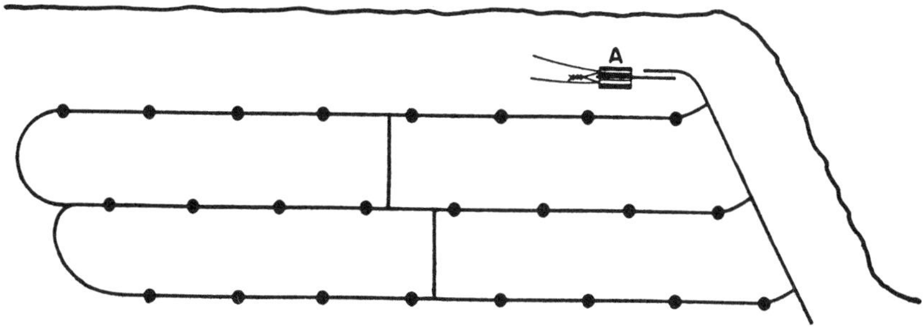

Figure 60. Three rows of holes connected by "primacord" (After DuPont Co., 1967).

Millisecond delays were quite widely employed prior to the introduction of millisecond connectors. These short-interval delays are attached, on the surface, to each detonating cord downline at the individual holes or to groups of holes by means of a trunkline. With M.S. delays and the absence of a trunkline, noise is reduced; however, precautions must be taken because of the large amount of electrical wiring required.

Millisecond connectors are tied directly into the detonating cord trunklines. Unlike M.S. delays, M.S. connectors come in a variety of delay intervals, 5, 9, 17, and 25 milliseconds. Figures 61 through 65 show sample configurations with M.S. connector placement marked by an "X". Special attention should be drawn to Figure 65 because it shows a configuration for horizontal drilling.

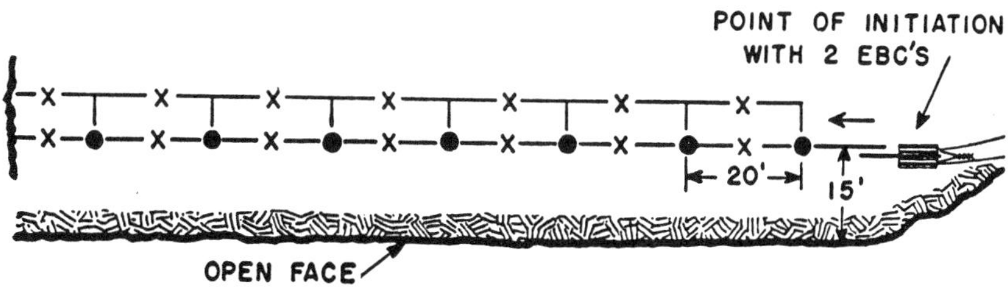

Figure 61. Single row shot fired in sequence. "Primacord" M.S. connectors inserted at points "X" (After DuPont Co., 1967).

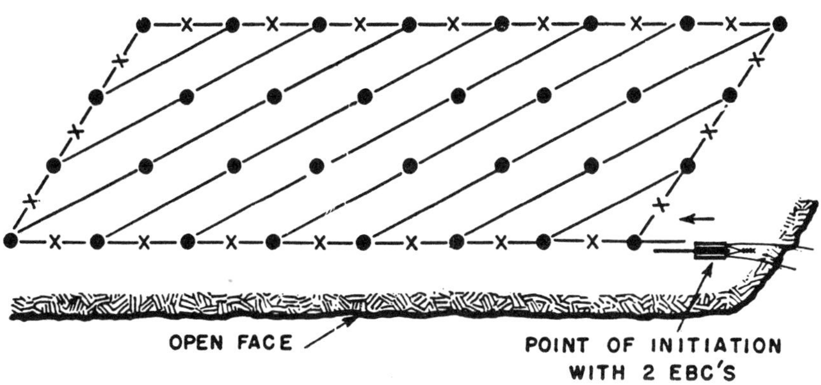

Figure 62. Multiple row shot with holes and rows fired in sequence from one end (After DuPont Co., 1967).

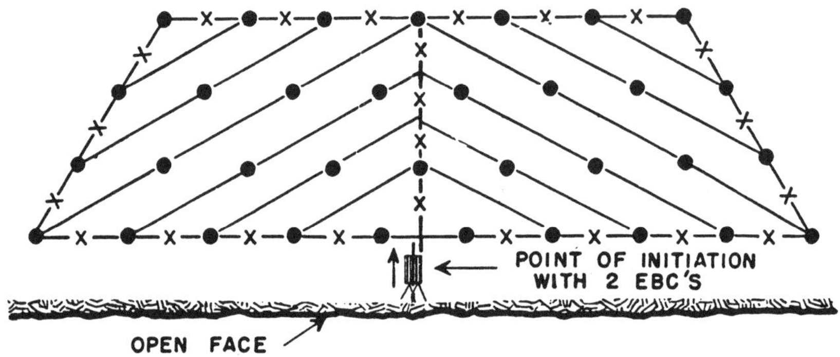

Figure 63. Multiple row shot in sequence initiated at center front (After DuPont Co., 1967).

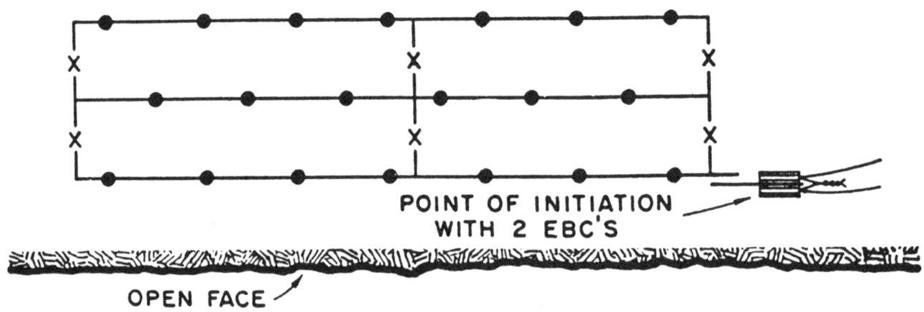

Figure 64. Multiple row shot entire row in sequence (After DuPont Co., 1967).

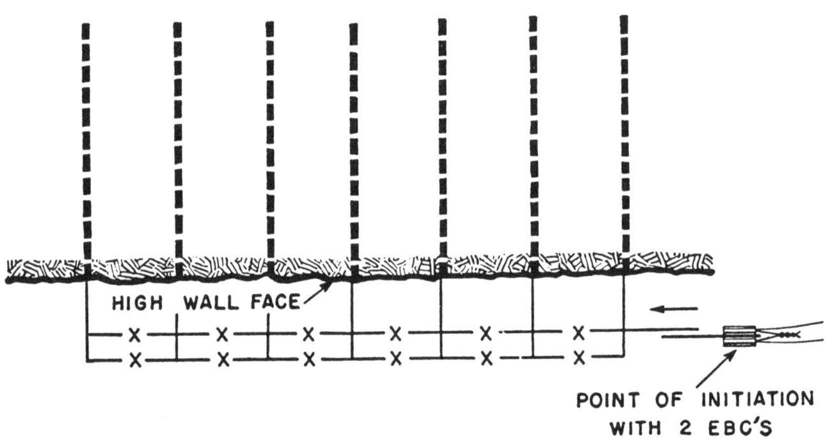

Figure 65. Single row horizontal hole shot in bituminous stripping (After DuPont Co., 1967).

A primer is usually the combination of an explosive cartridge and a detonating cap. By detonation, it sets off the remainder of the charge. The cartridge is placed at one end of the charge with the embedded detonator pointing towards the charge, not away from it. Detonating cord can also be used as a primer. The following precautions should be taken to assure the proper application of primers (DuPont Co., 1967):

1. The blasting cap should be securely inserted into the primer cartridge (deeply into the center) to guarantee that it cannot be pulled out.
2. The wires of electric firing devices or fuses should not be subject to harmful strains.
3. When necessary, the primer should be water resistant.
4. A dynamite punch should be used to make the hole for insertion of the blasting caps or detonating cord.
5. Since detonating cord will initiate every cartridge it comes in contact with, care should be taken that it is attached to the first cartridge loaded into the hole.
6. Primer assemblies should be made just before the holes are to be loaded and at the blast area, to reduce the exposure time to hazards such as fire, impact, or electricity.

A booster is an explosive, usually encased in a small metal shell, to improve the performance of another explosive. They are, quite often, placed at intervals along the borehole to reinforce the detonation of the column of blasting agent. Boosters differ from primers in that they do not contain an initiating device. Since boosters are placed between the detonators and blasting agents, they are essential for the ignition of cap insensitive blasting agents.

An explosive is a compound or mixture of compounds which undergoes a rapid, self-propagating exothermic reaction. This reaction is induced by either heat, impact, or shock. When the reaction moves through the explosive at a faster rate than the speed of sound in the unreacted explosive, it is called a detonation. If it is slower, it is called a deflagration. In any event, the products of the reaction, usually gases, exert tremendous pressure during expansion. Since the resultant condition created by the shock front is also dependent upon both detonation velocity and the sonic velocity in the rock, the reaction should be designed so that those velocities are equal.

Figure 66 shows a standard cartridge of explosive in the process of detonating. The right side is undetonated explosive, the left is expanding byproducts of the reaction, and the center represents the area where the explosion, a chemical reaction, is taking place. The explosion is a rapid oxidation reaction in which all elements necessary for the reaction are present in the explosive. The reaction takes place in a very definite zone though, in most cases, the zone is not as wide as is indicated in the figure. The actual width of the zone varies with a coarseness of the explosive particles, from small fractions of a millimeter in the fine-grained high explosives, to several centimeter in bulk, granular blasting agents such as ammonium nitrate/fuel oil (ANFO) mixtures.

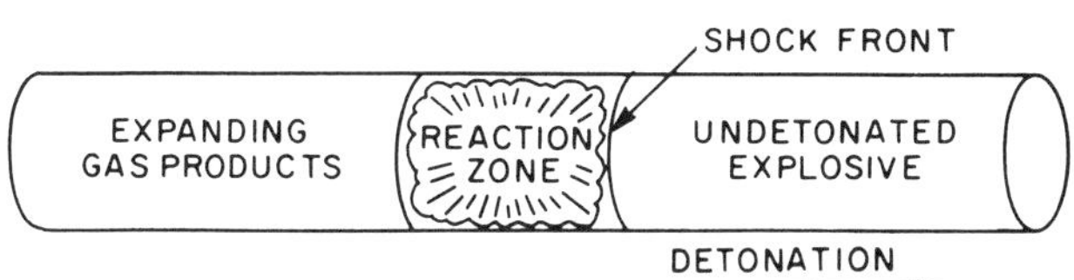

Figure 66. A cartridge of explosive undergoing the process of detonation (After Boddorff, 1974).

Just as rock properties, such as density, porosity, strength, energy absorption, etc., affect fragmentation, there are various explosive properties which also affect the outcome of the fragmentation process. Among these are detonation velocity, density, detonation pressure, borehole presure, and water resistance.

Detonation velocity is the speed at which the reaction front moves forward through a cylindrical charge. The velocity ranges between 1676.4 m (5500 ft) and 7620 m (25000 ft) per second. Faster velocities are appropriate in strong rock where a shattering effect is desirable. Slower velocities are applicable where less demand is on rock shattering, whereas a heaving effect is warranted. The following are steps that can be taken in the field to increase detonation velocity (Dick, 1973):

1. Use a larger charge diameter.
2. Increase density.
3. Decrease particle size.
4. Provide good confinement in the borehole.
5. Use a large initiator or primer.

Density refers to the ratio of explosive density to that of water (specific gravity). The ratio for commercial products varies between 0.5 and 1.7. By relating the specific gravity of the explosive to the borehole diameter, the amount of explosive per foot of borehole can be determined. This is particularly useful when matching blasting requirements to the economics of blasting. Figure 67 shows a nomograph to aid in this calculation.

Detonation pressure is the pressure of the detonation wave propagating through the explosive column. It is measured at the rear of the reaction zone (Chapmen-Jouget plane). Detonation pressure can be calculated from the following equation:

$$P = 4.18 \times 10^{-7}\ DC^2/(1 + 0.80D)$$

where,

P = detonation pressure in kb (1 kb (kilobar) = 100×10^7 pascals)
D = specific gravity
C = detonation velocity in meter per second

Figure 68 also shows the relationship by use of a nomograph. Although the detonation pressure range for commercial explosives varies between 5 and 150 kb, the higher pressures are applicable where a greater shock wave is needed, such as in competent rock.

Borehole pressure is that which is exerted perpendicular to the detonation pressure. Since it is exerted on the borehole walls, it is important in displacing, as well as breaking, the rock. Borehole pressures range from 10 to 60 Kilobar in commercial explosives, where borehole pressures common to ANFO are greater than detonation pressures. In high explosives, however, the converse usually hold true.

Water resistance is the ability of an explosive to withstand exposure to water without losing its sensitivity or efficiency. Gelatin dynamites and slurries have good resistance characteristics. In between these two extremes are the nongelatinized high explosives.

For surface mining applications, there are various types of explosives on the market. However, they are categorized into three groups: 1) high explosives, 2) dry blasting agents, and 3) slurries or water gels. Of all these varieties, dry blasting agents are the most common in coal applications because of cost considerations and safety, and dry blasting agents are cheaper than other forms of explosives and need a primer for detonation. Ammonium nitrate fuel oil, which has become the most common dry blasting agent, accounts for over 75 percent of the nation's consumption. Fifty percent of this figure, 453.6 x 10^6 Kg (10^9 lbs), is used annually for the surface mining of coal (Condon and Snodgrass, 1974). Due to its overwhelming acceptance, further discussion in this section will be dedicated to dry blasting agents.

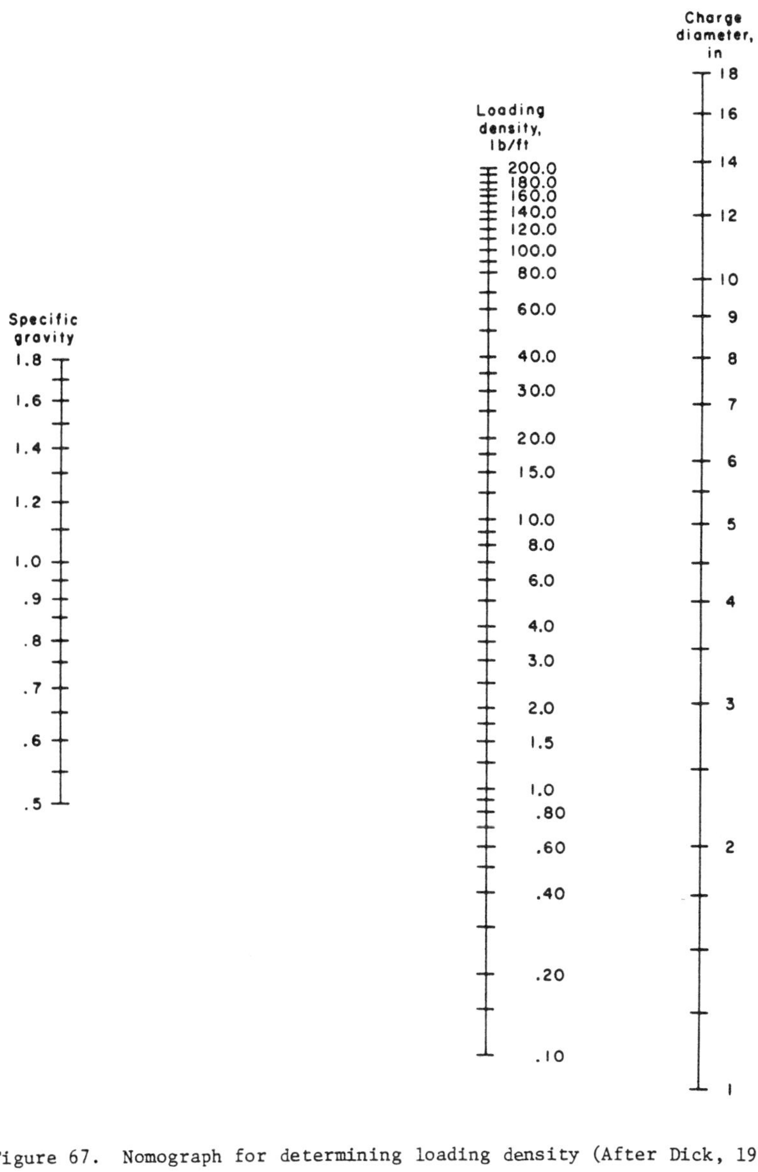

Figure 67. Nomograph for determining loading density (After Dick, 1973).

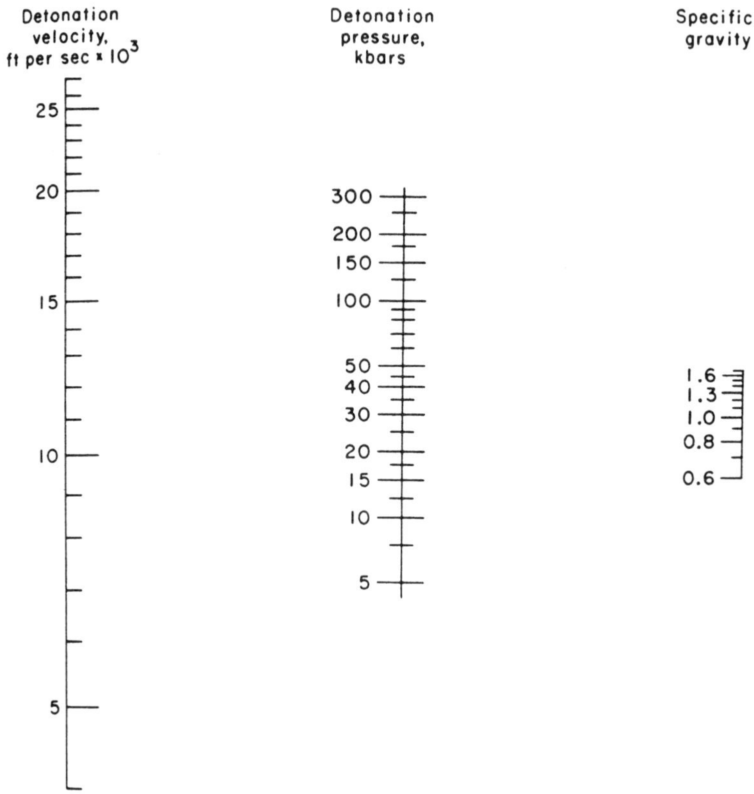

Figure 68. Nomograph for finding detonation pressure (After Dick, 1973).

A blasting agent is a mixture whose individual components are not classified as explosives and contains a fuel as an oxidizer. Ammonium nitrate fuel oil, though widely used, has only been commercially available since the early 1950's. It was discovered when detonation was observed after mixing ammonium nitrate prills with a certain proportion of a hydrocarbon fuel (typically, 3.78 x 10^3 cu m (4 qt) of oil to each 45.36 Kg (100 lbs.) ammonium nitrate). This led to the adoption of ANFO as a blasting agent with 60 percent of the effectiveness of dynamite with a cost, presently, of approximately $0.22 per Kg. Even though it represents an added cost, aluminum chips (5 to 30 percent by volume) are sometimes added to the mixture to increase explosive output. Ammonium nitrate fuel oil usually comes to the mine site in one of four manners:

1. As separate ingredients in bulk form for on-site mixing.
2. Premixed in bulk.
3. In paper or polyethylene bags.
4. In cartridges.

Because ANFO is usually mixed on site, there are numerous parameters which must be understood to maximize detonation velocities. Condon and Snodgrass (1974) list them as follows: fuel oil content, confinement, blasthole diameter, density, prill particle sizing, water content, critical diameter, borehole coupling (the percentage of the borehole diameter which is filled with explosives), temperature, and priming system. They examined the impact of primer type and borehole diameter on detonation velocity at two large surface mines in Indiana, since these variables are the easiest to control by mine operators. They concluded:

1. Larger diameter boreholes provide for you more efficient detonation of ANFO and greater energy per pound of powder.
2. The same steady-state velocities were achieved with different types of primers.

Although the application of ANFO is relatively safe, there are a number of precautions which must be taken into account (Coal Age, 1972):

1. Ammonium nitrate and fuel oil should be mixed in special plants under closely controlled conditions.
2. Locate mixing plants in an isolated area and build them with fireproof materials.
3. Mixed material should be removed from the plant as quickly as possible.
4. No welding or cutting should be done in the plant until cleared of explosive materials.
5. Primers, detonating cord, and caps should be transported in separate vehicles from mixed nitrate and oil. They should be brought together only at the point of use.

Design of Blasting Rounds

Figures 69 and 70 show some of the characteristics and parameters used in the design of blasting rounds.

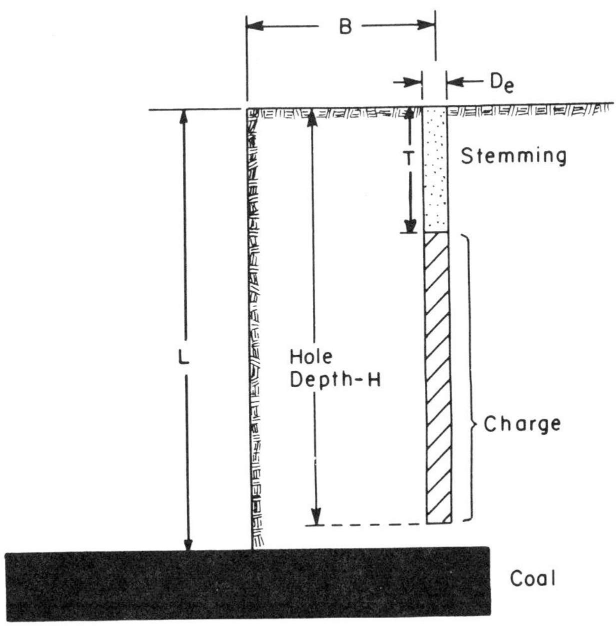

Figure 69. Bench cross section view showing D_e, B, H, T, and L (After Pugliese, 1972).

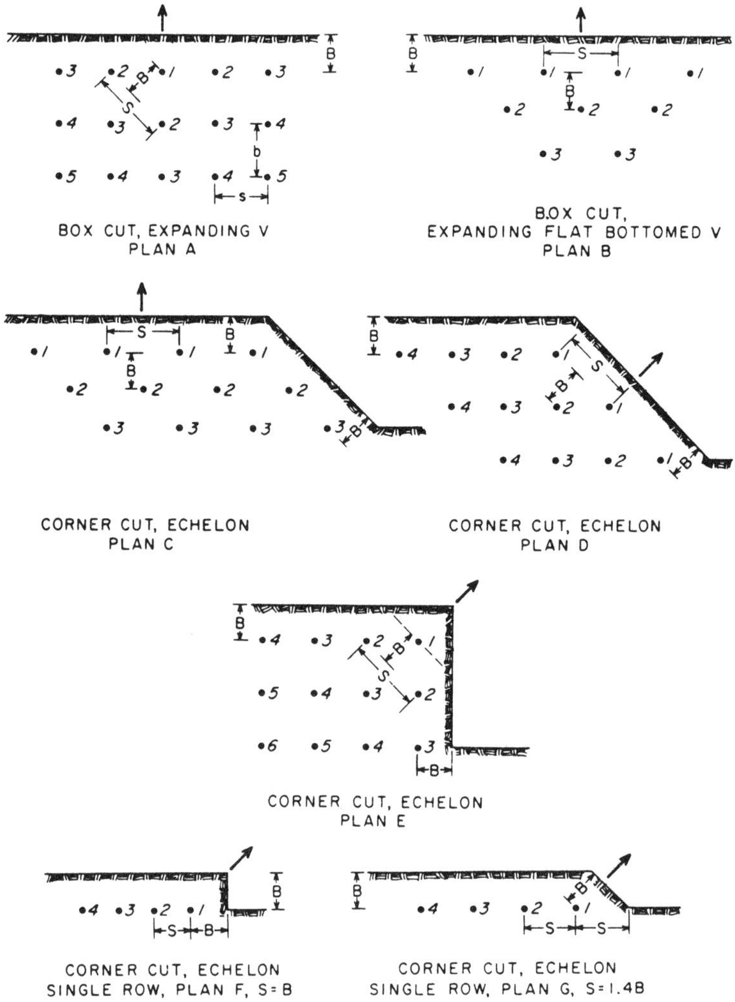

Figure 70. Generalized blasting patterns showing B, S, b, and s. (Numbers indicate firing sequence) (After Pugliese, 1972).

Figure 69 shows those dimensions which are considered for a single hole, and Figure 70 for a total round including:

- D_e = Diameter of the explosive in the borehole
- B = Burden, or distance from the charge to the nearest free face
- S = Spacing, or distance between two holes
- H = Length of hole
- J = Subdrilling length, or the length the hole is drilled below the floor of the working pit
- T = The portion of the hole that does not contain explosive, or collar length
- L = Bench height
- b = Spacing between rows of holes as measured perpendicular to the free face
- s = Separation between adjacent holes in a given row

These abbreviations are used by both Pugliese (1972) and Ash (1970). As both authors point out, b and s are terms used by operators in referring to the drilling grid, and should not be confused with B and S, the burden and spacing which are measured with reference to free faces.

Ash (1963) and Pugliese (1972) concur in the suggestion of five ratios for use in the design of rounds. Assuming that all measurements of length and diameter are expressed in the same units (cm, m, ft, or in.), the ratios relate the dimensions to be designed through the following set of equations:

$$B = K_B D_e \; ; \; K_B = \text{Burden ratio} \tag{1}$$

$$S = K_S B \; ; \; K_S = \text{Spacing ratio} \tag{2}$$

$$H = K_H B \; ; \; K_H = \text{Hole length ratio} \tag{3}$$

$$J = K_J B \; ; \; K_J = \text{Subdrilling ratio} \tag{4}$$

$$T = K_T B \; ; \; K_T = \text{Collar distance ratio} \tag{5}$$

From these relationships, all of the critical dimensions of blasting round design can be calculated if appropriate values for the various K ratios are selected. The work of Pugliese and Ash is primarily based on studies in limestone quarrying. While limestone quarrying is conceded as different from overburden removal in surface mining of coal, the relationships still hold and the K values which they have developed and assumed are valid for first assumptions. As is the case with any empirically based design criteria, empirical relationships from literature are usually accepted as a first approximation only, and succeedingly more exact relationships can be developed as actual practice of these design steps is followed.

For calculation of the burden, given a particular D_e, both authors found the following ratios appropriate for limestone and dolomite:

K_B = 30, under average conditions

K_B = 25, for low-density explosives such as ammonium nitrate fuel oil (ANFO)

K_B = 35, for high density slurry explosives and gelatins

The actual K_B ratios are related also to the density of the rock. Limestone and dolomite have a density of approximately 2.7 grams per cubic centimeter. Lower values of K_B should be used for rocks of greater density than 2.7 (e.g., 2.85 and up) or for lower density (e.g., 2.55 and less).

The calculation of spacing is based on K_S values between 1.8 and 2 for simultaneously initiated holes in a given row. Most favorable conditions are with staggered drill patterns and for rows in which all charges are not detonated simultaneously. If exact timing can be achieved, simultaneously fired holes can be designed using K_S values of 3 to 5. These charges must be sufficiently long and exactly timed to allow the proper enhancement of stress effects.

Spacing is also determined with respect to height and burden relationships, specifically the ratio H/B. Ash (1970) has shown that the relationships

$$S \approx (B\ H)^{1/2} \qquad \text{for } 2B \leqq H < 4B$$

$$S \approx 2B \qquad \text{for } 4B \leqq H$$

will be sufficient. In this situation, the K_S of the 1.8 to 2 is satisfactory, although lower K values can be assigned with H/B less than 3.

If delays are sequenced in a given row, the K_S should range between 1 and 1.2. Square drilling patterns are preferred, and rock will expand in a wedge or in a 45 degree pattern to the original open face (as in Figure 70, Plan A). Considering the ratio of s and b, where b = 1.4B for the regular case, the design relationship

$$S = K_S\ b = K_S\ (1.4B)$$

is appropriate where K_S = 1 to 1.2.

The value of the ratio K_S may be adjusted between 1 and 2 to meet local conditions, particularly the period of delay between charges. Large ratios create spacing so much greater than the burden that the vertical face craters and lumps are left on the pit floor. Small values of K_S, less than 1, create excessive breakage, premature breakage between holes, and progressive loss (rarefaction) from the hole, creating slabs, boulders, and toe problems. The nature of these particular relationships is such that they assume ideal energy balance between charges, so further experimentation with the balance of individual holes may be indicated if such problems are not resolved by varying the K_S ratio.

Hole length calculations are generally based on K_H ratios of 1.5 to 4. Pugliese (1972) cites 2.6 as the typical K_H value. The depth should almost always be greater than or equal to the burden to reduce overbreak and cratering tendencies. Conversely, high values, that is, greater than 4, may create underbreak or bootlegging if the hole is single-primed. Staged priming may allow lengths greater than those indicated by K_H values of 4. No definite limit to K_H can be established without a definite examination of the cratering properties of the specific material being blasted.

The subdrilling length, which is applicable only to quarries and not to coal mines, is generally indicated by K_J values greater than 0.2. Pugliese (1972) suggests 0.3 as preferable to insure a full face and quarry floor. In the case of overburden drilling, however, care is to be taken to limit fracturing of the coal seam and subsequent unnecessary removal of coal with the overburden. In some instances, negative K_J values may be applied. Common practice is to drill to the coal and then backfill by 1 to 1.3 m (3 to 4 feet).

The collar distance can be calculated by using a K_T value of 0.7. The uncharged portion of the hole is usually filled with stemming material and is left uncharged to prevent airblast and promote gas confinement within the borehole. In solid rock, Pugliese (1972) suggests that values of K_T less than 1 can lead to cratering, backbreak, and violence, particularly with collar priming.

The various calculations are presented in worksheet form in Table 24. Once the original face and the shape of the cut are selected, the geometrical relationships necessary to work with can be determined from the plan layouts of the cuts. For the given explosive type, blasthole diameter and rock density, the burden is calculated from equation 1. Equation 2 is then solved for the spacing. The hole length can then be checked according to equation 3 and the timing, delay, and sequencing constraints which the operator has established. Finally, the subdrilling length (equation 4), if necessary, can be calculated and the collar distance determined (equation 5).

TABLE 24. WORKSHEET FOR DESIGN OF BLASTING ROUNDS (AFTER PUGLIESE, 1972)

D_e = __________ in; Cut __________; Desired rock progression __________

$B = K_B D_e/12$ = __________ ft for $K_B \simeq 30$ (average)

$B = K_B D_e/12$ = __________ ft for $K_B \simeq$ ___ (alternative) (Eq. 1)

For staggered pattern, simultaneous timing

$S = K_S B$ = __________ ft for $K_S \simeq 2$ (average)

$S = K_S B$ = __________ ft for $K_S \simeq 1.8$ (alternative) (Eq. 2)

For square pattern, sequence timing

$s = K_S\, b = K_S\,(1.4\ B)$ = __________ ft for $K_S \simeq 1$ (average)

$s = K_S\, b = K_S\,(1.4\ B)$ = __________ ft for $K_S \simeq 1.2$ (alternative)

$H = K_H B$ = __________ ft for $K_H \simeq 2.6$ (average)

$H = K_H B$ = __________ ft for $K_H \simeq$ ___ (alternative) (Eq. 3)

$H_{Min} = K_H B$ = __________ ft for $K_H \simeq 1.5$ (minimum)

$H_{Max} = K_H B$ = __________ ft for $K_H \simeq 4$ (maximum)

$J = K_J B$ = __________ ft for $K_J \simeq 0.3$ (average)

$J = K_J B$ = __________ ft for $K_J \simeq$ ___ (alternative) (Eq. 4)

$T = K_T B$ = __________ ft for $K_T \simeq 0.7$ (average)

$T = K_T B$ = __________ ft for $K_T \simeq$ ___ (alternative) (Eq. 5)

$L \simeq H - J$ = __________ ft (average or alternative)

$L_{Min} \simeq H_{Min} - J$ = __________ ft (minimum)

$L_{Max} \simeq H_{Max} - J$ = __________ ft (maximum)

Blasting Pattern Problems

In the previous section, the design of blasting rounds was viewed from the standpoint of optimum fragmentation of solid rock. However, from a practical standpoint, there are many in situ properties of the overburden and the surrounding environment which also have a great impact on proper design. The major problems which result from these conditions are ineffective fragmentation, vibration, and noise.

Jointed ground complicates the blasting pattern layout. This is due to the fact that the joints act as avenues for the escape of explosive energy. This weakening of the explosive's effectiveness is costly and requires a design to limit the amount of escape. Since blasted material fragments along its structural pattern, it is natural to design the pattern to incorporate this tendency. For example, Figure 71 shows a typical condition in a sedimentary deposit. A rhombic structural pattern is shown with jointing angles at 75 degrees and 105 degrees. If the direction of the blast is designed to be out from the lesser angle (i.e., either N or S), overbreak, coarse fragmentation, and excessive ground vibrations are usually experienced. However, if the blast is designed to be out from the open angle (i.e., W), better fragmentation, a uniform particle size, and a straighter face will result. Therefore, in jointed ground, blasts should be directed out of the open angles, wherever possible.

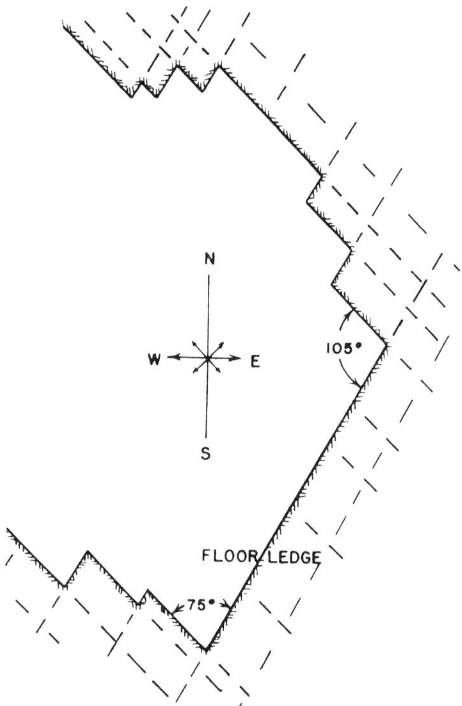

Figure 71. Representative sedimentary vertical jointing system (After Ash, 1968).

Ground vibration is an important consideration because of public demands. Damage to surface structures, which is visible evidence of blasting vibration effects, requires constant monitoring of blasting practices. With this in mind, Andrews (1975) developed the following timing guidelines:

1. The minimum delay between holes or corresponding decks in adjacent holes should be 1 millisecond per ft (ms/ft) of burden.
2. The minimum delay between rows should be greater than the average delay between holes in the same row, to allow sufficient forward and lateral relief for every hole.

Figure 72 portrays the results which led to the first recommendation. In this study (Bergmann, et al, 1974), a low delay ratio produced a large fragment size. However, all the curves tended to flatten out as the delay approached 1 ms/ft of burden. This implies that a greater delay than 1 ms/ft of burden produces little change in fragmentation. The second guideline was established by the comparison shown in Figure 73. If the delay time between rows is equal to the time between holes in the same row (Figure 73), high lateral confinement and poor fragmentation results. In Figure 73, however, a delay time between rows that is twice that between holes in the same row produces a jagged internal free face. This provides both forward and lateral relief for better breakage and reduced ground vibration.

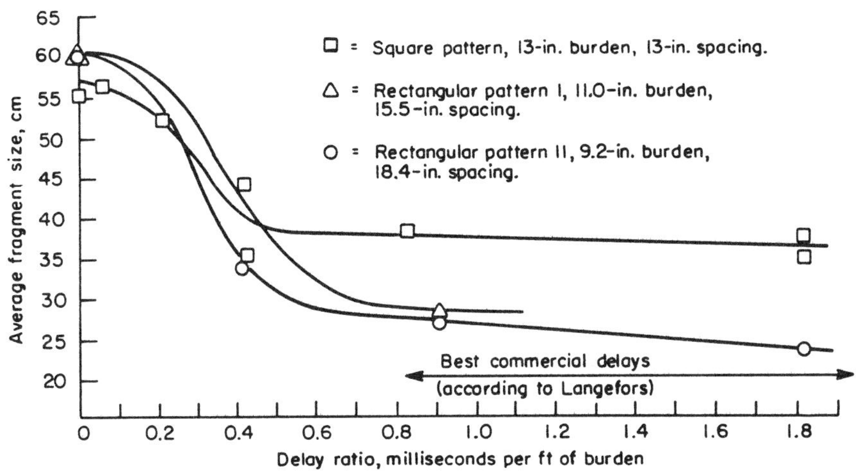

Figure 72. Effect of delay time between shotholes on average fragment size. The same powder factor was used for all shots (After Andrews, 1975).

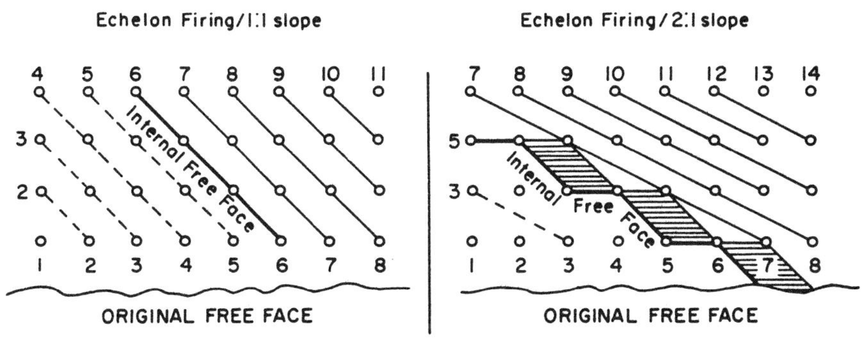

Figure 73. Comparison of multiple row patterns using echelon firing on 1:1 and 2:1 slopes (After Andrews, 1975).

Since any excessive energy from a detonation travels outward through ground, the intensity of the ground vibration is also a function of the distance traveled by the shock wave (i.e., there is a decrease in intensity with an increase in distance). Studies in this area (Nicholls, et al, 1971) reveal that damage to structures by these seismic waves is caused by particle velocity, rather than the displacement or acceleration, of the ground under a structure. Particle velocity is the speed at which the individual unit particles of the ground move as the shock wave is propagated. Figures 74 and 75 show some of the results of the aforementioned studies. Figure 74 shows data for the conclusion reached by the U.S. Bureau of Mines that major damage occurred above a velocity of 19.30 cm/sec (7.6 inch/sec), while minor damage occurred above 13.71 cm/sec (5.4 inch/sec). The Bureau then separated its findings into damage and safe zones, with 5.08 cm/sec (2.0 in/sec) set as the maximum value for the safe zone. Ninety-four percent of the minor damage occurred above this figure, which proved to be statistically acceptable. The value of 5.08 cm/sec (2.0 in/sec) was to be monitored in all three of the orthogonal components of the ground next to the structure.

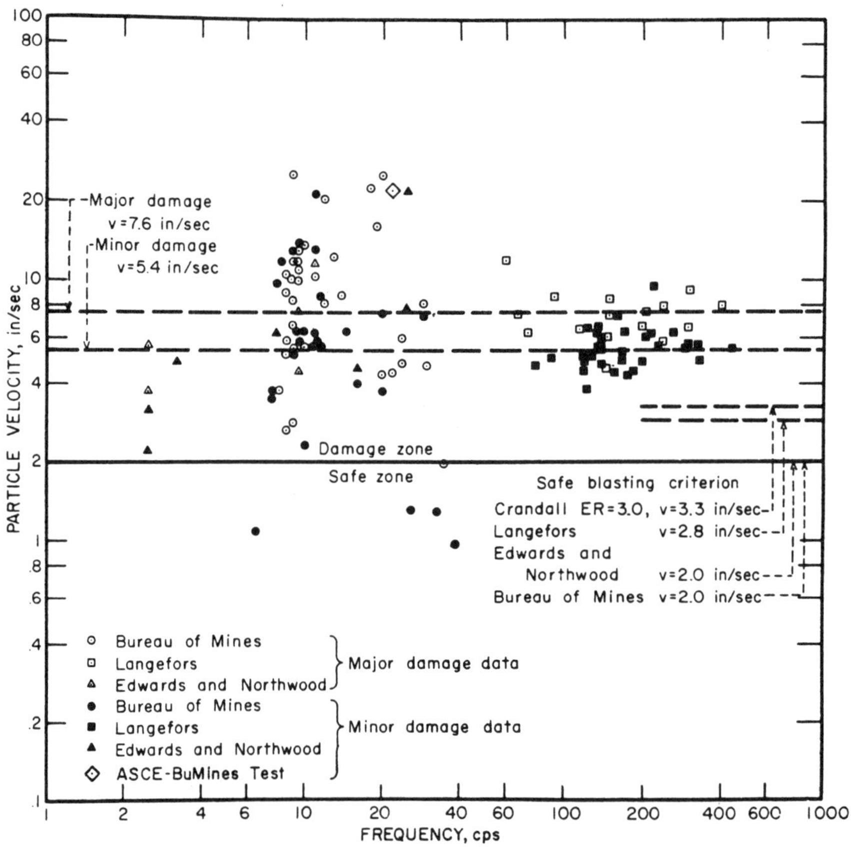

Figure 74. Particle velocity versus frequency with recommended safe blasting criterion (After Nicholls, et al., 1971).

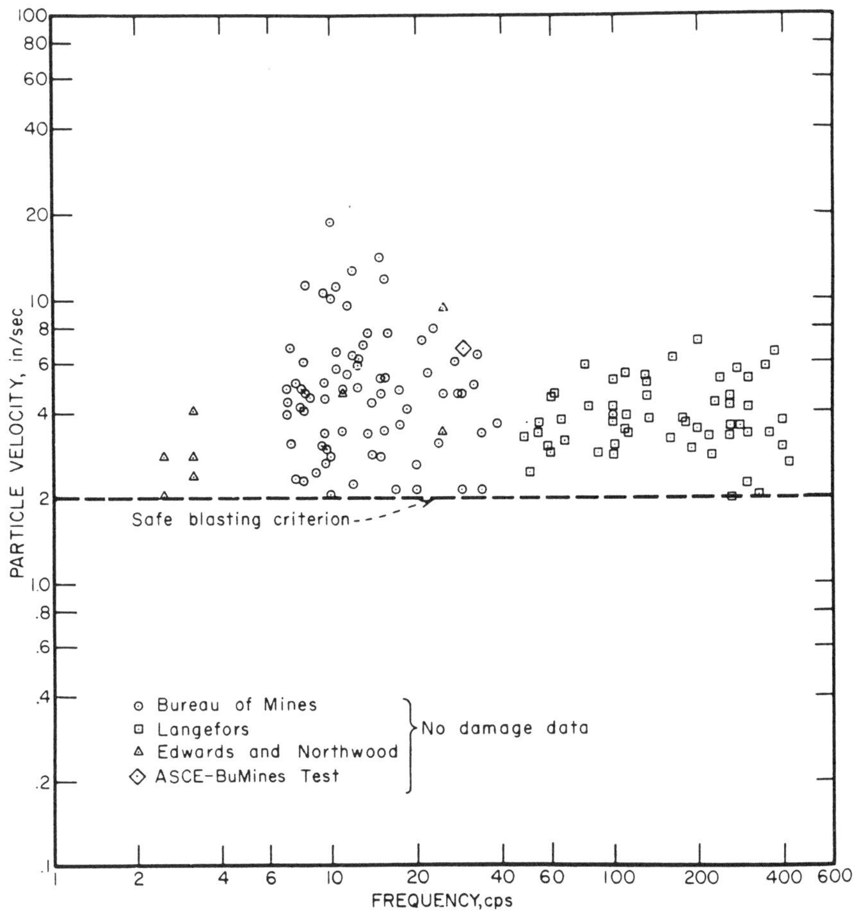

Figure 75. Particle velocity versus frequency for no damage data (After Nicholls, et al., 1971).

Figure 75 shows, however, that the value is not truly predictive insofar as damage is concerned. It shows that, in some instances, no damage was observed where the particle velocity was greatly in excess of the safe blasting criterion. However, the U.S. Bureau of Mines (USBM) merely feels that damage occurrence increases or decreases as the particle velocity increases or decreases with respect to the value of 5.08 cm (2.0 inch). Therefore, a safe blasting limit of 5.08 cm (2.0 inches) per second for the particle velocity, as measured from any of three mutually perpendicular directions in the ground next to the structure in question, has been used (Grim and Hill, 1974). The maximum peak particle velocity is under reevaluation (see note on Table 25).

For the control of ground vibrations from blasting, a simple formula has been devised relating the blasting charge and proximity to structures (Grim and Hill, 1974). It is given by the equation: $W = (D/50)^2$. In this relationship, W equals the weight of explosives (in pounds) which are detonated at any one instant, and D equals the distance (in feet) from the nearest structure. Instantaneous detonation, in this case, is considered to be all detonation within 8 milliseconds of each other. According to the Federal standards, the formula is revised as $W = (D/60)^2$. The effect of this revision is to reduce the weight of explosive that can be detonated for the same distance, (D). Table 25 and Figure 76 summarize the application of this means of vibration control. Other recommended or practiced blasting patterns are provided in Figures 77 and 78 (McGraw Hill Inc., 1970).

TABLE 25. MAXIMUM EXPLOSIVE CHARGES USING SCALED DISTANCE FORMULA

Distance to nearest residence building or other structure (ft)	Maximum explosive charge (to be detonated) (lb)
300	25
350	34
400	44
500	69
600	100
700	136
800	178
900	225
1,000	278
1,100	336
1,200	400
1,300	469
1,400	544
1,500	625
1,600	711
1,700	803
1,800	900
1,900	1,002
2,000	1,111
2,500	1,736
3,000	2,500
3,500	3,403
4,000	4,444
4,500	5,625
5,000	6,944

Note: In all blasting operations, except as otherwise authorized, the maximum peak particle velocity shall not exceed 1 inch per second at the location of any dwelling, public building, school, church, or commercial or institutional building. Peak particle velocities shall be recorded in 3 mutually perpendicular directions. The maximum peak particle velocity shall be the largest of any of the three measurements. The maximum peak particle velocity allowed may be reduced if it is determined that a lower standard is required because of density of population or land use, age or type of structure, geology or hydrology of the area, frequency of blasts, or other factors.

The reduction in noise is beneficial, not because noise creates property damage but because its reduction lessens the social problems faced by operators. Most complaints received by surface mine operators start, initially, because of noise. Although vibrations cause visible physical damage, noise has more of a psychological effect on the nearby population. Since that effect is so difficult to predict, measures must be taken to control the situation as

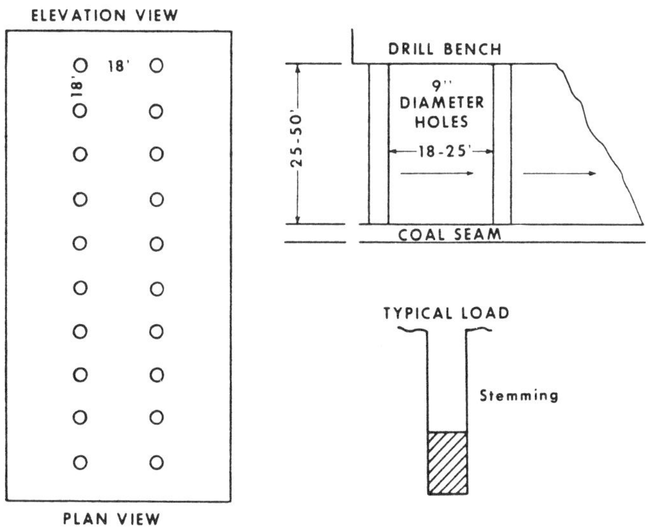

Distance to nearest dwelling = approximately 5000 feet.
Amount of explosive per hole = 340 lb.

Figure 76. Proposed blasting plan (example) (After Grim and Hill, 1974).

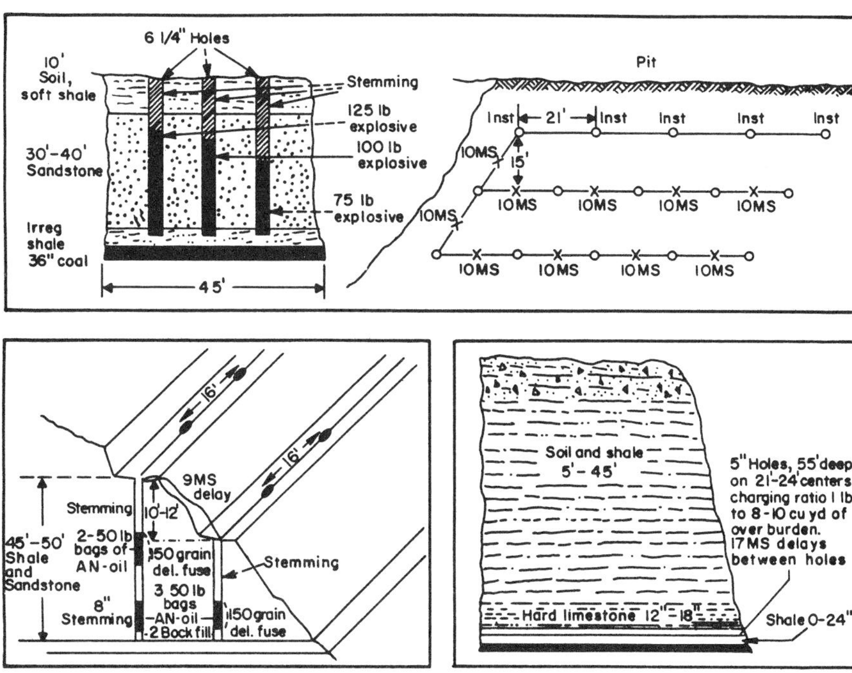

Figure 77. Some commonly used drilling and blasting patterns (McGraw-Hill, Inc., 1970).

best as possible. As far as noise control is concerned, solutions are generally twofold: 1) attention to blasting techniques and 2) proper evaluation of weather conditions.

Andrews (1975) concluded that, to minimize air blast intensity, the average rate of blast progression along the free face should be less than the velocity of sound in air. If t is the delay time between holes, the ratio d/t, when compared to c, is of extreme importance. When the ratio d/t is less than c, separate shocks radiate in all directions. The resulting air blast behind the face is, therefore, normally weaker and less noisy. When the ratio is equal to or greater than c, a strong air blast is produced due to super-position.

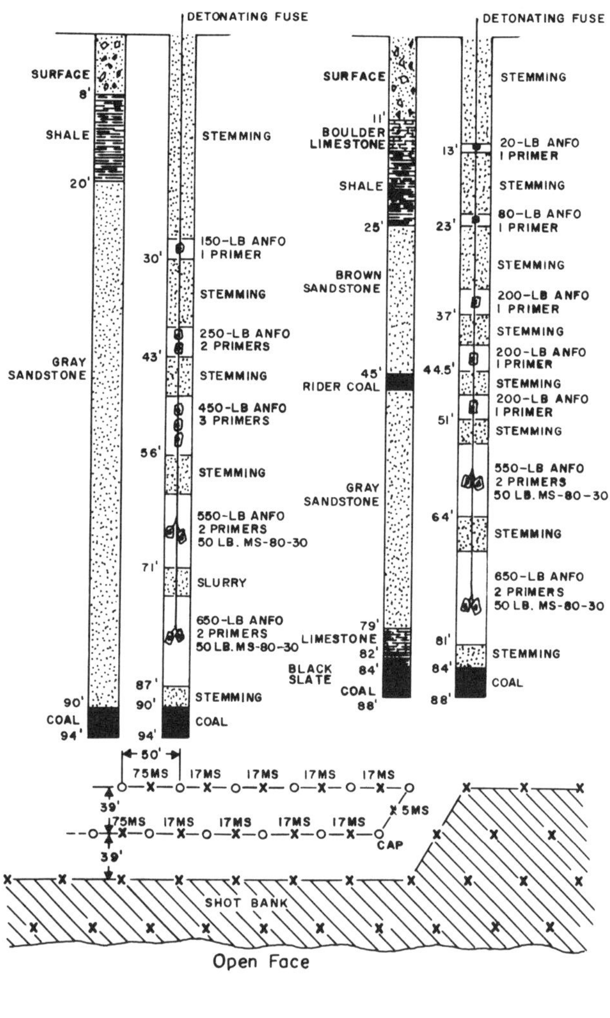

Figure 78. Metalized slurry is used to improve blasting results with ANFO (McGraw-Hill, Inc., 1970).

On-site noise control can also be practiced with detonating cord. When cord is used for trunk lines, covering it with 25.40 cm (10 in) of dirt can aid in noise abatement (McGraw Hill Inc., 1970). Further, a new form of low-energy detonating cord has been developed. One-hundred-fifty feet of this cord produces no more noise than 5.08 cm (2 in) of conventional cord (McGraw Hill Inc., 1970).

Weather can also affect noise propagation. Blasting should, when possible, be avoided on foggy, hazy, or smoky days as these conditions are prevalent during temperature inversions. Sound propagation is increased due to these inversions. The following weather conditions, when noted by mine operators, indicate unfavorable conditions for noise control (McGraw Hill Inc., 1970):

1. Relatively high atmospheric pressure that remains static more than 25 hours.
2. Wide daily temperature variations.
3. Poor visibility and light winds early in the morning, followed by cloudiness.
4. Clear, somewhat hazy days with little wind and fairly constant temperatures.

SUMMARY

This chapter has briefly reviewed the fragmentation practices in surface coal mining. Vertical rotary blast hole drills are the most common for overburden drilling. Ammonium Nitrate Fuel Oil is the explosive that is widely used. It is common in area mining for the drilling and blasting unit operation to be in advance of stripping by one or more cuts. In contour mining, where small loaders load onto trucks, the size of the blasted overburden becomes important. Therefore, the powder factor in contour operations is very low (varying from 1.68 cu m per Kg (1 cu yd per lb) of ANFO to 4.21 cu m per Kg (2.5 cu yd per lb) of ANFO). Important environmental considerations on which blasting designs must be based are noise, damage to surface structures, and protection from fly rocks.

Surface Mine Haulroad Design

INTRODUCTION

For the purposes of this report, coal mine haulroads are defined as roads constructed and/or used by a mine operator, which end at the pit or the bench. These roads may cover large areas of ground, and, according to a U.S. Department of the Interior report (1967), constitute approximately 10 percent of the total area directly disturbed by the surface mining operation.

According to the Surface Coal Mining and Reclamation Operations, Permanent Regulatory Program (Public Law 95-87, Title 30, Chapter VII, Subchapter A), road means a surface right-of-way for purposes of travel by land vehicles used in coal exploration or surface coal mining and reclamation operations.

A road consists of the entire area within the right-of-way, including the roadbed, shoulders, parking and side area, approaches, structures, ditches, surface, and such contiguous appendages as are necessary for the total structure. The term includes access and haulroads constructed, used, reconstructed, improved, or maintained for use in coal exploration or surface coal mining and reclamation operations, including use by coal-hauling vehicles leading to transfer, processing, or storage areas. The term does not include pioneer or construction roadways used for part of the road construction procedure and promptly replaced by a Class I, Class II, or Class III road located in the identical right-of-way as the pioneer or construction roadway.

The term also excludes any roadway within the immediate mining pit area.

(a) Class I Road means a road that is utilized for transportation of coal.

(b) Class II Road means any road, other than a Class I Road, planned to be used over a 6-month period or longer.

(c) Class III Road means any road, other than a Class I Road, planned to be used over a period of less than 6 months.

Subchapter K of the above regulatory program, specifically sections 816.150 to 816.175, details the general considerations, location, design and construction, drainage, surfacing, maintenance and restoration of the Class I, II and III roads. Movement of coal, equipment and personnel within the mine plan area may require, other than roads, railroad loops, spurs, sidings, surface conveyor systems, and other transportation facilities. General standards are set forth in section 816.810 to ensure the minimization of the adverse effects to hydrology, fish and wildlife and their habitats, and public and private property as a result of the design, construction and utilization of these transportation facilities. It is not the purpose here to detail all the provisions of the regulations but a brief summary of the important requirements for Class I roads will be provided.

CLASS I ROADS

General

Class I roads should be located, insofar as possible, on ridges or on the most stable available slopes to minimize erosion. No part of this road may be located in the channel of an intermittent or perennial stream, and excepting for approved stream fords, all stream crossings shall be made using bridges, culverts or other structures which are designed, constructed and maintained as per the provisions of the law. All Class I roads shall be removed and the land affected regraded and revegetated unless the retention of the road is approved as part of the post mining land use or as being necessary to control erosion.

Design and Construction

The design and construction of Class I roads shall be approved by a registered qualified professional engineer and shall incorporate the demand for mobility and travel efficiency, based on geometric criteria (vertical alignment, horizontal alignment, road cuts, and road embankments) appropriate for the volume of traffic and the weight and speed of vehicles that will be used for coal haulage. The overall grade shall not exceed 10 percent, and the maximum pitch grade shall not exceed 15 percent. Considerable importance has been attached to ensuring that adequate drainage facilities are included in the design and construction of these roads. Specifically, ditches, culverts, and bridges must be provided so as to minimize 1) the erosion from the roadway, 2) the siltation of adjacent streams, 3) the adverse impact on fish and wildlife habitat and 4) the instability of the roadway to protect the welfare of the public. In general, the water control system shall be designed to safely pass the peak run-off from a 10-year, 24-hour precipitation event or a greater event if required by the regulatory authority.

Maintenance

Class I roads should be maintained so as to meet the approved design standards at all times. Maintenance will include repairs to the road surface, minor reconstruction of road segments as necessary, brush removal and revegetation. Roads that have been damaged by floods or earthquakes will not be used until the damaged areas have been reconstructed.

OTHER TRANSPORTATION FACILITIES

These facilities shall be designed, constructed, and maintained, and the area restored to 1) prevent, to the extent possible, damage to fish, wildlife, and other environmental values, 2) prevent contribution of suspended solids into streamflow or runoff outside the permit area, 3) control water quality and quantity degradation, 4) minimize erosion, siltation, and air pollution, and 5) prevent damage to public and private property.

This chapter will briefly describe the steps necessary for the design of haulroads, since a well-designed haulroad will minimize the problems enunciated before.

SURFACE MINE HAULROAD

The distinguishing feature that sets a mine haulroad apart from any other road is the duration of its usefulness. In general, most roads are designed with a specific lifetime in mind or with perpetual existence assumed. Haulroads may be in use for a period of days to many years, depending upon the mine plan and other related factors. Therefore, it is important to know the estimated time a proposed haulroad will be in use. The use of very expensive surfacing material on a short-duration haulroad would not be justified if the road could be maintained with an additional road grader for less total cost. In fact, in most surface coal mines, haulroads are maintained in mined-out pits and through spoil banks for considerable distances before the permanent section of the haulroad is reached. For this reason, the total mine plan should be considered when planning haulroads.

HAULROAD DESIGN

The design principles used for haulroads differ little from that for any other road; however, the roads for many large surface mines today have to be designed to carry loads far in excess of the loads found on such highways as the interstate system. With the introduction of larger equipment, such as off-highway trucks, there presently seems to be no weight limit with trucks weighing 355.6×10^3 Kg (350 tons) now in existence. Unlike the interstate highway system which limits the loads on its roads, haulroads may be subjected to larger loads simply because the mine operator is forced to purchase larger equipment in order to remain competitive. The ability to increase the bearing capacity of hualroads with minimum cost is a design feature that warrants some thought during the initial design stages.

Often, the road-building material is a mine product, crushed overburden and/or bottom ash. Therefore, the construction of subbase should be on the basis of the results of the soil survey and the design of traffic loads. In area mining, since production from a mining site lasts for several years, the topography is generally gentle, the hauling equipment used is usually large compared to that used in contour mines, and the roads are usually better designed. In contour mines, the roads are long and usually have to be laid on severe grades. Water seepage from the hillsides adds to the maintenance problem. It is also not uncommon for haulroads to be maintained through spoil piles and old pits. Even without these, the fugitive dust problem with vehicular traffic demands that the roads be wetted and maintained in good condition to keep ahead of grading and surfacing problems. Regular checking of ditches and culverts for drainage and incorporation of $CaCl_2$ or NaCl into the road surface will add to the mechanical stability of the road surface. Adequate facilities for watering and water trucks must be available in the mines.

When a haulroad is abandoned, steps must be taken to minimize erosion and establish a vegetative cover. For complete abandonment, culverts and other structures are removed, and the natural drainage pattern is restored. Side ditches should be obliterated, and properly spaced grade dips or water bars should be constructed to handle roadway cross drainage. A water bar must be placed at the head of all pitch grades, regardless of spacing. All road surfaces must be ripped, treated with soil amendments, seeded with grasses, legumes, and trees, and mulched. Seeding will help stabilize the abandoned roads, provide food for wildlife, and improve the aesthetics.

ALIGNMENT

In general terms, alignment refers to the design of the geometric elements of the haulroad. Such things as horizontal and vertical curves, site and stopping distance, superelevation, and maximum grade all come under this general term. Any standard text on highway construction can be referred to for more information (Meyer, 1969). Specifically, with regard to surface mines, a recent study by Skelly and Loy (1976) and the U.S. Bureau of Mines Information Circular 8758 (1977) are useful references. Murray (1978) provides a computer program to calculate the best alignment of a road (i.e., road grade) for a given topographical profile, taking into consideration the truck capacity and speed-rimpull characteristics.

Superelevation refers to the sloping of the road toward the center of the curve. Superelevation must be provided on curves to reduce the side friction between the tires and the road. It is recommended that the superelevation should never be less than 0.20 meter per meter because most road surfaces are crowned that much to provide drainage of the road from side to side. Because of the danger of sliding off the inside of a steeply superelevated highway when traveling slower than the design speed, the superelevations in Table 26 are recommended.

TABLE 26. RELATIONSHIP BETWEEN SPEED AND RADIUS OF A CURVE AND THE SUPERELEVATION RECOMMENDED

	SPEED/MPH							
RADIUS/FT	10	15	20	25	30	35	40	50
50	0.04	0.04						
100	0.04	0.04	0.04					
150	0.04	0.04	0.04	0.05				
250	0.04	0.04	0.04	0.04	0.06			
300	0.04	0.04	0.04	0.04	0.05	0.06		
600	0.04	0.04	0.04	0.04	0.04	0.05	0.06	
1000	0.04	0.04	0.04	0.04	0.04	0.04	0.05	0.05

Horizontal curves should be well laid out. When horizontal curves are excessively sharp, the road surface should be widened to account for the added width of passing vehicles (Figure 79).

It would be impractical to build a haulroad with an abrupt change in grade. When negotiating changing grades (i.e., vertical curves), one must consider two other aspects of vertical design, namely sight distance and stopping distance. It is recommended that a vertical curve should never be less than 30.48 meter (100 feet) in length.

By keeping haulroads at their minimum slopes, cycle times will be increased and wear and tear on the equipment will be reduced. Although it would be impractical to design haulroads so all vehicles could operate at their maximum speed at all times, some compromise between percent grade and cycle time can be reached.

The stopping distance (SD) for various size vehicles can be computed using the following formula:

$$SD = [1/2\ gt^2 \cdot \text{Sin}\ \Theta + V_o t] + [\frac{(gt\ \text{Sin}\ \Theta + V_o)^2}{2g(U\ MIN - SIN\ \Theta)}$$

where

SD = Stopping distance in meter.
g = Acceleration of gravity, 9.81 m/sec^2 (32.2 ft/sec^2)

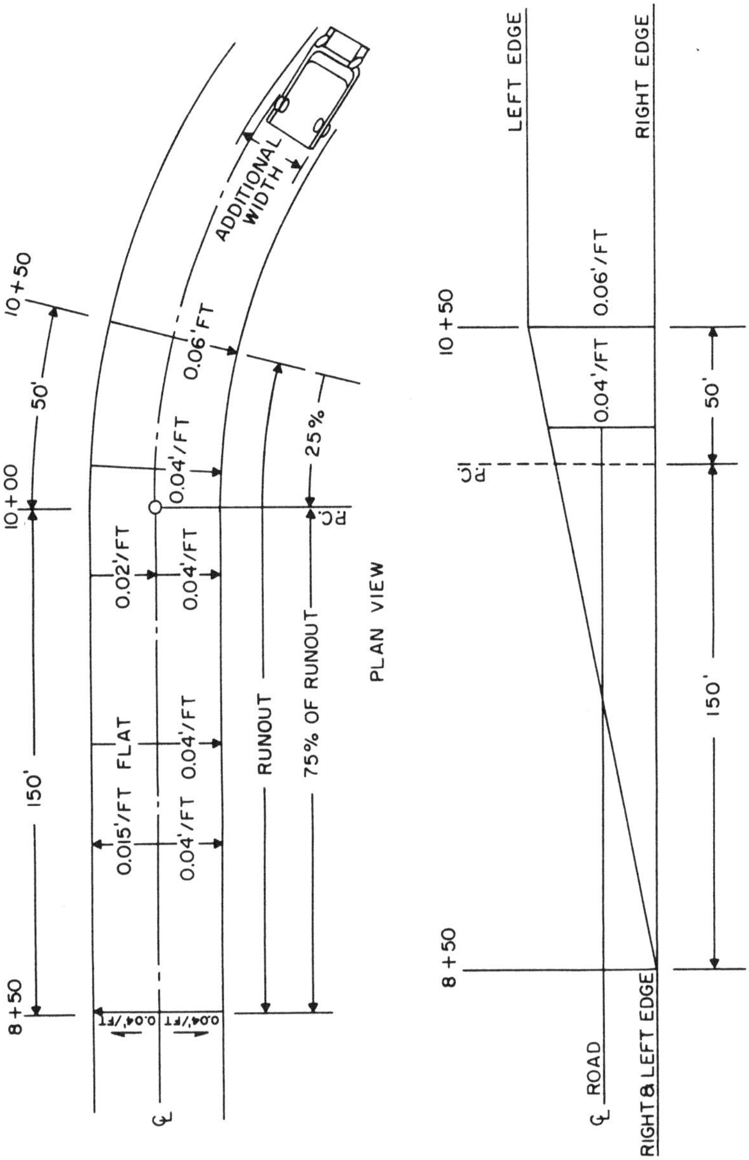

Figure 79. Superelevation runout considerations.

t = Reaction time of the driver
Θ = Angle of descent in degrees
V_o = Initial speed before breaking
U MIN = Coefficient of friction between break shoe and drum

Sight distance is the extent of peripheral area visible to the vehicle operator. As stated earlier, sight distance should never be less than stopping distance. The easiest and possibly the most used method of determining sight distance is to graphically construct the vertical curve, and then from the known stopping distance, see if the sight distance is adequate.

ROADWAY BEARING CAPACITY

Once the geometric parameters of the haulroad have been established and the route of the road known, it will be necessary to determine the load-bearing capacity of the various materials used in constructing the road.

Most soil materials are subject to three types of deformation, namely elastic, consolidation, and plastic (Figure 80). The elastic deformation will rebound to its original shape when the load is removed. The consolidation deformation will not rebound, but leave the loaded area denser than the unloaded area. Plastic deformation occurs when the water and air in the pores of the soil combine with loading forces to displace the roadway material (see Figure 80).

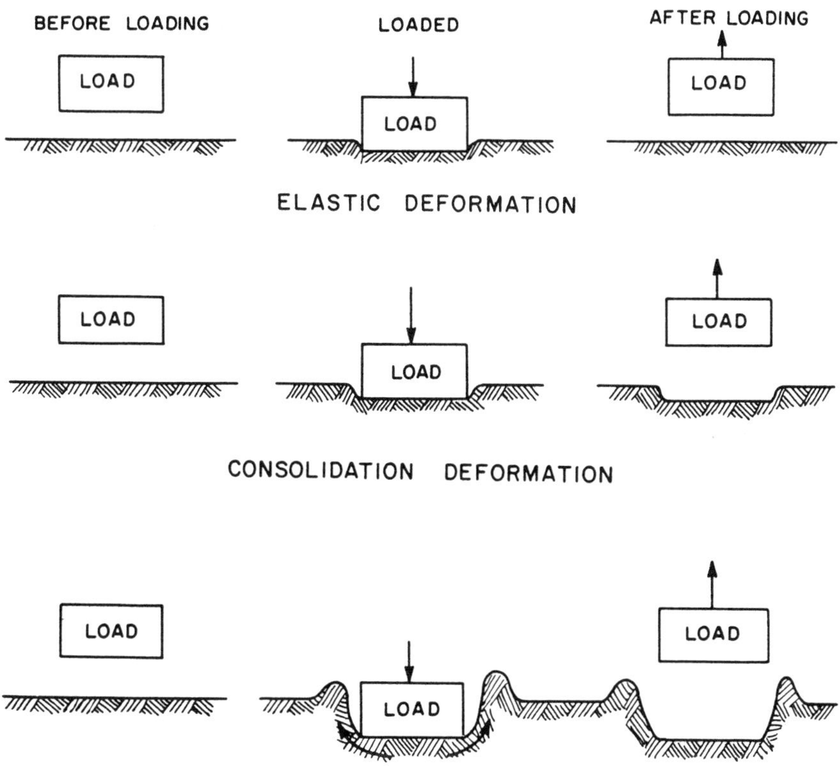

Figure 80. Plastic deformation.

PLASTIC DEFORMATION

Plastic deformation is the most damaging type of deformation to the haulroad surface, and is the deformation that the road must be designed to withstand.

The initial step in this design is to determine the loads that the road is expected to withstand. This is done by computing the wheel loads for the heaviest vehicle that will use the road. From the manufacturer's specifications, the axle load is divided by the number of tires on the axle to determine wheel load. If dual wheels are on the axle, the wheel load must be increased by 20 percent to account for the overlapping stresses (see Figure 81).

For example, suppose there is an axle loading of 45360 Kg (100,000 lbs) on the rear axle of a truck with dual wheels. It is known that the wheel load is 45360 ÷ 4 or 11340 Kg (25,000 lbs), but because of the dual wheels, one must add 20 percent, or in this case, 2236 Kg (5,000 lbs) to the wheel load to account for the stress overlap under the wheels. The road will now be designed to withstand wheel loads of 13608 Kg (30,000 lbs).

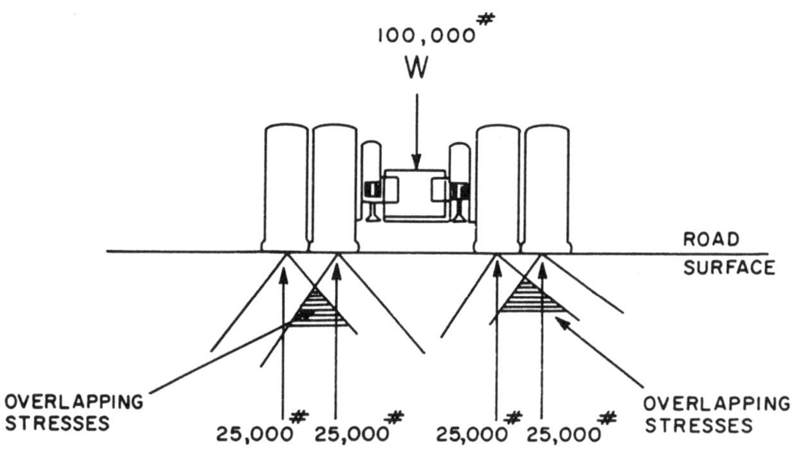

Figure 81. Dual wheels on an axle.

The wheel loading of today's off-highway vehicles is seldom less than 78122.7 Kg per square meter (16,000 lb per square foot), which is the approximate bearing capacity of soft rock. Table 27 shows the approximate bearing capacity of various materials encountered in haulroad construction. As indicated by Table 27 few materials can support large vehicles' wheel loads in their natural state. For this reason, designers must compute the amounts of additional materials that must be placed over the original surface material in order to provide the desired bearing capacity. One of the most widely used methods for achieving this is an empirical method developed by the California Department of Highways in 1942. This is the California Bearing Ratio (CBR) which is the ratio of the load carrying capacity of the test soil to that of crushed rock. The curves shown in Figure 82 were developed for designing road surfaces with known CBR. It should be emphasized that the CBR for any material can be determined by taking a sample of the material to a testing lab. If the CBR is to be assumed, some conservatism should be employed. The use of the chart, Figure 82, is as follows. If the wheel loading is 195306 Kg per square

TABLE 27. BEARING CAPACITIES OF MATERIALS

MATERIAL	BEARING CAPACITY LBS/FT^2
HARD SOUND ROCK	120,000
MEDIUM HARD ROCK	80,000
HARD PAN OVERLYING ROCK	24,000
COMPACT GRAVEL	20,000
SOFT ROCK	16,000
LOOSE AND SANDY GRAVEL	12,000
HARD DRY CLAY	10,000
LOOSE COARSE SAND	8,000
LOOSE FINE SAND	4,000

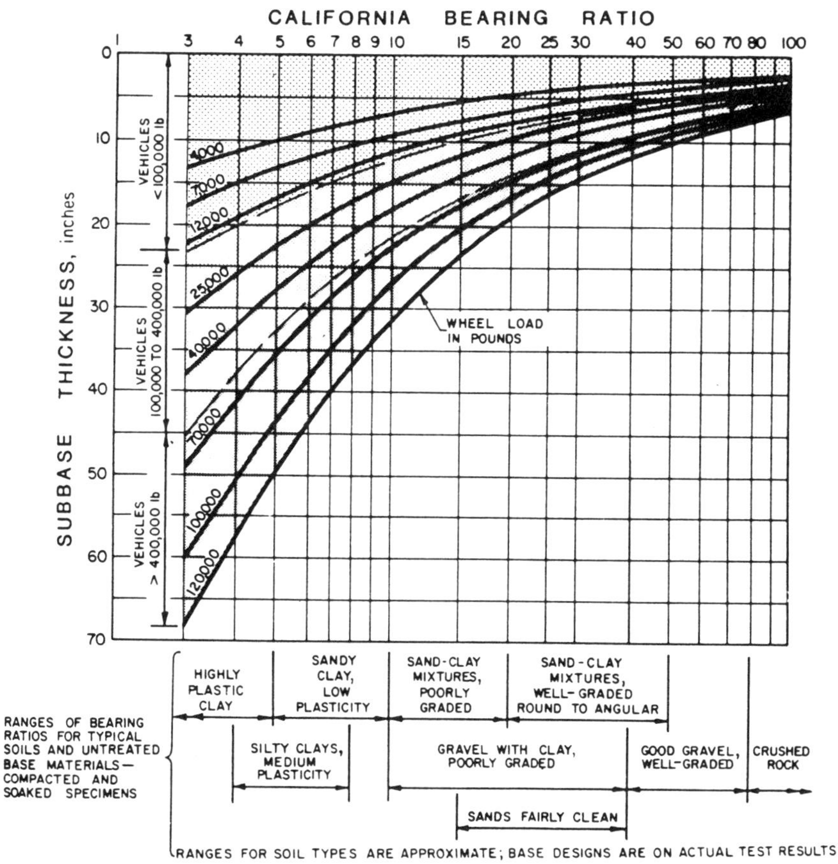

Figure 82. CBR curves.

meter (40,000 lb per square foot) and it is desired to build a road over silty clay with a CBR of 4, it can be seen from the chart that the road surface will be required to be 81 cm (32 in) above the original soil (see Figure 83). This will allow the 18144 Kg (40,000 lb) load which is distributed over possibly 0.0929 square meter (1 square foot) at the surface to be distributed over about 3.716 square meter (40 square ft) at a depth of 81 cm (32 inches). A gravel and clay mixture can be used as material with a CBR of 25. From the chart, it is clear that the top of the subgrade must be 25.4 cm (10 inches) from the surface. Crushed stone base with a CBR of 80 can be planned. Crushed stone must be 12.7 cm (5 inches) from the top, as can be seen from the chart.

This leaves 12.7 cm (5 inches) of material still not accounted for between the surfacing and the top of the base. This would normally be filled with a similar base material with a CBR of 80 or some paved surface such as blacktop or concrete.

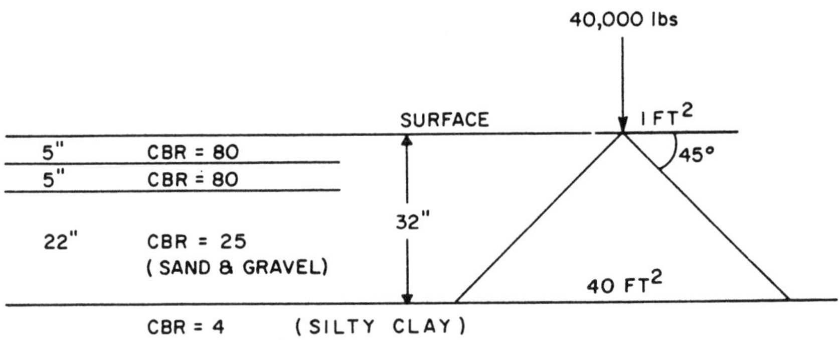

Figure 83. Haulroad materials and layer thicknesses.

Materials other than crushed stone for haulroad surfaces have been employed at some mines; however, the advantage of such surfaces is difficult to justify economically. When roads are paved, the surface must be protected when running tracked vehicles such as bulldozers on the road. Spillage is always a problem on haulroads and requires the use of a grader for cleaning, even when the road is paved. The grader is used to continually dress and smooth a crushed stone surface so the advantage of paving is unwarranted. The major advantages of surfacing are the reduction of dust during dry weather and the excellent drainage provided during wet weather.

EMBANKMENTS

When designing embankments upon which the load bearing subgrade and base materials are to be placed, it is necessary to understand some characteristics of the various soils to be used. There have been several methods of soil classification of which two are used extensively, namely the American Association of State Highway Officials' (AASHO) method shown in Table 28 and the Unified Soil Classification. The AASHO method seems to be more popular, possibly because it has been used for a longer period of time. There are two other characteristics of cohesive soils that play an important part in describing soils, liquid limit, and plastic limit. When a soil mixture passes from a viscous liquid state to that of a plastic soil, the moisture content at which this transition occurs is called the Liquid Limit (LL). With further reduction in the moisture content the material becomes

TABLE 28. CLASSIFICATION OF HIGHWAY SUBGRADE MATERIALS

General classification	Granular materials (35% or less passing No. 200)							Silt-clay materials (more than 35% passing No. 200)			
	A-1			A-2							A-7
Group classification	A-1-a	A-1-b	A-3	A-2-4	A-2-5	A-2-6	A-2-7	A-4	A-5	A-6	A-7-5, A-7-6
Sieve analysis, percentage passing:											
No. 10	50 max.										
No. 40	30 max.	50 max.	51 min.								
No. 200	15 max.	25 max.	10 max.	35 max.	35 max.	35 max.	35 max.	36 min.	36 min.	36 min.	36 min.
Characteristics of fraction passing No. 40:											
Liquid limit				40 max.	41 min.	40 max.	41 min.	40 max.	41 min.	40 max.	41min.*
Plasticity index	6 max.		NP	10 max.	10 max.	11 min.	11 min.	10 max.	10 max.	11 min.	11 min.
Usual types of significant constituent materials	Stone fragments, gravel, and sand		Fine Sand	Silty or clayey gravel and sand				Silty soils		Clayey soils	
General rating as subgrade	Excellent to good							Fair to poor			

Classification procedure: With required test data available, proceed from left to right on above chart, and correct group will be found by process of elimination. The first group from the left into which the test data will fit is the correct classification.

*Plasticity index of A-7-5 subgroup is equal to or less than LL minus 30. Plasticity index of A-7-6 subgrade is greater than LL minus 30.

more stiff and less plastic until it finally reaches a semi-solid state. The moisture content at this point is known as the Plastic Limit (PL). The numerical difference between the LL and the PL is known as the Plasticity Index (PI). Granular soils have a PI = 0. Table 28 shows the various PI for the different classifications.

Any embankment is made up of three components, soil, air, and water. The relationship between moisture content and density is particularly important. All soils have a moisture content at which the soil is at its maximum density. If embankments are placed with the soil moisture content not at this optimum value, the embankment cannot be fully compacted. This optimum moisture content can be determined with a test known as the AASHO density or Proctor density test. Examples of different soil types and their moisture density relationship is shown in Figure 84.

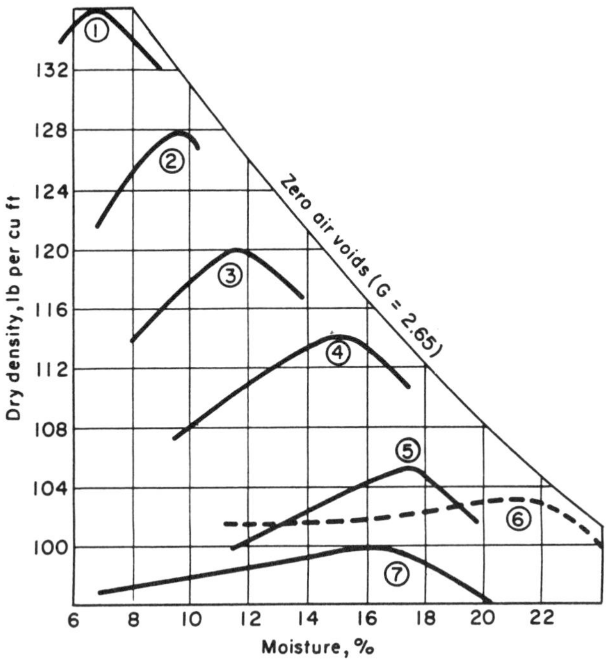

Soil Texture and Plasticity Data

No.	Description	%Sand	%Silt	%Clay	L.L.	P.I.
1	Well-graded loamy sand	88	10	2	16	NP
2	Well-graded sandy loam	72	15	13	16	0
3	Med-graded sandy loam	73	9	18	22	4
4	Lean sandy silty clay	32	33	35	28	9
5	Loessial silt	5	85	10	26	2
6	Heavy clay	6	22	72	67	40
7	Very poorly graded sand	94	← 6 →		NP	NP

Moisture-density relationships for seven soils each compacted by the AASHO standard method

Figure 84. Relationship between moisture content and density.

The following tables are included as a guide in selecting embankment material. Table 29 is a selection of soil on the basis of anticipated embankment performance, and Table 30 lists the recommended minimum requirements for compaction of embankments.

When using rock as an embankment material, the size of the individual rock pieces should not be greater than 12 inches in any dimension, and the rock embankment should be placed in layers not to exceed two feet. The top 0.91 m to 1.52 meter (3 to 5 feet) of rock embankments should be placed in soil. The final 30.48 cm (12 in) of embankment should be of material with a dry density of at least 1633.7 Kg/meter cube (102 lb/ft^3) at optimum moisture. When placing earth embankments, the layers should not exceed 22.8 cm (9 inches) loose before compaction for fine-grained cohesive soil and 30.48 cm (12 inches) for granular cohesionless soil.

The final step before actually building the haulroad is to design adequate drainage to keep the road from deteriorating.

DRAINAGE

Once the haulroad has been designed, it is necessary to provide the road with ditches and culverts of sufficient size to remove all surface water before it can damage the road embankments and surface.

If rainfall were applied at a constant rate to an impervious area, the runoff from this area would eventually equal the rate of rainfall. The time required to reach this equilibrium is known as the time of concentration (t_c). This value is the basis for a method of estimating runoff known as the Rational Method. Although there are many methods employed for estimating runoff, this method is simple and does provide reasonable results for small drainage areas which are usually encountered in haulroads. The Rational Method can be reduced to the following empirical formula:

$$Q = K\ i\ A \qquad (1)$$

where

Q = Runoff in cu ft/sec
K = Runoff coefficient, a function of drainage area (Table 31)
i = Intensity of rainfall, in/hour for a duration equal to the time of concentration t_c
A = Drainage area in acres

The time of concentration can be calculated from the following formula:

$$t_c = \frac{60\ (11.9\ L^3)^{0.385}}{H} \text{ mins} \qquad (2)$$

where

L = Length of channel from headwater of watershed to the end of the channel (miles)
H = The fall in the channel in ft

Once the time of concentration has been established the intensity i can be calculated from the intensity frequency curve shown in Figure 85. This type of curve can be constructed for any location. Each area will have its own unique set of rainfall criteria. The curve shown in Figure 85 is for southern Ohio.

TABLE 29. GENERAL GUIDE TO SELECTION OF SOILS ON BASIS OF ANTICIPATED EMBANKMENT PERFORMANCE

HRB classification	Visual description	Max. dry-weight range, lb per cu ft	Optimum Moisture range, %	Anticipated embankment performance
A-1-a A-1-b	Granular materials	115-142	7-15	Good to excellent
A-2-4 A-2-5 A-2-6 A-2-7	Granular materials with soil	110-135	9-18	Fair to excellent
A-3 A-3a*	Fine sand and sand	110-115	9-15	Fair to good
A-4a* A-4b*	Sandy silts and silts	95-130	10-20	Poor to good
A-5	Elastic silts and clays	85-100	20-35	Unsatisfactory
A-6a* A-6b*	Silt-clay	95-120	10-30	Poor to good
A-7-5	Elastic silty clay	85-100	20-35	Unsatisfactory
A-7-6	Clay	90-115	15-30	Poor to fair

* Ohio modification.
SOURCE: Ohio Department of Highways.

TABLE 30. RECOMMENDED MINIMUM REQUIREMENTS FOR COMPACTION OF EMBANKMENTS

Class of soil (AASHO M 145-49)	Condition of exposure					
	Condition 1 (not subject to inundation)			Condition 2 (subject to periods of inundation)		
	Height of fill, ft	Slope	Compaction (% of AASHO max. D)	Height of fill, ft	Slope	Compaction (% of AASHO max. density)
A-1	Not critical	1-1/2 to 1	95+	Not critical	2 to 1	95
A-3	Not critical	1-1/2 to 1	100+	Not critical	2 to 1	100+
A-2-4	Less than 50	2 to 1	95+	Less than 10	3 to 1	95
A-2-5	Less than 50			10 to 50		95 to 100
A-4	Less than 50	2 to 1	95+	Less than 50	3 to 1	95 to 100
A-5	Less than 50					
A-6	Less than 50	2 to 1	90-95*	Less than 50	3 to 1	95 to 100
A-7						

Remarks: Recommendations for condition 2 depend upon height of fills. Higher fills of the order of 35 to 50 ft should be compacted to 100 per cent, at least for part of fills subject to periods of inundation. Unusual soils which have low resistance to shear deformation should be analyzed by soil-mechanics methods to determine permissible slopes and minimum compaction densities.

*The lower values of minimum requirements will hold only for low fills of the order of 10 to 15 ft or less and for roads not subject to inundation nor carrying large volumes of very heavy loads.

SOURCE: Highway Research Board.

TABLE 31. RUNOFF COEFFICIENTS

Type of Drainage Area	Coefficient of Runoff, K
Concrete or bituminous pavements	0.8-0.9
Gravel roadways, open	0.4-0.6
Bare earth (higher values for steep slopes)	0.2-0.8
Turf meadows	0.1-0.4
Cultivated fields	0.2-0.4
Forested areas	0.1-0.2

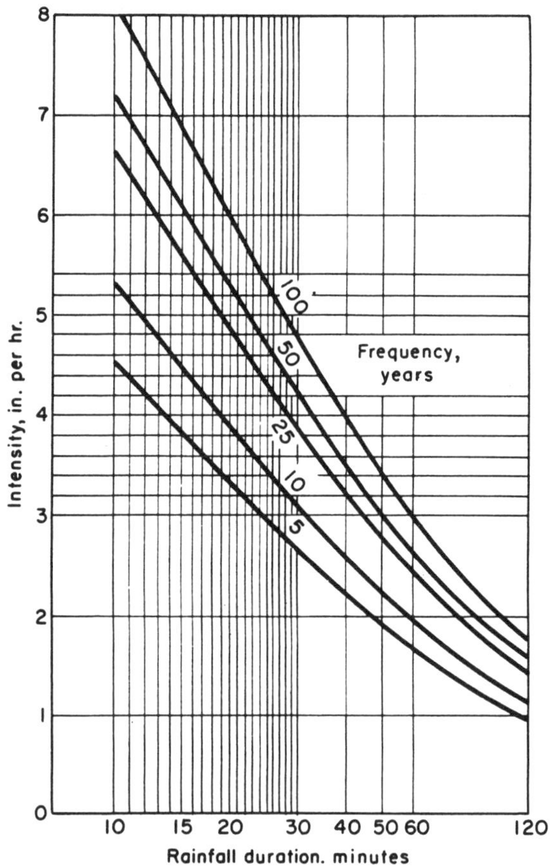

Figure 85. Rainfall intensity, duration, and frequency for the southern one-third of Ohio.

The use of the Rational Method is illustrated for a drainage area adjacent to a haulroad located in southern Ohio (Figure 86). The hatched area of 102 acres will all drain to a new culvert pipe that is to be installed. The values of L, H and K are shown in Figure 86.

$$t_c = \left(\frac{60\ (11.9\ L^3)}{H}\right)^{0.385} = \left(\frac{60\ (11.9 \times 0.75^3)}{10}\right)^{0.385} = 46 \text{ mins}$$

From Figure 85, for a 10-year frequency at a t_c of 46 minutes, the value for i is 2.3 inches per hour. Therefore, the value of Q can now be calculated.

$$Q = K\ i\ A$$

$$= 0.2 \times 2.3 \times 102 = 47 \text{ cu ft/sec}$$

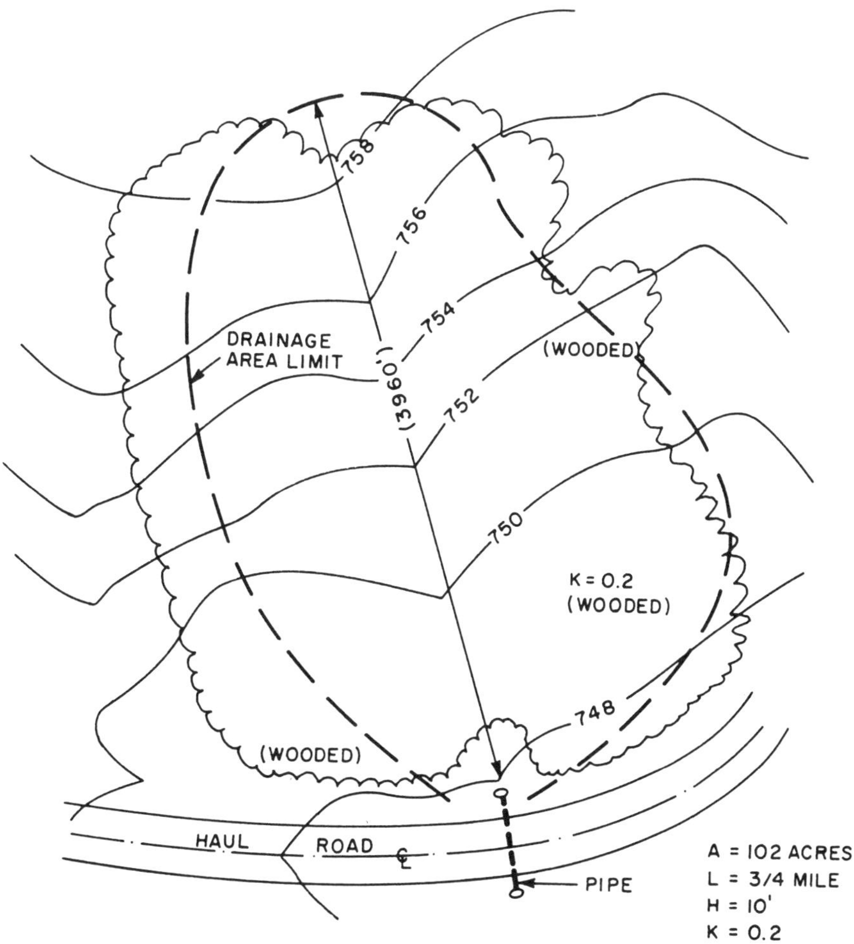

Figure 86. Drainage area adjacent to a haulroad.

There are other more elaborate empirical methods for computing runoff. Regardless of the method, all ditches and culverts must be designed to accommodate the expected runoff.

DITCHES

The use of V ditches for controlling and channeling runoff is economical. As the velocity of water increases with slope, it is advisable to keep all V ditches as flat as possible and yet allow the water to run. Several anti-erosion steps are to be taken within the ditch as the percent slope increases. Rock lined channels are required when slopes exceed 8 percent. The rock lining controls erosion and helps dissipate the velocity energy of the water.

CULVERTS

The size of culverts necessary to take the runoff must be adequate. The most common type of culvert is corregated metal pipe, mainly because of its ease of handling and low initial cost.

When placing pipe culverts in areas of traffic, it is advisable to have a minimum of 60 cm (2 feet) of cover over the pipes; where vehicles of 90718.47 Kg (100 tons) or greater cross the culvert regularly 91 cm (3 feet) would be advisable. When installing pipe culverts the major concern should be to provide lateral confinement to the pipe. The common practice is to excavate a trench 60 cm (2 feet) wider than the outside diameter of the pipe, place the pipe, then backfill in 15 cm (6-inch) maximum compacted layers.

Pipes should never be placed under an embankment prior to placing the embankment. The embankment should be placed first, and then a pipe trench should be excavated through the embankment. The graph shown in Figure 87 shows the size of pipe required for various flow rates. It may sometimes be advisable to place a pipe smaller than that required to take the maximum flow and let the water pond on the upstream side of the culvert. As an example, suppose the flow rate is 0.2832 cu m/sec (10 cu ft/sec). From the graph it can be seen that a pipe larger than 53 cm (21 inches) in diameter will be required. Therefore, a 60 cm (24-inch) diameter pipe is selected. However, if the water could be allowed to pond a maximum of 90 cm (3 ft), one can use a 38 cm (15-in) diameter pipe. Even 60 cm (2 ft) ponding would allow the use of a 45 cm (18-in) diameter pipe.

SUMMARY

This chapter has briefly reviewed the design of surface coal mine haulroads, and the environmental problems associated with them. Good haulroad design for coal mines can hardly be over-emphasized in view of the very heavy loads. Considerations from highway construction, logging roads, etc., can be transferred wherever applicable.

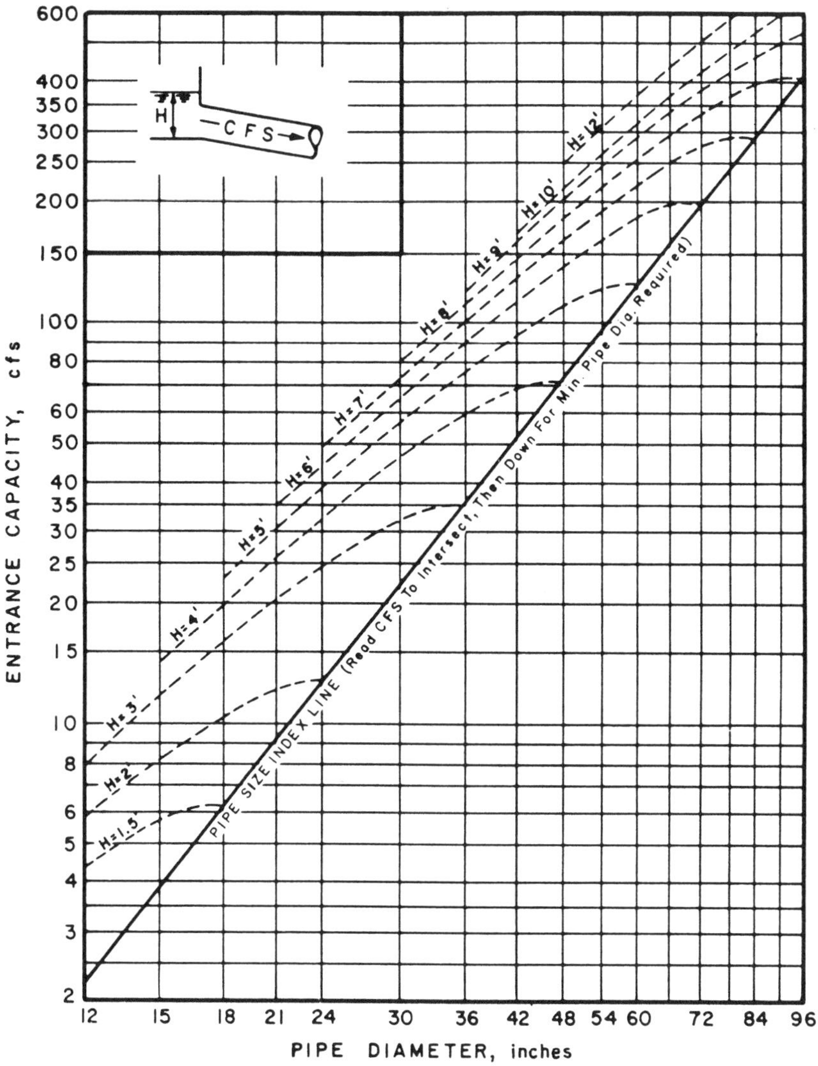

Figure 87. Pipe culvert capacity graph.

Slope Stability

INTRODUCTION

The earth's crust, particularly that portion termed continental, contains a variety of rock types which are arrayed in a wide display of geologic structures and subjected to a great variety of stresses. Some of these structures are very complex, such as overturned and faulted folds, others, such as undisturbed sediments, are very simple. The alteration of ground stability by the introduction of an excavated hole will depend upon the extent of this complexity plus a variety of other factors. The size of the excavation complexity of geologic structure, presence of hydrostatic pressure, and presence of tectonic stresses can all affect the stability of the slopes surrounding an excavation. Luckily, when consideration is given only to coal mines within the regions of interest to this report, the factors simplify themselves greatly. Thus, this slope stability section discusses only the structures of concern to coal mines: highwalls, low-walls, storage piles, and to some extent, impoundments. In this section, the word rock is used generally to cover all crustal material that is found naturally. For the purpose of mechanical analysis, soil is merely rock with a low or absent tensile strength.

The section is divided into three major parts. The first part, the introduction, defines structures of concern, gives a qualitative description of the elements of stability, and ends with a list of helpful references. The middle part, on the theory of stability, gives a quantitative description of the stability elements. The final part of the section gives practical suggestions for constructing a safe slope or for remedying a failing one. The mine engineer should read all three sections, while the operating manager may be content to skip the middle section.

Generally speaking, coal is overlain by sediments possessing relatively simple and flat-lying structure. Although these strata may contain some faults and jointing, the principal mechanical discontinuity is that created by bedding. Since there is but little tectonic disturbance, the stresses in the overburden are primarily due to body weight, that is, the pull of gravity on the rock itself. Therefore, an analysis of slope stability in these coal mines generally requires knowledge of the excavation geometry, rock type and strength, density of bedding, strength across bedding, and the amount and pressure of included groundwater. Many of these data should be determined by careful examination of exploration results, before mining begins.

Structures of Concern

Man-made slopes in surface coal mines abound; generally, however, they may be classified into four distinct types. Thus, the structures of concern are highwalls, overburden or spoil piles, waste or refuse piles, and impoundments.

Highwalls--As explained elsewhere in this report, highwall heights are usually constrained by economic considerations, with the dimensional limitations of the excavator sometimes acting as a bound. In other words, highwall heights are defined, through stripping ratios, by the price of the coal and the cost of removing it. However, in dragline mining particularly, a maximum digging depth of 50 to 65 m (150 to 200 ft), depending on machine size, is imposed. Rarely does the mechanical stability of the highwall itself enter into these constraints. Generally speaking, this situation exists because coal surface mine slopes are stable with little hydrostatic head in the sediments, and because draglines mine with such speed that they can remove an incipiently failing slope before it becomes catastrophic. However, several specific items enter into the present consideration of highwall stability.

1. No mine is ever totally free from wall failure. Although small, from a mechanical viewpoint, falls of rocks can be disastrous to personnel in the pits. Some major failures have occurred; these can be extremely disruptive to stripping shovels and coal loading machinery working on the pit floor, as well as equipment located on the surface within the shear plane of the failure zone.
2. In times of expensive capital goods, lost time on excavators due to the need to clean up slope failures can be very costly. Human injury and fatality costs are impossible to assess realistically.
3. High-priced coal is leading to ever greater stripping ratios and, hence, higher highwalls. Some wall heights may soon reach the limits of equilibrium.
4. Multiple benching, either to accommodate multiple seams or to allow for more flexible excavation of deep seams, puts large machines on small working platforms. There is limited room available on an intermediate bench to which a machine can be moved if a slope is failing. Although items 1, 2, and 3 refer to limited-height dragline benches, depths of 304.8 m (1000 feet) or more can be achieved by truck-shovel combinations on multiple benches.

Overburden Piles--Spoil piles, often called low-walls, present another version of the stability problem. Mechanically, the material is excavated in discrete particles; it is more homogeneous than when in situ and possesses little or no tensile strength. Since its strength is dependent upon confinement, rather than using a convenient measure of strength such as unconfined compressive strength, more complex measures such as yield criteria and the angle of internal friction are required. Problems in existing mines point to situations which will worsen with increasing depth of mines.

1. Planned stacking space may be inadequate if the swell factor changes at different sites, or if it has been underestimated. To prevent rehandling or covering of coal when this occurs, the temptation is to steepen the slope of the spoil pile. While this steeper slope may be temporarily stable, changes from increased loading or changed weather conditions may cause rapid failure.
2. Mining around an inside curve causes increased amounts of spoil to be deposited in one place. Again, the temptation is to steepen the slope. This is because the fixed casting radius and heights of most stripping machines prevent a safe spreading of the excess material.
3. Maintaining access ramps through spoil causes excess material to be stacked. As with curves, the temptation is to steepen slopes.
4. The presence of water can materially alter a slope's stability. Slopes constructed on a dry day may fail, either by surface erosion or by massive translation, on a wet day. Obviously, the predominant rock type will have a major control on this occurrence.

Sandstone blocks, with little recompaction and a great deal of interparticle porosity, will drain quickly and well and remain stable during wet conditions. On the other hand, poorly-cemented sandstones or shales, particularly shales with water-sensitive clays such as bentonite or montmorillonite, may well fail during wet periods.

Storage piles--Although preparation-waste or refuse piles are not peculiar to surface mines, they represent a surface structure which can fail. Inasmuch as their instability may evidence itself years or even decades after the pile was begun, as with the Aberfan disaster in South Wales, great care must be taken in their design and construction. Similar piles that fall into this category are those for the temporary storage of topsoil and for the permanent storage of excess overburden in head-of-hollow or valley fills.

Impoundments--Because of the additional problems created by the storage of water, a separate category is given for those waste piles used as impoundments. The dam at Buffalo Creek, which failed before it was completed, is testimony to the need for consideration of and good design for impoundments.

Elements of Stability

Although a more explicit review of the theory of stability is presented later, a review of the elements of stability is in order in this introductory material. These elements are often grouped, conveniently, into sliding forces and resisting forces; that is those elements which tend to cause failure and those which tend to prevent it.

Sliding Forces--There are two principal sources of sliding forces. The first is the force of gravity on the rock itself or on any equipment resting on the slope; in other words, this force is due to body weight of the slope and the weight of any equipment on top of the slope. The second force is that due to any water contained within the material of the slope. Frequently, groundwater pressure is the major contributor to slope failure. A third force, rarely seen in the relatively undisturbed strata in the coal fields, is that of residual stress remaining from mountain-building activity.

Resisting Forces--Those forces that resist failure are unconfined tensile and shear strengths and the shear strength resulting from internal friction. Since the weakest part of an overall slope is often at bedding or jointing planes, the unconfined strengths of concern are those measured across these discontinuities. In addition to its pressure, water can serve to diminish the strength or friction of some rock types. Thus, where clays are present in the rock, water can be doubly damaging to stability.

Sliding Surfaces--The above forces act along pre-existing surfaces such as bedding planes or joints. In slopes of homogeneous material, either solid without pre-existing fractures or granular such as in spoil piles, the forces will act along a curved surface that develops in such a way as to minimize the expenditure of energy during failure. In granular material with no internal friction, this arc will be circular; where friction exists, the arc takes the form of a logarithmic spiral. Apparently, the use of only circles, and not spirals, in the analysis of slopes does not materially affect the prediction of failure. The same cannot be said for the substitution of a plane, particularly when the slopes deviate much from the vertical.

References

Rather than reference each statement in this section, a list of useful references is given. These references encompass the statements made here, but also provide the design engineer or consultant with a useful library.

"Slope Stability" is discussed in Section 12.1 of Surface Mining [Pfleider, 1968]; essentially the same analysis is presented in Section 7.1 of the SME Mining Engineering Handbook [Cummins, 1973]. Rock Mechanics and the Design of Structures in Rock [Obert and Duvall, 1967] gives background work theory, but has little specific slope stability work.

A recently-released design manual, Engineering and Design Manual: Coal Refuse Disposal Facilities [MESA, undated], is one that no surface-mine designer should be without. It gives a thorough review of slope stability, including extensive references, before embarking on the design of refuse disposal facilities, both piles and impoundments.

Finally, slope stability is discussed in most soil mechanics textbooks. Because the authors did extensive research in stability, a favored reference is Soil Mechanics in Engineering Practice [Terzaghi and Peck, 1967]. One other text, used by this author is Principles of Soil Mechanics [Scott, 1963]. Several other papers, referred to in other parts of this chapter, are also listed.

THEORY OF STABILITY

Mechanics

The theories of fundamental mechanics are, in their most comprehensive forms, complex and cumbersome. An attempt will be made here to extract only that material needed to understand subsequent analyses.

It is reasonably well understood in mechanics that relationships among the three principal stresses can be expressed which define a material's yield limit. When the stresses are less than the defined relationship, the material is stable. When the stresses reach their limit, or when conditions attempt to push these stresses beyond this limit, the material responds by yielding. This yielding prevents the stresses from exceeding the set limit. At the yield point, further applications of stress on the material result in further deformation but not in a stress rise within the material. Although yield is sometimes called failure, it should not be confused with rupture or collapse. Yield means deformation beyond the elastic limit; the physical manifestation of yield may take a variety of forms from crystal deformation to intergranular slippage to the creation of visible cracks and failure zones. Since failure is energy conservative, microscopic evidences of incipient failure will occur within predictable planes or zones. Further applications of stress will cause the micro-cracks to coalesce into one large crack. Even in granular material, high differential deformation will occur in narrow zones that surround undisturbed blocks. This can be seen in failing slopes, where large blocks will move as one piece along a failure surface.

Although the subject of yield criteria is extensive and yet evolving, particularly for three-dimensional analyses of anisotropic material, the theory for a two-dimensional analysis such as the slip-circle analysis, to be given later, is not that involved. In a general soil, the criterion for stability, and thus failure, was given by Coulomb in the year 1773 as

$$|\tau| \leq \sigma \tan\phi + c \tag{1}$$

That is, the shearing stress (τ) on a section of isotropic, cohesive soil must be less than the cohesion (c) plus the product of the normal stress (σ) on that same section with the coefficient of internal friction ($\tan\phi$). Failure occurs when the shear stress equals the cohesion plus this value of the normal stress. The geometry of this situation is shown in Figure 88.

In soils, cohesion can be thought of as unconfined tensile strength; in rock, however, the tests for shear strength give a better value for cohesion than do those for tensile strength. Since it is difficult, if not impossible to test soils directly for their tensile strength, cohesion is usually calculated from a series of shear tests.

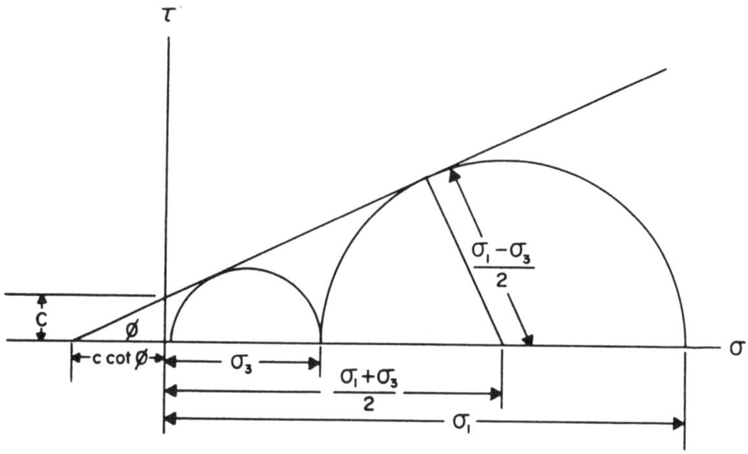

Figure 88. Coulomb's failure surface.

Mohr adapted this yield condition and represented it by means of a graphical plot of principal stresses upon stress co-ordinates, where the normal stress, σ, is the abscissa and the shear stress, τ, is the ordinate. In Figure 88, σ_1, the major, and σ_3, the minor, are the principal stresses, c is the cohesion, c $\cot\phi$ is the tensile strength, and ϕ is the angle, called the angle of internal friction, made by the envelope surrounding and tangent to the stress circles and the abscissa. The envelope describes the limiting yield conditions of the principal stresses and satisfies the equality of equation (1); therefore a circle that is tangential to the envelope represents a stressed soil sample that is yielding, and a circle that lies within the envelope represents a stressed soil sample that is rigid. Under the stated conditions, it is impossible for the stresses to be arranged such that a Mohr circle intersects the envelope. In terms of the principal stresses, failure occurs when

$$2c\cos\phi + (\sigma_1 + \sigma_3)\sin\phi = \sigma_1 - \sigma_3. \tag{2}$$

Considering cohesionless soil, c $\cos\phi$ must equal zero and equation (2) reduces to

$$\sigma_1/\sigma_3 = (1 + \sin\phi)/(1 - \sin\phi). \tag{3}$$

Equation (3) represents the maximum value of σ_1, for any σ_3, and any ϕ; however, if Figure 88 is examined, it is seen that a circle can be drawn to the left at σ_3 so that σ_1 is smaller than σ_3. This is the minimum value of σ_1 and is represented by

$$\sigma_1/\sigma_3 = (1 - \sin\phi)/(1 + \sin\phi). \tag{4}$$

The ratios of equations (3) and (4) are recognized as defining Rankine's lateral earth-pressure coefficients. The major principal stress (σ_1) divided by the minor principal stress (σ_3) of equation (3) is called the passive coefficient (K_p) and the same relation in equation (4) is the active coefficient (K_a).

It is known that when the limiting conditions of equilibrium are reached, the soil will fail along preferred orientations. In plane conditions, these orientations are called sliplines and form angles of $1/4\pi \pm 1/2\phi$ with the principal or algebraically greater stress direction. Only in homogeneous and isotropic material can it be shown that stress sliplines are coincident with strain characteristics.

The implication of a positive value for ϕ in a material is that there is an increase in shear strength for each increase in normal stress. Put differently, a confined soil will be stronger than an unconfined soil. The weight of a structure will provide the confinement necessary to create sufficient foundation strength to support that structure, presuming, of course, that yield is not exceeded. A zero value for ϕ, as seen in cohesive materials such as clays and most metals, means that, derived from Equation (1),

$$|\tau| \leq c \tag{5}$$

and, in terms of Equation (2), failure occurs when

$$|\sigma_1 - \sigma_3| = c \tag{6}$$

A purely cohesive substance is weakened, not strengthened, by confining pressures. Equation (6) is known as the yield criterion of Tresca, and is often applied in metal plasticity.

Demonstration of the notion of yield criteria and establishment of values for c and ϕ is done with a shear tester. Either a direct-shear (shear box) or triaxial tester may be used. The shear box is preferred when drained tests are to be performed and where strengths across predetermined sliding surfaces are desired. The triaxial tester is preferred for homogeneous material which is allowed to form its own failure plane. The triaxial tester can also be arranged to manipulate pore pressure. Thus, the rise of pore pressure due to deformation can be measured or the effect of increased pore pressure (effective stress: see the subsequent section on effect of water) can be simulated. Both tests are standard and can be obtained commercially.

Effect of Geology

In the previous discussion on mechanics, reference is made to homogeneity and isotropy. However, earth materials, especially undisturbed overburden, are unlikely to be uniform. Therefore, some consideration must be given to the effect of geology. Namely, strengths and angles of internal friction vary among the rock types; also, within a single rock type these variables may change according to direction of measurement. Finally, the presence of discontinuities, such as bedding planes, faults, and joints, may reduce unconfined strengths to zero. Since friction exists between any two surfaces, angles of internal friction are often unaffected by discontinuities. This statement is modified by the fact, that many discontinuities, particularly faults, are filled with gouge or other clay-like material which acts as a lubricant and reduces the angle of internal friction.

The effect of geology, therefore, is to take the homogeneous system discussed previously and impose upon it predetermined sliding surfaces and highly variable strength factors. The mine engineer must analyze each slope according to both its geometric and geologic factors. A purely homogeneous and cohesionless material can be analyzed on geometric factors only. For instance, in an undisturbed highwall the sliding surface will be strongly influenced by the strength across bedding planes. If, as is often the case, the contact between the coal and its immediate superjacent stratum (roof) has low or even zero strength, and if the wall fails, it probably will be by sliding (translation) rather than by rotation. Should rotation occur, it will undoubtedly intersect this contact plane. Since coal itself is a weak foundation material, it would be improbable to find a rotational failure which involved the strata below the coal. That is, it is unlikely to find base-circle failures. Similarly, the presence of faults or joints (cleat) will help to define the sliding surface.

Looking closely at the rocks which compose the slope, it is seen that a high degree of variation can occur within a single geologic material. The best example is shale, which may have a shear strength of over 1000 pounds per square inch normal to bedding and a strength near zero parallel to bedding. Similar variations can occur in the angle of internal friction. This angle, ranging between 30 and 40 degrees for most soil materials, is a direct measure of surface roughness and particle shape and is closely correlated to mineralogy. A smooth spherical particle will have a ϕ of around 20 degrees, while an angular substance can have a ϕ as high as 45 degrees. Angles of internal friction over 45 degrees probably represent interlocking particles rather than true friction. Some soil materials, those that contain clay and most notably the shales, can have dramatic alterations to ϕ with the addition of a small percentage of water. Sand and gravel, derived from quartz or unaltered silicate rocks, will demonstrate cohesionless characteristics (c near zero, ϕ between 30 and 40 degrees), while clays and clay stones will demonstrate either cohesionless or cohesive (a measurable c, ϕ near zero) behavior, depending upon the amount of water present. Obviously, shale particles, with little or no cohesive strength, can lose all strength when suddenly wetted.

Narrowing from a general discussion to the specifics of surface coal mines, certain simplifying conditions are seen to exist. Undisturbed strata are flat, and maximum shear strength is parallel to the vertical; as a corollary, maximum tensile strength is normal to the vertical. This is a relatively favorable orientation for slope stability. However, of greatest significance is that most slopes of interest in coal mines are composed of broken materials. Spoil piles, refuse piles, and certainly, the blasted portion of the highwall can be characterized as cohesionless soils. Because the dragline so often sits on blasted material, the last is often of great concern.

Effect of Water

In the preceeding paragraphs, mention was made of some simplifying conditions; it is now appropriate to introduce one important, complicating condition. The presence of water in a slope can reduce its overall stability or even induce catastrophic failure. As discussed in other portions of the background reports, water enters a slope as precipitation, as surface drainage, or as groundwater. Succinctly, water can increase body weight, which is a sliding force, reduce normal stress which contributes to the resisting force, and, for some minerals, reduce ϕ which is tantamount to lubrication and thus a decrease of resisting force. The following comments will give some guidelines on how water affects stability.

Water's Weight--All earth material contains holes or empty spaces termed voids. Even solid rock contains cracks and fissures of interparticular spaces in sediments. When the rock contains a large volume of voids, it is termed porous. When the voids are connected so that a fluid may flow through the rock, it is termed permeable. The ratio of void volume to solid volume is appropriately termed the void ratio. The ratio of void volume to total volume is called porosity. These voids, whether in rock or in soil, contain either fluid or gas. Commonly the fluid is water, but may be petroleum liquids; the gas is usually air, but can be hydrocarbon. For these purposes, concern lies only with air and water. When the entire void space is filled with water, the soil is called saturated. Percent saturation is a ratio of water volume to total void volume.

Similar to rock, soils may be permeable or impermeable, porous or dense, according to their makeup. Soils with particle sizes of sand or larger (greater than one millimeter) are usually porous and permeable. A high degree of fines or clays will decrease a soil's permeability and increase its density. This is because the fines can occupy the void spaces between the larger particles. Naturally, for one kind of particle, the densest soil will be well-graded, that is possessing particle sizes from the finest to the coarsest. Distinction is made between particle density, which is a function solely of mineralogy, and bulk density, which is an overall measure and is connected to both particle density and void ratio.

Concern lies with the fact that water can add weight to a slope by filling voids. If a rock, whose density is 2.65 grams/cm^3 (165 lbs/ft^3), has a void ratio of 20 percent, then its bulk density is 2.21 grams/cm^3 (138 lbs/ft^3). At saturation, its density is 2.38 grams/cm^3 (148 lbs/ft^3) which is an 8 percent increase. However, 20 percent void ratio equals 17 percent porosity and is quite low for soils. For a soil, with a 50 percent void ratio (33 percent porosity) and the same particle density, bulk density would be only 1.89 grams/cm^3 (118 lbs/ft^3). Saturation gives a bulk density of 2.18 (136) or a percentage increase of 15. Since a soil is much more likely to saturate than intact rock, this weight change should be of concern to exposed spoil piles, refuse banks, and blasted highwalls.

Water pressure--When the voids of a soil become saturated, it is possible for the pore water to develop a pressure which is measurably greater than atmospheric. This pressure is dependent only upon the pore water head and not upon its volume. Hence, even a low porosity material can develop a pore pressure. For simplicity, assume that the water is not flowing and that consideration is given only to the static head. The development of pore-water pressure can lead to the most disastrous situations. These range from improbable and feared quicksand to the collapse of the Buffalo Creek Impoundments on February 26, 1972, which caused an estimated death toll of 118 people.

The net effect of pore pressure is to reduce directly the normal stress equation (1). The notion of effective stress ($\bar{\sigma}$) must be introduced,

$$\bar{\sigma} = \sigma - p_w \tag{7}$$

Equation (7) says that the normal stress on a surface should be the stress due to the body weight of the substance plus any imposed loads, such as the weight of a dragline, less the pore-water pressure (p_w). This reduced quantity is the effective stress. Equation (1) can be rewritten

$$|\tau| \leq \bar{\sigma} \tan\phi + c. \tag{8}$$

The natural and logical question is what happens when $p_w = \sigma$? Obviously,

$\bar{\sigma} = 0$, $\bar{\sigma} \tan\phi = 0$ and the substance will behave as a liquid if c is inconsequential. Since liquids have no shear strength, the effect of having a zero value for effective stress is to cause a zero shear strength and to cause seemingly liquid behavior in a bulk solid. This process, called liquefaction, can happen precipitately fast, since saturation and then pressurization of the pores can occur with the addition of only a very small quantity of water. Indeed, during earthquakes, such as at Hebgen Lake, Montana in 1959, where 28 people were killed by a landslide, liquefaction can occur due to the consolidation of the soil from the quake's vibration and the accompanying reduction in porosity, and hence, a rapid rise in pore pressure.

Mine designers must guard against pore water pressure in all surface mine structures. Areas of concern are joint fractures in a highwall which intersect the surface and can intercept surface run-off; blasted areas of a highwall which collect run-off; fill areas (head-of-hollow or valley) and spoil piles which saturate due to precipitation, surface run-off, or groundwater flow; and, most specifically, impoundment structures and their foundation zones. Water interception structures such as diversion ditches, free-flowing (French) drains where appropriate, and in the extreme, dewatering wells should be built to keep these structures drained. Pore-water pressure can be measured by placing packers and a piezometer in specially drilled wells. Even without precise measurement of pressure, its presence can be presumed if there is a water seep or flow from a slope's surface.

In closing, it should be noted that the so-called lubricating effect of water is often the liquefaction of narrow failure zones. Liquefaction can occur when a porous layer is entrapped by impermeable layers. While this is not often a problem in undisturbed terrain, it immediately becomes so when an excavation enhances the instability of a constrained and pressurized layer. Liquefaction can also occur within fault or joint zones, particularly if they are coincident with failure zones predicted by mechanics theory.

DESIGN OF SAFE SLOPES

The following discussion draws from those preceeding, but can, for ease of reading be taken on its own. It starts with some computational schemes--as it were, theoretical analyses--for factor of safety in a standing slope. It then proceeds into some practical or field considerations for slope stability. The mine designer is warned against taking the simplest approach if it is not appropriate for his conditions. Evaluate the site and select the approach that encompasses the given conditions.

Factor of Safety, First Considerations

Generally speaking, the mechanical stability of any substance can be found if the stresses within that substance and the substance's strength (stresses that the substance can withstand at the point of failure) are known. When failure, in the sense of yield, occurs at a point, such as on an unconfined surface, where strain may develop, then failure may progress to the point of complete mechanical rupture. Because of confinement, it should be remembered that yield failure need not necessarily lead to rupture. By the use of appropriate yield criteria and computers to handle the iterative computations, it is possible to perform such a total stress analysis on a slope.

Because of the complexity of these total solutions, and because of an inability to perform the iterative computations before the widespread introduction of digital computers, a series of limit-equilibrium-analyses were devised. Some of these analyses are still appropriate for the field engineer. Basically, these analyses compare those force components inducing a slide against those resisting this slide. To make the problem tractable, a rupture or slip surface has to be assumed. The ratio of resisting forces to sliding forces along any slip surface is the factor of safety for that surface. This is expressed in equation (9),

$$\text{Factor of Safety} = \frac{\Sigma \text{ Resisting Moments}}{\Sigma \text{ Sliding Moments}}, \tag{9}$$

where the moments are the forces multiplied by the radius. Because slip surfaces are assumed, a series of analyses using differently-located slip surfaces is performed in order to find the minimum factor of safety for any slope.

Sliding forces are usually due to weight and water pressure. The weight of the slope material, its contained water, and any surface loads due to structures or machines, resolved to the components parallel to the slip surface, are the prime cause for sliding. Any unbalanced pore pressure can also add to sliding. Other, more limited, sliding forces might be due to residual ground stresses or formation pressure resulting from expansion of wetted clays or shales.

Resisting forces would be from cohesive strength across the sliding surface and from friction. Obviously, friction forces are a function of the normal component of weight and the coefficient of friction. Resistance may also be due to toe loadings, such as those that occur in base slip circles.

The slip surface itself is often thought to be circular, although other curves such as a logarithmic spiral are possible. Irregular slip surfaces develop where discontinuities such as faults and joints exist in the ground. Often, slip surfaces are a mixture of smooth curves and connecting straight lines. This can be visualized if pre-existing geology is imposed on the theoretical circle of Figure 88. If one of the circles were to intercept a vertical joint which had been opened due to freezing, then the upper part of the slip surface would be defined by the joint rather than a continuation of the arc. Figure 89 can also be used to locate the critical circle (lowest factor of safety) for cohesionless slopes ($\phi = 0$). The variable D is the depth to a firm foundation and is expressed as a function of the slope height (H). Since the coal is assumed to be the solid foundation, most surface coal mine failures will be toe failures and D = H. It should be emphasized that this simplifying condition does not exist for impoundments, refuse piles, and fill areas that are built on soil or poor foundation materials.

Slip-circle Analyses

A calculation of a slope's factor of safety requires some manipulative method to simplify the determination of forces. This calculation is delineated in the paragraphs which follow and the steps to follow are listed in Table 32. Once a rupture surface on which failure will occur is assumed, the slope is divided into slices and each slice is assumed to have the chord, defined by its sides, as its base rather than the intercepted arc. Ten or fifteen such slices give almost as accurate a solution as using the arc itself. Five or six such slices often suffice, for the straight-based slice method gives a more conservative factor of safety than a circle-based slice. The method of slices is shown in Figures 90 and 91. Figure 90 is an overview of the method and represents the two-dimensional plane normal to the slope face. One assumption in this method is that the third dimension has no effect on the analysis. This assumption is called plane strain. Figure 91 shows the force on an individual slice.

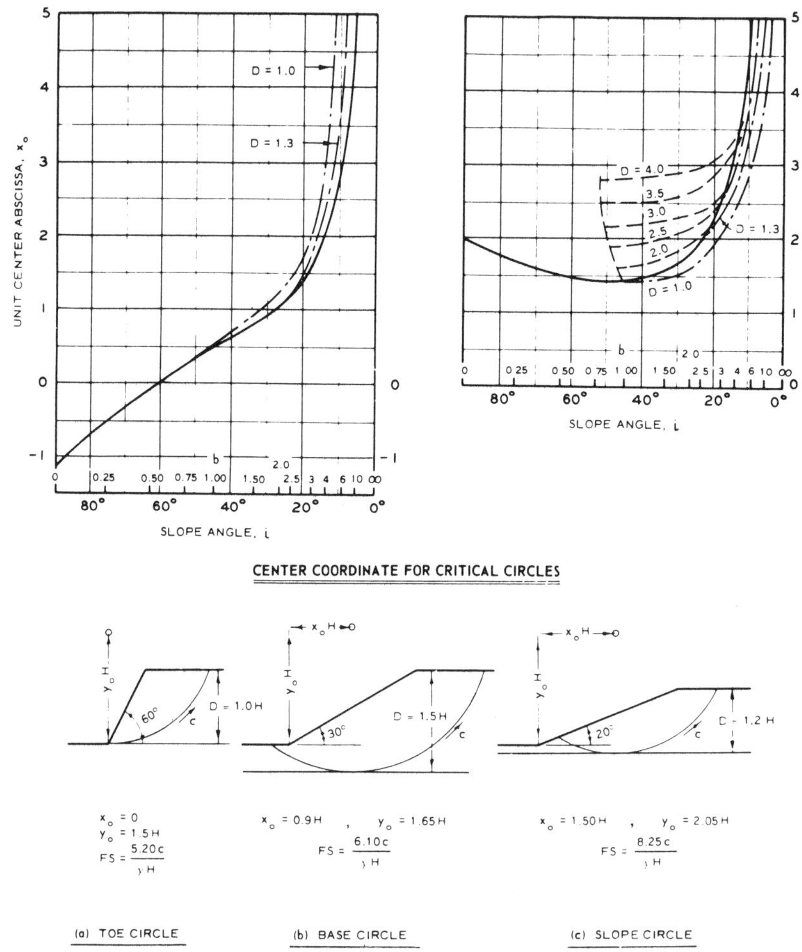

Figure 89. Critical slip circles for purely cohesive slopes, ϕ=0.

Looking at the idealized slice, it is seen that the sliding force is the component of the total weight, W_t (W_t = weight of solid, W_s + weight of water, W_w + weight of overlying equipment or structures, W_e) which acts parallel to the base or

$$F_s = W_t \sin \alpha, \qquad (10)$$

where α is the angle between the normal and the vertical. For the purposes of calculation, the slice can be considered to be one foot in depth; this allows the multiplication of the slice area by its assigned density to obtain W_t.

TABLE 32. STEPS IN COMPUTING SLOPE SAFETY FACTORS

1. Draw a true cross-sectional profile of the slope to be analyzed. If the location of bedding, faulting, jointing, or other mechanical discontinuities is known, then place these on the cross-section.

2. Impose a slip circle on the slope. For a first trial, consult Figure 87, and use the following configuration.

a) Highwall over coal: toe circle.
b) Overburden or refuse pile on solid foundation: slope circle or toe circle.
c) Overburden or refuse pile on weak foundations: base circle.
d) A bench in homogeneous material (highwall and foundation are in the same material): base circle.

3. If groundwater is known to be present, impose the phreatic surface (groundwater table) on the slope.

4. Divide the slope into equal-thickness slices as shown in Figure 90. Determine the weight, including groundwater of each slice. Assign an angle of internal friction and a value of cohesion for each slice base. Be conservative and consider worst-case conditions such as in wet or wintery weather. Measure α for each slice.

5. Determine or assign pore-water pressures, if any, for each slice.

6. Calculate the sliding force according to equations (10) and the resisting force with equation (11) for each slice.

7. Summarize the forces and determine the factor of safety according to equation (12).

8. Repeat this procedure starting with step 2. However, alter the position of the slip circle. At this point, experience has been gained with the analysis, it is wise to try other slope slip surfaces. Instead of a complete circle, for instance, a partial circle truncated by fault or joint cutoffs can be tried. If a plane sliding surface is observed in the slope face, then this should be used also as the base for slices. Continue repeating the analysis until confident that the lowest factor of safety has been calculated.

9. If this factor of safety is less than 1.5 for a working slope, or 4.0 for a permanent slope, then it is necessary to redesign the slope, altering either slope height or overall slope angle or both.

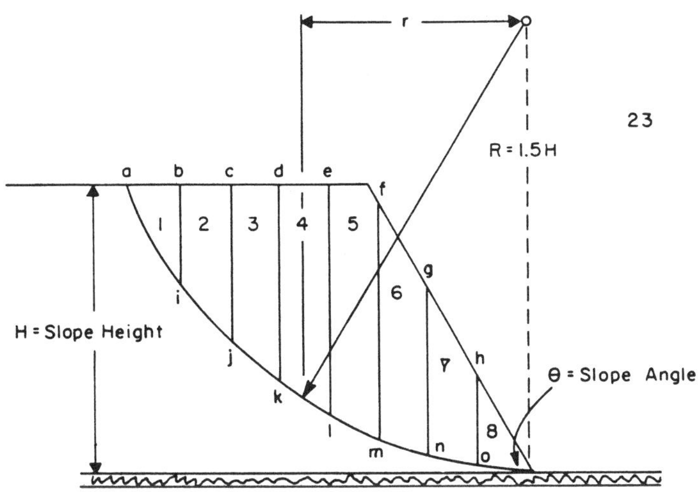

Figure 90. Slices on a toe circle for use in slip-circle analysis of slope stability.

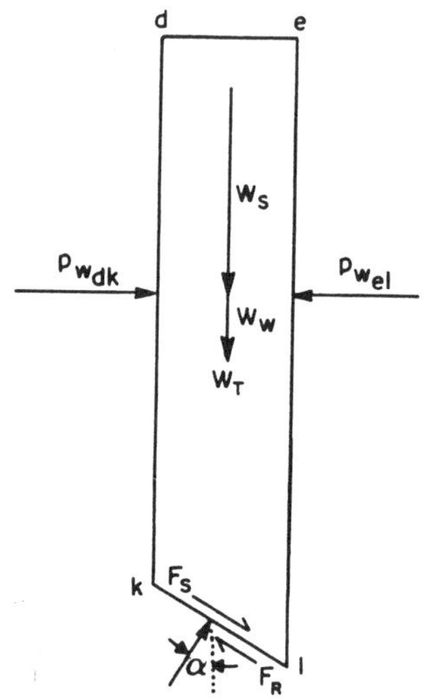

Figure 91. The forces on slice 5.

The resisting forces are the combination of shear strength due to friction and cohesion as defined in equation (8). It should be noted that the method of slices compares forces, not stresses, and that the equations in this section refer to the forces on a slice, not stresses (force per unit area). Since the friction or resisting force is a function of the effective normal force on the sliding surface, equation (11) can be written,

$$F_R = (W_t - \overline{Pw}_{kl}\ L_{kl})\ \cos\alpha\ \tan\phi_{kl} + c'L_{kl} \qquad (11)$$

where W_t is the total slice weight, $\overline{Pw}_{kl}$ is the pore-water pressure acting on line kl, L_{kl} is the length of line kl, ϕ_{kl} is the angle of internal friction along kl, and c' is the effective cohesion along the line kl. Effective cohesion refers to the cohesion that exists when wetted. If there is no groundwater and the slope is dry, then c' = c. Substituting in equation (11), using moments instead of forces, gives the analysis for any one circle.

$$\text{Factor of Safety} = \frac{R\Sigma\ (W_t - \overline{Pw}L)\cos\alpha\ \tan\phi + R\Sigma c'L}{R\Sigma\ W_t\ \sin\alpha} \qquad (12)$$

An equivalent value for the denominator is

$$\text{Sliding Moments} = \Sigma W_t\ r = R\Sigma\ W_t\ \sin\alpha.$$

When performing the analysis, it is often helpful to tabularize the variables. It must be remembered that the analysis should be repeated with different circles until a minimum factor of safety is found. Both the circle's radius and the circle's center should be varied. If it is impossible to obtain a factor of safety greater than that required by regulation, then the mine engineer must consider changing the slope itself by altering either H or Θ or both.

Factor of Safety, Additional Considerations

The analysis just presented has several simplifying conditions implicit in it. However, the method of slices will accommodate a more complex approach. For instance, the weight of a slice was given as if the density were uniform throughout the slice. It is relatively easy to impose layered geology over Figure 90 and calculate W_t as the aggregation of the weight of each layer present in the slice.

No mention was made of the net effect of pore water pressure on the sides of the slice and its contribution to sliding. Returning to Figure 91, the average pore-water pressure acting on side dk (Pw_{dk}) times the length of dk that is pressurized (the pore-water pressure need not act along the entire side) minus the similar force on side el gives an additional component of sliding. Converting this force to a moment and adding it to the denominator has the effect of lowering the factor of safety. This calculation for side pressure is not precise because it assumes that the side pressures act on the same horizontal plane and do not create a moment couple. A computerized analysis of slices has been developed to handle these unbalanced forces, but its use goes beyond the level of this report.

Additionally, the analysis given assumes that each slice stands alone and shares no strength with its neighboring slice. This assumption is valid for broken material but not for intact geologic material. The more sophisticated computer analysis (Morgenstern and Price, 1965) mentioned above is capable of handling this additional strength variable. The effect of including this side strength is to increase the safety factor so that ignoring it tends to be a conservative move. Finally, the analysis ignores residual ground stresses due to mountain-building activities. It also ignores poro-water pressures due to the movement of groundwater.

The U.S. Bureau of Mines has published a report (Wang et al, 1972), RI 7685, which describes a computer program for slope stability that combines a total stress analysis with a limiting equilibrium analysis. Basically, the program determines the most likely sliding surface by using a finite element stress analysis. It then determines the factor of safety for that surface. For the computer-oriented engineer, this approach gives the most realistic and rapid solution to stability determinations.

It should be noted and emphasized that the method of slices is not synonymous with slip-circle analysis. Although, in the example in Figures 90 and 91, slices were used with a slip circle, the method of slices is equally appropriate for any sliding surface. For a plane surface, such as overburden stacked on the outslope of a contour mine, slice analysis can be done, but forces are used instead of moments. Equally, slices can be used with a sliding surface determined from existing geology rather than with an assumed slip circle. If the presumed sliding surface is based on existing conditions, only one factor-of-safety analysis need be performed. As a precaution, sliding surfaces other than the obvious one should be analyzed to be certain that the lowest factor of safety for the slope is known.

Constructing a Safe Slope

In review, the following elements, when varied, contribute to or diminish slope stability: slope geometry (height and angle); the physical properties of the slope material (weight, angle of internal friction, and cohesion); geology (structure and presence of discontinuities); the presence of water (its added weight, pore-water pressure which diminishes shear strength, and the possibility of liquefaction); and the strength of the foundation soil or rock which underlies the slope. Slope stability is obtained by varying those factors in the control of the engineer in response to weaknesses or, if lucky, strengths over which there are no control. Thus, a safe slope is one in which the properties of the slope are known and the slope's height, angle, and method of construction are controlled in response to these properties. The next few paragraphs deal with those elements of stability that can be controlled realistically in coal mines.

Slope Height--In many cases the height of a slope is fixed by premining conditions. However, heights can be reduced by use of top benches or pull back of spoil piles. In multiple-seam mining, it may be necessary to remove an upper seam and its overburden completely in order to limit eventual highwall heights. Concern should be felt when an analysis shows that the factor of safety is below 1.5. Obvious signs of impending failure (see section on monitoring) should lead to immediate corrective action. However, it must be borne in mind that slope stability results from a combination of factors. Since slope height and slope angle both affect stability, it is better, for permanent design changes, to change the slope angle. Nonetheless, the following conditions should be watched.

1. Continuing incipient failure of a dragline bench (tension cracks and slumps) may be an indication that the bench is too high. Since other factors such as dragline productivity through cycle time reduction may be improved, an advance bench may be beneficial. This advance bench, cut by mobile machinery or dragline chopdown, lowers the working highwall.

2. Spoil piles which are allowed to become too high should be watched and repaired. A particular problem is the height of spoil on an inside curve in area mining. Because the dragline continues to dump at relatively the same point as it moves around the curve, the pile builds up above normal height. The excess material should be pulled back by mobile equipment because slope failures would cover valuable coal. The same type of problem exists where a haulroad ramp goes into the pit through the spoil. The spoil at either side of the ramp is higher than normal. Slope failure here could close the haulroad.

3. Refuse piles or overburden storage areas should never be allowed to go above their design height. Regulation may mandate maximum size of these piles; but even if not regulated, their permanence demands a cautious approach to factor of safety.

Slope Angle--Many of the permanent remedies for failing slopes involve reducing the slope's overall angle. Most slope angle reduction methods have the effect of increasing the horizontal dimension of the mine. Thus, a reduction in highwall angle will require an increase in the casting radius of a dragline. A reduction in slope angle of overburden piles, either stacked on a slope or in fills, will require more land area to dispose equal amounts. The following methods elaborate on slope angle reduction.

1. Benches cut into a slope have the effect of reducing a slope's angle. As each bench's width is increased, or as its individual height is reduced, the overall slope angle is decreased. Thus, the use of benches, regularly spaced, is an important way of maintaining stability in deep-haul pits. As stated before, many strip mines use at least one bench, the advance dragline bench, and its design and location should consider stability factors. Multiple-seam coal strip mines give the opportunity to use multiple benches for increased slope control.

2. Even though it is not needed for overall slope reduction, all high-

walls which have personnel or vehicles under them should have a safety bench to catch falling rock. Even the most stable slopes will have some loose rock on its face. Loosening can be due to prior blasting or to freezing and weathering. Leaving a safety bench of some five meters (15 feet) in width, no more than one third of the way up the slope, is a good way to safely catch falling rock. In this case, the safety bench is most effective for steep slopes because the falling rocks have little horizontal movement when falling and do not bounce or bound over the safety bench. In the most protected cases, such as public highways, a chain-link fence is provided to catch falling rock.

3. Sometimes a bench is built up at the base of an unstable slope. This transported material increases toe stability, and decreases the overall slope angle. Such built up toes can be used for unstable refuse or overburden fill piles. If the transported material is taken from the top of a slope and placed at its toe, both slope angle and slope height are reduced.

4. The twenty-degree limitation (Public Law 95-87) on out-slope stacking and the current requirement for approximate original contour, have greatly reduced the concern for slope stability of spoil on sloping foundations.

Stabilization of Slope and Foundation--Until now, the physical properties of the slope material have been taken as a given. However, when constructing an overburden or refuse pile, there are many steps that can be taken, other than changing the pile geometry, which enhance the pile's stability. These steps improve the foundation and strengthen the fill.

1. The fill should not go on top of vegetative debris such as brush, shrubs, branches, or general forest matter. Vegetation rots or burns and eventually looses all strength which can create settlement and piping; rotting debris is slippery and can act as a lubricant. Finally, rotting vegetable matter can lower effluent water quality below acceptable limits.

2. Top soil is too valuable to bury with refuse or spoil. Additionally, because of its organic, clay, and water content it makes a poor foundation. Material piled on top soil can fail in the base-circle mode. Top soil should always be stripped from a pile site.

3. Some sub-soil may also be poor foundation material, but yet be too thick to be stripped. If the pile cannot be built elsewhere, some steps can be taken to improve the bearing capacity of the subsoil. Compaction of the soil will improve it, but maximum compaction occurs only at optimum water content. Dry soils may have to be wetted, or, conversely, wet soils dried. Wetting is by spraying, but drying depends both upon the weather and mechanical exposure (plowing, scarifying, etc.) of the soil to the sun. Proctor compaction tests on the soil with varying degrees of contained water will identify the optimum water content. Sometimes a poor soil can be improved by adding crushed stone. In all cases, good drainage must be provided. When the foundation material is less strong than the pile's, more sophisticated computation of bearing stability should be undertaken than has been described here. A soil mechanics consultant is suggested.

4. It is possible to increase the density and cohesion and decrease the porosity and permeability of cohesive soils and rocks that are put into piles by placing them in horizontal layers and compacting these layers as they are placed. Although compaction will have little effect on purely cohesionless material, most refuse and strip mine overburden has a high proportion of shale, mudstone, or other clay-containing materials. These cohesive materials should be dumped on a horizontal surface and then spread into two-to four-feet thick layers. The layer thickness limitation is so

that the weight of the spreading machine will be effective throughout the layer. Layers that are too thick will not be compacted as the stresses from the machine's weight will be too diffuse. High-bearing load spreaders such as rubber-tired dozers or even spreader-compactors should be used in favor of low-bearing machines such as track-type dozers. This form of construction is imperative for refuse piles and strongly suggested for overburden fills where the overburden could be classed as cohesive in nature.

Layering and compacting has decided advantages in reducing spontaneous combustion in refuse piles. It also reduces water infiltration and associated pollution from these piles. Finally, it is a good way of reducing the potential man-made sliding surfaces within a pile. By insuring that all layers are horizontal and by locking one layer to another through compaction, the natural stability of flat-lying sediments is emulated.

Drainage--Generally speaking, a permanently-stable slope is one that does not contain excessive water. Mine slopes, either in the highwall or in the overburden, should be protected from surface water by diversion ditches. Built-up slopes, such as refuse piles or valley fills, especially if they are on sloping ground, should have foundation drains. Existing slopes that show signs of saturation—seepage can be partially dewatered by horizontal wells or drains that are drilled into the base of the slope. These are particularly effective if they intercept vertical fractures or joints. Although good drainage is of paramount importance, it is a complex subject that is handled well in other manuals.

Other Stabilizing Actions--There are a variety of ways to stabilize a slope that have not been discussed. These include cable or steel-bar reinforcement, grouting, retaining walls, and so forth. Most of these methods are expensive and not often used in mining. However, there may be occasions when they are required. For instance, if the slope above a major haulroad or other permanent facility, such as a tipple, begins to move, it may be necessary to stabilize it by some complicated means, such as tie rods; there is no room or no time to take the simpler steps that have been described. Again, a consultant should be hired.

For convenience, a variety of slope stabilizing methods are listed in Table 33 (Kimball, undated). Both methods that have been described and those that have only been alluded to are included.

Monitoring

The final portion of this section deals with the monitoring of existing slopes to determine their stability. This monitoring can range from casually collected visual clues to the most intensive observations by scientific equipment. Three categories are examined: visual clues, field surveying, and special slope monitoring equipment.

Visual Clues--Frequently, incipient slope failures are spotted by recognition of surface clues by knowledgeable and responsible mine workers. These people, who should be taught to recognize the signs of impending failure, are encouraged to report their concerns immediately. Optimally, corrective action can be taken to repair the slope; at a minimum, personnel and equipment can be removed from the dangerous area.

Water in a bank is a potential hazard. Therefore water seepage from the face of a bank or from the toe of a bank, unless from a designed drain point, is an early warning sign of problems.

Cracking is a sure sign of slope movement and potential massive failure. Look for tension cracks at the upper part of the slide. If failure is soon to occur, the downhill side of the tension crack will be lower than the

TABLE 33. SUMMARY OF METHODS FOR PREVENTION AND CORRECTION OF LANDSLIDES

Effect on Stability of Landslide	Method of Treatment	General Use		Frequency of Successful Use[1]			Position of Treatment on Landslide[2]	Best Application and Limitations
		Prevention	Correction	Fall	Slide	Flow		
Not affected	I. Advoidance Methods A. Relocation	X	X	2	2	2	Outside slide limits	Most positive method if alternate location economical
Reduces shearing stresses	II. Excavation[3] A. Removal of head	X	X	N	1	N	Top and head	Deep masses of cohesive material
	B. Flattening of slopes	X	X	1	1	1	Above road or structure	Bedrock: also extensive masses of cohesive material
	C. Benching of slopes	X	X	1	1	1	Above road or structure	where little material is removed at toe
	D. Removal of all unstable material	X	X	2	2	2	Entire slide	Relatively small shallow masses of moving material
Reduces shearing stresses and increases shear resistance	III. Drainage: A. Surface: 1. Surface ditches	X	X	1	1	1	Above crown	Essential for all types
	2. Slope	X	X	3	3	3	Surface of moving mass	Rock facing or previous blanket to control seepage

(continued)

TABLE 33. (continued)

Effect on Stability of Landslide	Method of Treatment	General Use		Frequency of Successful Use[1]			Position of Treatment on Landslide[2]	Best Application and Limitations
		Prevention	Correction	Fall	Slide	Flow		
Reduces shearing stresses and increases shear resistance (cont'd)	3. Regrading surface	X	X	1	1	1	Surface of moving mass	Beneficial for types
	4. Sealing cracks	X	X	2	2	2	Entire, crown to toe	Beneficial for all types
	5. Sealing joint planes and fissures	X	X	3	3	N	Entire, crown to toe	Applicable to rock formation
	B. Subdrainage							
	1. Horizontal drains	X	X	N	2	2	Located to intercept and remove subsurface water	Deep extensive soil mass where ground water exists
	2. Drainage trenches	X	X	N	1	3		Relatively shallow soil mass with groundwater present
	3. Tunnels	X	X	N	3	N		

(continued)

TABLE 33. (continued)

Effect on Stability of Landslide	Method of Treatment	General Use		Frequency of Successful Use[1]			Position of Treatment on Landslide[2]	Best Application and Limitations
		Prevention	Correction	Fall	Slide	Flow		
Increases shearing resistance	IV. Restraining Structures:							
	A. Buttresses at foot							
	1. Rock fill	X	X	N	1	1	Toe and foot	Bedrock or firm soil at reasonable depth
	2. Earth fill	X	X	N	1	1	Toe and foot	Counterweight at toe provides additional resistance
	B. Cribs or retaining walls	X	X	3	3	3	Foot	Relatively small moving mass or where removal of support is negligible
	C. Piling:							
	1. Fixed at slip surface		X	N	3	N	Foot	Shearing resistance at slip surface increased by force required to shear or bend piles

(continued)

TABLE 33. (continued)

Effect on Stability of Landslide	Method of Treatment	General Use		Frequency of Successful Use[1]			Position of Treatment on Landslide[2]	Best Application and Limitations
		Prevention	Correction	Fall	Slide	Flow		
Increases shearing resistance (cont'd)	2. Not fixed at slip surface		X	N	3	N	Foot	
	D. Dowels in rock	X	X	3	3	N	Above road or structure	Rock layers fixed together with dowels
	E. Tie-rodding slopes	X	X	3	3	N	Above road or structure	Weak slope retained by barrier which in turn is anchored to solid formation
Primarily increases shearing resistance	V. Miscellaneous Methods							
	A. Hardening of slide mass:							
	1. Cementation or chemical treatment							
	(a) At foot		X	3	3	3	Toe and Foot	Non-cohesive soils
	(b) Entire slide mass		X	N	3	N	Entire slide mass	Non-cohesive soils to prevent movement

(continued)

TABLE 33. (continued)

Effect on Stability of Landslide	Method of Treatment	General Use		Frequency of Successful Use[1]			Position of Treatment on Landslide[2]	Best Application and Limitations
		Pre-ven-tion	Cor-rec-tion	Fall	Slide	Flow		
Primarily increases shearing resistance (cont'd)								temporarily in relatively large moving mass
	2. Freezing	X		N	3	3	Entire	
	3. Electro-osmosis	X		N	3	3	Entire	Effects hardening of soil by reducing moisture content
	B. Blasting		X	N	3	3	Lower half of landslide	Relatively shallow cohesive mass underlain by bedrock. Slip surface disrupted; blasting may also permit water to drain out of slide mass
	C. Partial removal of slide at toe			N	N	N	Foot and Toe	Temporary expedient only; usually decreases stability of slide

1 1 = frequently; 3 = rarely; N = not considered applicable

2 Relative to moving or potentially moving mass. 3 Exclusive of drainage methods.

uphill side. En echelon shear cracks will appear at the sides of the slumping material, and there will be a zone of uplift or bulging at the base of the slump. The uplift zone, as it spreads, can also show signs of tension cracks. Unlike the crest cracks which are normal to the slide direction, the toe cracks may well be oriented parallel to the slide direction.

If vegetation covers a slope, then these cracks may not be seen readily. However, leaning trees or bare patches of soil, called soil stretching, may indicate movement. Terracettes, which look like regular cattle traces around a hillside, are signs of soil creep. Other indicators of movement are surface features that were installed vertically, or in a row, or both, and that are now out of alignment. Fence lines, stone walls, roads, pole lines, or pipelines that are out of alignment toward the downhill slope may be indications of creep or slumping.

Field Surveying--Actual displacement of sliding can be measured by a variety of field surveying techniques. The precision of measurement is limited only to that of the measurement system itself; there are no peculiarities, in other words, that reduce a method's effectiveness. The selection of any particular method will depend upon the conditions on the slope and the equipment available to the engineer. Some of the more useful methods are listed.

1. The vertical descent of permanent markers can be obtained by running a line of levels from a stable zone across the slump to the next stable zone. Short lengths of pipe or concrete are sunk into the soil; periodic checks on their level gives their vertical movement. This survey is quick, cheap, and possible with normal mine equipment. It does require a man to cross the slump zone and does not give horizontal displacement.

2. Several surveying methods which use a transit can improve leveling. Again, permanent markers are installed. To get total displacement, vertical angles are taken as well as horizontal angles from a measured base line. Although triangulation computation is more complicated, nobody is required to traverse the slump. Should the slump be stable enough to cross, however, then stadia measurements can reduce computational work. Although a high-order transit traverse can be done across a slump, triangulation is preferred.

3. Electronic Distance Measurers (EDMs), particularly those that embody a transit for angular measurement, substitute well for transits and expedite the survey. However, the targets are not disposable, and thus, a worker is required to cross the zone. Modern EDM's are reliable, replicable, fast and easy to use. They are, however, relatively expensive and subject to deterioration if not used often.

4. Photogrammetry is an excellent way to observe ground motion. The camera records the position of permanent markers; displacements are scaled from the photograph by measuring distances between markers in the slump and one on stable ground. Although aerial photography gives horizontal displacements more precisely than vertical, oblique stereo photography will give both. Actually, photogrammetry does not require an airplane and good results can be obtained with a tripod-mounted camera.

Special Monitoring Equipment--If more sophisticated (although not necessarily more accurate) measurement of slope failure is desired, there is a variety of instruments capable of giving displacements, strains, tilts, or record of acoustic emissions. Home-made extensometers and the most sophisticated telemetric monitors have been used on moving slopes. Some of the reasons for using this equipment over field surveying are continuous output, unattended performance, information from depth, or scientific faddishness. A slope stability consultant will be glad to advise whether any of this equipment is appropriate. Frankly, because much of it is fragile and prone to failure from tampering, it should be used only when normal surveying appears to be inadequate.

Assessing Displacement Information--To close this section, it is necessary to suggest what can be done with monitoring information, without requiring a textbook on rock mechanics. It is suggested that one step is most important. Displacement data should be plotted against time so that total displacement from the beginning of measurement and incremental displacements can be observed. The mathematical slope of such a plot is the displacement rate. If the rate is increasing over time, then the mine slope is surely failing. If the rate is decreasing—the graph's slope is flattening—, then the mine slope is stabilizing.

By reasonable extrapolation of a displacement-time plot, predictions can be made of when a slope will fail disastrously. Although continuing but small displacements can be considered a failure, they are of no consequence to a working slope that may soon be removed or stabilized. On the other hand, continuing displacements in a permanent slope are signs of need for immediate, corrective action. If these small displacements get larger day by day, then a day will come when the displacement is total.

SUMMARY

This chapter has given a three-part analysis of slope stability. The introductory and concluding sections can be read independently to get a descriptive review of slope stability elements and the diagnosis of slope failure. If engineering calculations of safety factors are required, then the middle section should also be read. For the engineer who is deeply involved with stability analyses, further references are suggested.

Operating Considerations

INTRODUCTION

In judging the performance of stripping equipment, several material and equipment characteristics need to be taken into account. Given the best of conditions, a piece of equipment may not achieve the optimum level of production due to operational conditions. In practice, the mechanical efficiency of machines never equals one hundred percent, and designed production is rarely realized. Therefore, several correction factors must be applied to adjust equipment performance from a theoretical maximum to that realizable in practice.

CORRECTION FACTORS

The correction factors arise from machine design, material characteristics, and modes of operation.

Swing Factor

The superstructure on which the dragline and shovel revolve can swing 360° in either direction. The cycle time of the machines is affected by the swing angles. The factor that determines the effect of swing angle variations on cycle time, with 90° taken as the normal, is called the swing factor. The smaller the angle of swing, the less the cycle time and the more cycles of operation completed per hour. For example, an angular increase from 90° to 120° for a shovel will increase the cycle time approximately 10 percent (Tables 34 and 35).

Depth Factor

There is an optimum cutting depth for a dragline or a shovel. When the equipment operates at any depth other than this optimum, machine production decreases. The correction for operating at different depths is usually made in the cycle time calculation. Tables 34 and 35 incorporate the effect of depth.

Operating Time Per Hour (MPH)

This factor is applied to derive the machine working time from the theoretical 60 minutes per hour. For example, if a machine works 48 minutes every hour, the MPH factor is 0.80 (48/60) (Table 36).

Bucket-Fill Factor

The bucket-fill factor is sometimes referred to as carry factor or dipper factor for shovels and bucket factor for draglines. This factor expresses the actual quantity carried as a fraction of the nominal dipper or bucket capacity, and is a function of the digging conditions (Table 37).

TABLE 34. EFFECT OF DEPTH OF CUT AND ANGLE OF SWING ON SHOVEL CYCLE TIME (DREVDAHL, 1961)

Depth of cut in % of Optimum	Angle of Swing in Degrees						
	45°	60°	75°	90°	120°	150°	180°
40	.93	.89	.85	.80	.72	.65	.59
60	1.10	1.03	.96	.91	.81	.73	.66
80	1.22	1.12	1.04	.98	.86	.77	.69
100	1.26	1.16	1.07	1.00	.88	.79	.71
120	1.20	1.11	1.03	.97	.86	.77	.70
140	1.12	1.04	.97	.91	.81	.73	.66
160	1.03	.96	.90	.85	.75	.67	.62

Production at a given depth and angle of swing = Production at optimum depth and 90° swing x Factor for swing and depth

TABLE 35. EFFECT OF DEPTH OF CUT AND ANGLE OF SWING ON DRAGLINE CYCLE TIME

Depth of cut in % of Optimum	Angle of Swing in Degrees							
	30°	45°	60°	75°	90°	120°	150°	180°
20	1.06	.99	.94	.90	.87	.81	.75	.70
40	1.17	1.08	1.02	.97	.93	.85	.78	.72
60	1.24	1.13	1.06	1.01	.97	.88	.80	.74
80	1.29	1.17	1.09	1.04	.99	.90	.82	.76
100	1.32	1.19	1.11	1.05	1.00	.91	.83	.77
120	1.29	1.17	1.09	1.03	.985	.90	.82	.76
140	1.25	1.14	1.06	1.00	.96	.88	.81	.75
160	1.20	1.10	1.02	.97	.93	.85	.79	.73
180	1.15	1.05	.98	.94	.90	.82	.76	.71
200	1.10	1.00	.94	.90	.87	.79	.73	.69

Production at a given depth and angle of swing = Production at optimum depth and 90° swing x Factor for swing and depth

TABLE 36. MINUTES OF OPERATING TIME PER HOUR AND EQUIVALENT OPERATING EFFICIENCIES

Minutes per hour	Efficiencies, %
60	100
55	92
50	83
45	75
40	67
35	58
30	50
25	42
20	33
15	25

TABLE 37. FILL FACTOR (DREVDAHL, 1961)

Easy Digging	Medium Digging	Hard Digging	Rock
Shovel Dipper Factor 85% to 100%	Shovel Dipper Factor 80% to 90%	Shovel Dipper Factor 70% to 80%	Shovel Dipper Factor 40% to 60%
Loose, soft, free running materials Close lying, which will fill dipper or bucket to capacity and frequently provide heaped load. Overload compensates for swell of material	Harder materials that are not difficult to dig without blasting, but break up with bulkiness causing voids in dipper or bucket.	Materials requiring some breaking up by light blasting or shaking. More bulky and somewhat hard to penetrate, causing voids in dipper or bucket.	Blasted rock, hardpan and other bulky materials, which cause considerable voids in dipper or bucket and are difficult to penetrate.
Dry sand or small gravel. Moist sand or small gravel. Loam. Loose earth. Muck. Sandy clay. Loose clay gravel. Cinders or ashes. Bituminous coal. Very well blasted material.	Clay - wet or dry. Coarse gravel. Clay gravel, packed. Packed earth. Anthracite coal.	Well broken limestone, sand rock and other blasted rocks. Blasted shale. Ore formations (not of rock character) requiring some blasting. Heavy wet, sticky clay. Gravel with large boulders. Heavy, wet gumbo. Cemented gravel.	Hard tough shale. Limestone. Trap rock. Granite. Sandstone. Taconite. Conglomerate. Caliche rock. Any of these blasted to large pieces mixed with fines and dirt. Tough, rubbery clay that shaves from bank.

These conditions are classified as easy, medium, or hard digging. Since the values describing the conditions are relative, detailed field tests should be conducted to determine the fill factors.

Load and Swell Factors

The load factor (LF) relates the bank volume to the loose volume handled. This factor is particularly important in contracted overburden removal, since payments are based on either bank or loose volume. Mathematically, the load factor is equal to loose density divided by bank density (Table 38).

TABLE 38. MATERIAL WEIGHTS, PERCENTAGE, SWELL AND LOAD FACTORS (CATERPILLAR TRACTOR CO., 1975)

Weights shown are averages, rounded off to the nearest 50 pounds. Since actual weights are affected by moisture content, grain size, degree of compaction and other local conditions, tests must be made to determine exact material characteristics. The load factor and swell have been determined from the weights given, using the equation found on Page 190.

Material	lb/bcy	% Swell	lb/lcy	Load Factor
Bauxite	3200	33	2400	.75
Callche	3800	82	2100	.55
Cinders	1450	52	950	.66
Carnotite, Uranium Ore	3700	35	2750	.74
Clay - Natural Bed	3400	22	2800	.82
Dry excavated	3100	23	2500	.81
Wet excavated	3500	25	2800	.80
Clay & Gravel: Dry	2800	41	2000	.71
Wet	3100	11	2800	.90
Coal: Anthracite, raw	2700	35	2000	.74
washed	2500	35	1850	.74
Bituminous, raw	2150	35	1600	.74
washed	1900	35	1400	.74
Decomposed Rock:				
75% Rock 25% Earth	4700	43	3300	.70
50% Rock 50% Earth	3850	33	2900	.75
25% Rock 75% Earth	3300	25	2650	.80
Earth - Dry packed	3200	25	2550	.80
Wet excavated	3400	27	2700	.79
Loam	2600	23	2100	.81
Granite - Broken	4600	64	2800	.61
Gravel - Pitrun	3650	12	3250	.89
Dry	2850	12	2550	.89
Dry 1/4"-2" (6-51 mm)	3200	12	2850	.89
Wet 1/4"-2" (6-51 mm)	3800	12	3400	.89
Sand & Clay - Loose	3400	27	2700	.79
Compacted	—	--	4050	--
Gypsum - Broken	5350	75	3050	.57
Crushed	4700	75	2700	.57
Hematite, Iron ore	4900	18	4150	.85
Limestone - Broken	4400	69	2600	.59
Magnetite, Iron ore	5500	18	4700	.85
Pyrite, Iron ore	5100	18	4350	.85
Sandstone	4250	67	2550	.60
Sand - Dry, loose	2700	12	2400	.89
Damp	3200	12	2850	.89
Wet	3500	12	3100	.89
Sand & Gravel - Dry	3250	12	2900	.89
Wet	3750	10	3400	.91
Slag-Broken	4950	67	2950	.60
Stone - Crushed	4500	67	2700	.60
Taconite	7100-9450	75-72	4100-5400	.57-58
Top Soil	2300	43	1600	.70
Traprock - Broken	4400	49	2950	.67

It is expressed as a fraction. On the other hand, the swell factor (SF) is the percentage increase in volume of material from bank state to loose state. The relationship between load factor and swell factor is as follows:

$$LF = \left(\frac{100}{100 + SF}\right)$$

Other Factors

There are several other factors such as management and operator efficiencies and climatic conditions that affect equipment performance. These cannot be accurately quantified. Atkinson (1971) has divided both the job conditions and management practices into four categories each, from excellent to poor and has provided a matrix of values for various combinations of job conditions and management conditions (Table 39). Drevdahl (1961) has also presented a similar table of correction factors for combined job-management conditions.

For some earth-moving equipment, a more detailed breakdown of these factors is possible. For example, Caterpillar Tractor Company has provided correction factors for dozers for grade, visibility, type of materials, and mode of operation. Since values for these factors are specific to particular equipment, they are discussed and presented in the report along with the equipment production calculations.

TABLE 39. OPERATING EFFICIENCY FACTORS (ATKINSON, 1971)

JOB CONDITIONS ↓	MANAGEMENT CONDITIONS			
	EXCELLENT	GOOD	FAIR	POOR
Excellent	0.83	0.80	0.77	0.77
Good	0.76	0.73	0.70	0.64
Fair	0.72	0.69	0.66	0.60
Poor	0.63	0.61	0.59	0.54

CYCLE TIME

The cycle time of stripping equipment consists of two distinct components, the fixed and the variable. For a stripping shovel, crowding, dump, digging, and retract times are constant or fixed in every cycle. However, the swing and swing-back times can vary from cycle to cycle, and therefore, constitute the variable component of the cycle time. The relationship between production and cycle time is shown in Table 40.

Calculation of theoretical cycle times is rather involved. Manufacturers usually provide enough horsepower for various motions to maintain a uniform cycle time for comparable conditions. Every activity and motion of the machine is dependent on motor horsepower, machine weight, and the need for proper control. For example, when calculating the swingtime, one has to know the average translational inertia of the machine and the average torque from the speed-torque curve. Here, one must also take into consideration the swing-gear ratio. The acceleration and deceleration times, plus

TABLE 40. WALKING DRAGLINE MACHINE OUTPUT (LEARMONT, 1975)

$$\text{Bank Yards per Operating Hours} = \frac{3600}{\text{Cycle Time}} \times \text{Operating Delay Factor} \times \frac{\text{Efficiency Factor}}{\text{Swell Factor}} \times \text{Bucket Capacity}$$

	2570-W 335 Ft Boom, 38°300' Radius, 110 Cu Yd	2570-W 360 Ft Boom, 31-1/2°340' Radius, 90 Cu Yd
Case 1, 110 Ft Overburden		
Weighted Average Cycle Time, From Table 42	48.2	45.7
Operating Delay Factor, From Table 43	0.80	0.82
Efficiency Factor	0.85	0.85
Swell Factor	1.25	1.25
Bank Yards per Operating Hour	40.6 x Bucket Capacity = 4470 yd per hr	43.9 x Bucket Capacity = 3950 yd per hr
Case 2, 130 Ft Overburden		
Weighted Average Cycle Time, From Table 42	57.7	47.7
Operating Delay Factor	0.82	0.84
Efficiency Factor	0.85 Bank Material 0.95 Rehandled Material	0.85
Average Efficiency Factor	0.88	
Swell Factor	1.25	1.25
Bank Yards per Operating Hour (including 38% rehandle)	36.0 x Bucket Capacity = 3960 yd per hr	43.1 x Bucket Capacity = 3880 yd per hr
Productive Bank Yards per Operating Hour	26.1 x Bucket Capacity = 2870 yd per hr	

the continuous running time at full speed are also taken into account for this calculation. In industry literature, on the basis of machine parameters and design information, standard cycle times have been provided (Table 41). These ideal cycle times are subject to correction for any nonstandard application (chopping) or differences in swing angle, digging height, etc. The influence of these factors were discussed earlier. Cycle times can also be obtained through time studies in the field.

TABLE 41. APPROXIMATE CYCLES PER HOUR FOR STRIPPING SHOVELS AND DRAGLINES (ATKINSON, 1971)

Bucket or Dipper Size		Dragline	Shovel
Yd^3	m^3		
8-35	6-27	58	69
36-59	28-45	56	68
60-200	46-150	53	64

* These figures are based on a 90° swing for a shovel and a 120° swing for a dragline, which approximates most field conditions.

Decreases in cycle time can be achieved by increasing the motor horsepower. For example, maximizing the swing efficiency would not only require an increase in power, but also a reduction in inertia loads, resulting in only a modest reduction in cycle time. Swing power can be increased, but this does not yield a proportional reduction in cycle time. An increase of 50 percent in power, for example, reduces total cycle time of a dragline by only 10 percent. Such major increases in power will be reflected in high operating costs.

For draglines, the swing time is the most critical, although in some cases, hoisting time is important. In cases where the hoisting time makes up to about 40 percent of the total cycle time, an increase of 33 percent in power reduces the cycle time by 10 percent. This reduction in cycle time occurs only in the hoisting time portion of the cycle, unlike the swing-time reduction which spreads over the entire range of the total cycle. The average reduction in cycle time, due to decreases in hoisting time, in actuality, will be less than 10 percent.

The bucket-fill time appears fixed and is not affected by the size of the bucket, provided the machine operator is efficient.

It should be noted that a chopping operation increases the cycle time by as much as 40 to 50 percent. An average increase of 45 percent would be appropriate for calculation over the normal overall cycle time without extended swing or hoisting time (Table 42). Also, bench digging or extension increases the dragline cycle time by as much as 80 percent over the overall cycle time.

Operating Delay

In production calculations, not all the hourly minutes are parts of the cycles as there are inevitable delays that occur which take some of the operating time that could have been better utilized for useful cyclic operations. These are called operating delays (Table 43). Such delays can occur due to: 1.) machine movement, 2.) cleaning coal, and 3.) miscellaneous minor causes. An estimate of these delays for selected conditions is neces-

TABLE 42. WALKING DRAGLINE CYCLE TIME CALCULATION (LEARMONT, 1975)

	2570-W 335-Ft Boom, 38° 300' Radius 110 Cu Yd				2570-W 360-Ft Boom, 31-1/2° 340' Radius, 90 Cu Yd		
	Chop Down	Median Depth	Maximum Depth	Extend Bench	Chop Down	Median Depth	Maximum Depth
Case 1, 110 Ft Overburden							
Depth, Ft	20	45	90		20	45	90
Percent of Total	18	41	41		18	41	41
Average Swing Angle, Degree	120	45	90		105	40	80
Spot Bucket, Sec	3.0	3.0	3.0		3.0	3.0	3.0
Fill Bucket, Sec	10.2	10.2	10.2		9.5	9.5	9.5
Hoist Bucket, Sec	--	--	20.6		--	--	--
Swing Over, Sec	23.2	13.5	--		22.0	13.2	18.8
Swing Back, Sec	22.3	12.7	18.5		21.0	12.4	17.8
Total Time, Sec	58.7	39.4	52.3		55.5	38.1	49.1
Weighted Average Cycle Time, Sec		48.2				45.7	
Case 2, 130 Ft Overburden							
Depth, Ft	20	55	110		20	55	110
Percent of Total	11	30	30	29	15	43	42
Average Swing Angle, Degree	135	50	100	180	120	45	90
Spot Bucket, Sec	3.0	3.0	3.0	3.0	3.0	3.0	3.0
Fill Bucket, Sec	10.0	10.2	10.2	10.2	9.5	9.5	9.5
Hoist Bucket, Sec	--	--	24.4	--	--	--	--
Swing Over, Sec	25.1	14.2	--	30.8	23.9	14.1	20.1
Swing Back, Sec	24.2	13.4	19.8	29.8	22.9	13.2	19.1
Total Time, Sec	62.5	40.8	57.4	73.8	59.3	39.8	51.7
Weighted Average Cycle Time, Sec		57.7				47.7	

sary to arrive at actual production time. The actual cycle time in practice is a weighted average of the chopping, minimum, maximum, and optimum-depth cycle times (Table 42).

TABLE 43. WALKING DRAGLINE - OPERATING DELAYS (LEARMONT, 1975)

	2570-W 335-Ft Boom, 38° 300' Radius, 110 Cu Yd	2570-W 360-Ft Boom, 31-1/2° 340' Radius, 130 Cu Yd
Case 1, 110 Ft Overburden		
Approximate rate of face advance per hr, ft	10.8	8.8
Movement across face for a 30 ft advance, ft	80	80
Total dragline movement per hr. ft.	39.6	32.3
Move time per hour (1 min per 8 ft step including preparation), min	5.0	4.0
Clean coal time per hr, min	3.0	2.4
Miscellaneous delays per hr, min	4.0	4.0
Total operating delays per hr, min	12.0	10.4
Operating delay factor, %	80	82
Case 2, 130 Ft Overburden		
Approximate rate of face advance per hr, ft	6.6	7.4
Movement across face for a 30 ft advance, ft	160	80
Total dragline movement per hr, ft	41.8	27.1
Move time per hour (1 min per 8 ft step including preparation), min	5.2	3.4
Clean coal time per hr, min	1.8	2.1
Miscellaneous delays per hr, min	4.0	4.0
Total operating delays per hr, min	11.0	9.5
Operating delay factor, %	82	84

SURFACE MINE DESIGN

The topography of the area determines the stripping method to be employed. For example, mountainous areas call for mountaintop removal or other contour mining methods, while flat terrain may dictate simple overcasting by area techniques. Seam thickness, as well as multiple-seam occurrences, must also be considered in equipment and method selection. Seam pitch, characteristics of the coal, the amount of groundwater over the coal seam, and stability of the highwall are also important factors. Finally, information concerning faults within the deposit, limits of the deposit, property lines, roads, surrounding private properties, and local land-use regulations must be carefully evaluated before actual mine design.

The economic life of the mine is related to pit planning by the yearly average stripping ratio and other factors that could assure profitability of the venture. Early return on the investment is desired and often pit designs are planned with this in mind. Equipment selection has direct financial consequences on pit planning. It may be possible to achieve low operating and maintenance costs with high initial costs. On the other hand, low initial capital investment will call for expansion capital in the future.

Pit entrances should be centrally located so that haulage distance between the center of gravity of the pit area and the dump is near minimum. Dead heading can be avoided by construction of several pit entrances preferably perpendicular to the cut.

Environmental concerns about surface mining have been on the increase in the last decade. Mine planning should include plans for reclamation procedures prior to the commencement of mining. Regulations require that reclamation be accompanied by revegetation and burial of toxic material to avoid groundwater contamination. Acid mine drainage must be treated and plans for its treatment initiated prior to mining. To gain public and community acceptance, special consideration should be given to the immediate community in hiring unskilled labor. Job training plans should include preparing those hired for more skilled jobs and positions of responsibility.

ENGINEERING DESIGN AND PIT LAYOUT

Generally, equipment is selected to conform with an existing mine design or mines are designed based on the anticipated equipment to be employed. In either case, the equipment performance characteristics and operating conditions determine the actual production levels. The following is a set of definitions for use with the mathematical equations to be developed in this section. Since the use of English units is most common, the derivations here are in English units.

P = Production per month in tons
A = Reject fraction in run-off mine coal
U = Recovery fraction from coal in ground
COAW = Equivalent lb per cu yd of coal in the solid seam (Usually = 2000 lbs)
TT = Equivalent lb per ton of coal
1 ton = 2000 lb for short ton
1 ton = 2240 lb for long ton
1 ton = 1000 kg for metric ton
T = Thickness of coal seam in ft
H = Overburden height in ft
H1 = Depth of dragline working bench in ft
SMH = Scheduled monthly operating hours
CT = Cycle time in seconds
Bs = Bucket size for shovel in cu yd
Bd = Bucket size for dragline in cu yd
BF = Bucket fill factor (fraction)
OF = Operating efficiency (fraction)
W = Pit width in ft
SP = Swell of material (percentage usually 25% to 30%)
Φ = Angle of highwall in degrees
Θ = Angle of spoil in degrees
FACD = MUF factor for dragline in lb/MUF
FACS = MUF factor for shovel in lb/MUF
Rs = Shovel reach in ft
Rd = Dragline reach in ft
Rsd = Shovel dumping radius in ft
Rdd = Dragline dumping radius in ft
Bb = Bucket size of bulldozer in cu yd
Bl = Bucket size of loader in cu yd
Bc = Bucket size of scraper in cu yd
S = Speed in feet per minute (fpm)
S1 = Loaded speed in mph
S2 = Empty speed in mph
LF = Load factor (fraction)
MF = Material factor (fraction)

MAT = Maintenance factor (fraction)
CF = Carry factor (fraction)
MPH = Available minutes per hour (fraction)
BCYPH = Maximum production per hour in cu yd
HD = Haul distance in ft
TCT = Total cycle time in sec
VL = Loose volume in cu yd
VB = Bank volume in cu yd
HPS = Hours per shift
SPD = Shifts per day
DPW = Days per week
WPM = Weeks per month

Overburden Yardage to be Moved

Knowledge of the amount of overburden to be moved is the primary information on which the selection of a shovel or dragline is based. For a clean coal production of P tons per month, the acreage of coal to be exposed and the amount of overburden to be removed, calculate as follows:

P = Production per month in tons
A = Reject fraction in run-off-mine coal
U = Recovery fraction from coal in ground
COAW = Equivalent lb per cu yd of coal in the solid
T = Thickness of coal seam in ft
H = Overburden height in ft
ACP = Amount of run-of-mine coal to be mined in tons
ACE = Amount of coal to be exposed in tons
CYE = Cubic yards of coal to be exposed
SYE = Square yards of coal to be exposed
ACYE = The acreage of coal to be exposed
COB = Cubic yards of overburden to be removed

$$ACP = \frac{P}{1 - A}$$

$$ACE = \frac{ACP}{U}$$

$$CYE = \frac{(ACE)(TT)}{COAW}$$

$$SYE = \frac{(CYE)(3)}{T}$$

$$ACYE = \frac{SYE}{4840}$$

$$COB = \frac{(SYE)(H)}{3}$$

Calculation of Bucket Size

Knowing the amount of overburden to be removed each month, a suitable bucket size can be calculated.

SMH = Scheduled monthly hours
CT = Cycle time in sec
OF = Operating efficiency

BF = Bucket-fill factor
NTCPM = Number of theoretical cycles per month
TBC = Theoretical bucket capacity
NBSIZE = The actual bucket capacity after considering the swell percentage, the bucket-fill factor and the operating efficiency.

$$NTCPM = \frac{(SMH)(3600)}{CT}$$

$$TBC = \frac{COB}{NTCPM}$$

$$NBSIZE = \frac{TBC}{(OF)(BF)}$$

It is now a common practice among operators to restrict the bucket sizes within the range of 70 to 140 cubic yards. Therefore, if the calculations should lead to a large bucket size, it is advisable to use two or more machines in multiple pits to meet the production requirements.

Calculation of Reach

Since the primary equipment can be a shovel or a dragline, the formulas applicable to both machines are developed here.

1. Shovel Reach--

The usual operating practice with a shovel is to align one set of its track on the rib of coal adjacent to the top of the spoil. An imaginary plane passing through this location would contain the point "D" shown above the front crawler in Figure 92. Projection of a horizontal line from the

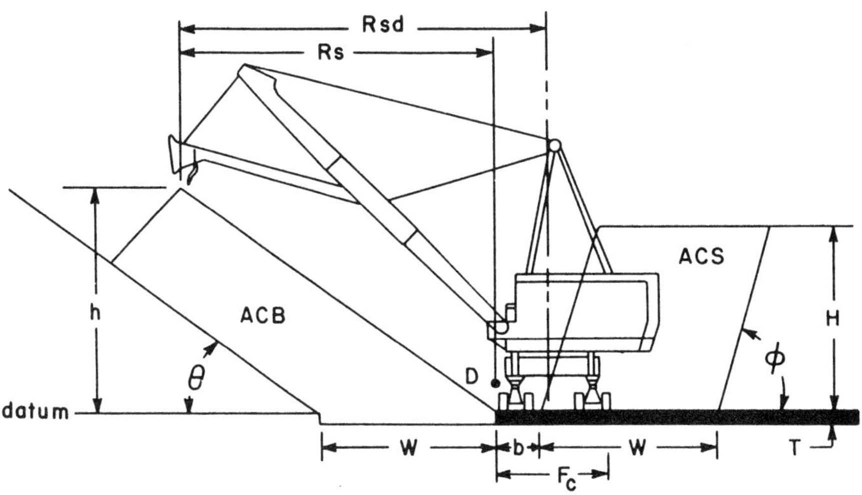

Figure 92. Shovel pit with coal fender (Stefanko, et al., 1973).

imaginary vertical plane to the point of discharge of the shovel dipper establishes the reach of the shovel. The reach is calculated as follows (Rumfelt, 1961). Attention is drawn to Figure 93.

ACS = Area cut in solid, sq ft
ACB = Area cut in broken (or spoil), sq ft
Φ = Angle of highwall, degrees
Θ = Angle of spoil, degrees
h = Height of spoil, ft
H = Height of overburden highwall, ft
T = Thickness of coal, ft
W = Width of pit, ft
b = Berm width, ft. This is assumed to be small and therefore is not included in the calculations.
F_c = Width across both crawlers in ft
Rs = Reach of shovel in ft
Rsd = Dumping radius of shovel in ft

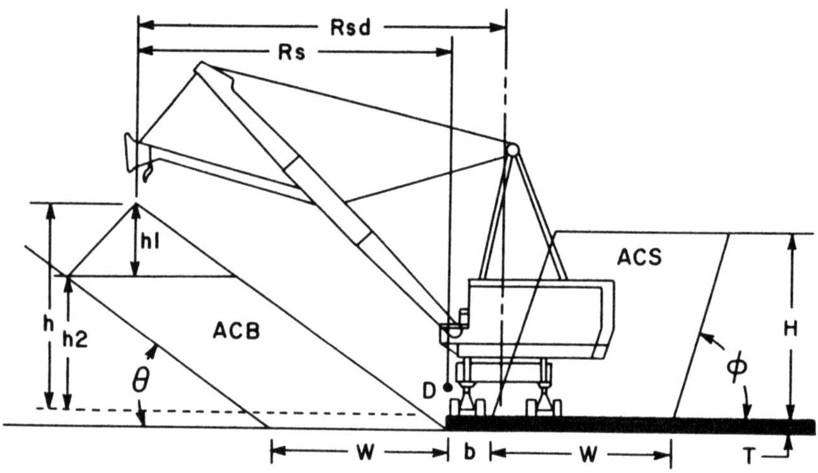

Figure 93. Shovel pit without coal fender (Stefanko, et al., 1973).

With reference to Figure 93

$$h_1 = (\frac{W}{2}) \text{ Tan } \Theta$$

$$h_2 = h - (\frac{W}{2}) \text{ Tan } \Theta$$

$$ACB = W [h - (\frac{W}{2}) \text{ Tan } \Theta] + (\frac{W}{2}) (\frac{W}{2}) \text{ Tan } \Theta + (W) (T)$$

$$= (W)(h) - (\frac{W^2}{2}) \text{ Tan } \Theta + (\frac{W^2}{4}) \text{ Tan } \Theta + (W)(T)$$

$$+ (W)(h) - (\frac{W^2}{4}) \text{ Tan } \Theta + (W)(T) \qquad (1)$$

The corresponding area in solid is given by,

$$ACS = (W)(H) \qquad (2)$$

Assuming that the length is the same in the spoil as in the solid, it can be deduced that

$$ACB = (W)(H)(1 + \frac{SP}{100}) \quad (3)$$

Therefore, equations (1) and (3) are the same, i.e.,

$$(W)(T) + (W)(h) - (\frac{W^2}{4}) \text{ Tan } \Theta = (W)(H)(1 + \frac{SP}{100})$$

and,

$$h = H(1 + \frac{SP}{100}) + (\frac{W}{4}) \text{ Tan } \Theta - (T)$$

Hence, from the geometry in Figure 93

$$Rs = h \text{ COT } \Theta$$

which yields

$$Rs = \frac{1}{\text{Tan } \Theta} [H(1 + \frac{SP}{100})] + (\frac{W}{4}) \text{ Tan } \Theta - (T)$$

Defining F_c = the width across both crawlers in feet, the dumping radius of the shovel is given by,

$$Rsd = Rs + .5F_c$$

2. Dragline Reach--

All definitions used in calculating shovel reach apply to the dragline case also. While the shovel operates on the top of the coal, the draglines work on a bench on the highwall or on top of the highwall itself. An imaginary vertical plane passing through the edge of the highwall would contain the point "D" shown to the left of the crawler in Figure 94. Projection of a horizontal line from the imaginary vertical plane to the point of discharge of the bucket content establishes the reach of the dragline. The reach is calculated as in the case of the shovel, with the following equations

Rd = Reach of dragline in ft
Rdd = Dumping radius of dragline in ft

$$h = H(1 + \frac{SP}{100}) + (\frac{W}{4}) \text{ Tan } \Theta - (T)$$

But the dragline reach as can be seen from Figure 93 is given by,

$$Rd = H \text{ COT } \Phi + h \text{ COT } \Theta$$

$$= \frac{H}{\text{Tan } \Phi} + \frac{1}{\text{Tan } \Theta} [H(1 + \frac{SP}{100}) + (\frac{W}{4}) \text{ Tan } \Theta - (T)]$$

Defining E_t = the outside diameter of the dragline tub, the dumping radius can be calculated as,

$$Rdd = Rd + .75E_t$$

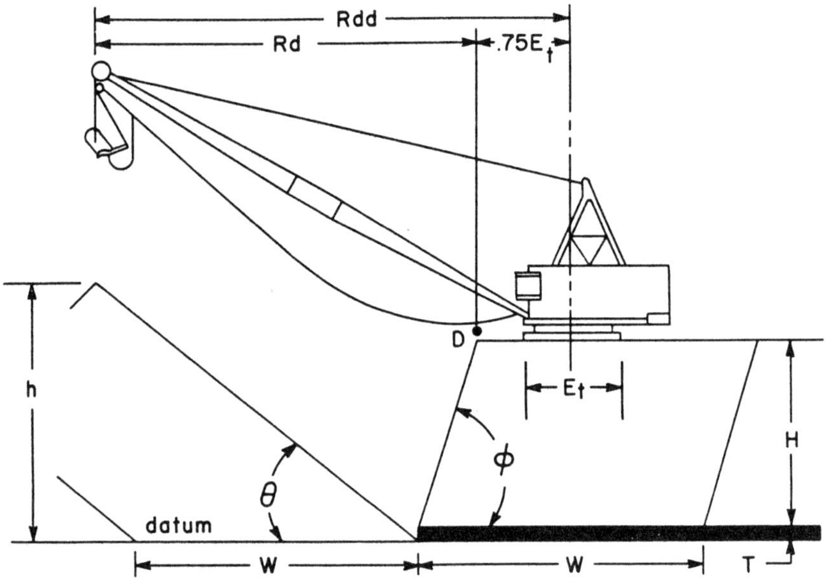

Figure 94. Dragline pit (Stefanko, et al., 1973).

3. Dragline with a Working Bench--

This case is illustrated in Figure 95. Here the dragline works on a bench at a depth H1 ft below the top of the highwall. This procedure may be practiced to increase the operating range of the dragline. The overburden to depth H1 is removed by some other equipment, and the reach calculation with consideration to bench is modified as follows:

$$Rd = \frac{H - H1}{\tan \Phi} + \frac{1}{\tan \Theta} \left[(H - H1)\left(1+ \frac{SP}{100}\right) + \frac{W}{4} \tan \Theta - T\right]$$

$$Rdd = Rd + .75E_t$$

However, if the dragline is used for cutting this bench also, the dragline bucket size should be larger for the same clean coal production.

Machine Weight Using the Maximum Usefulness Factor (MUF)

Maximum usefulness factor (MUF) is the measure of the capability of a machine to do a job. For a stripping machine, it is defined as follows:

MUF = Bucket size (cu yd) x Reach (ft)

Rumfelt (1961) investigated a large number of stripping machines and

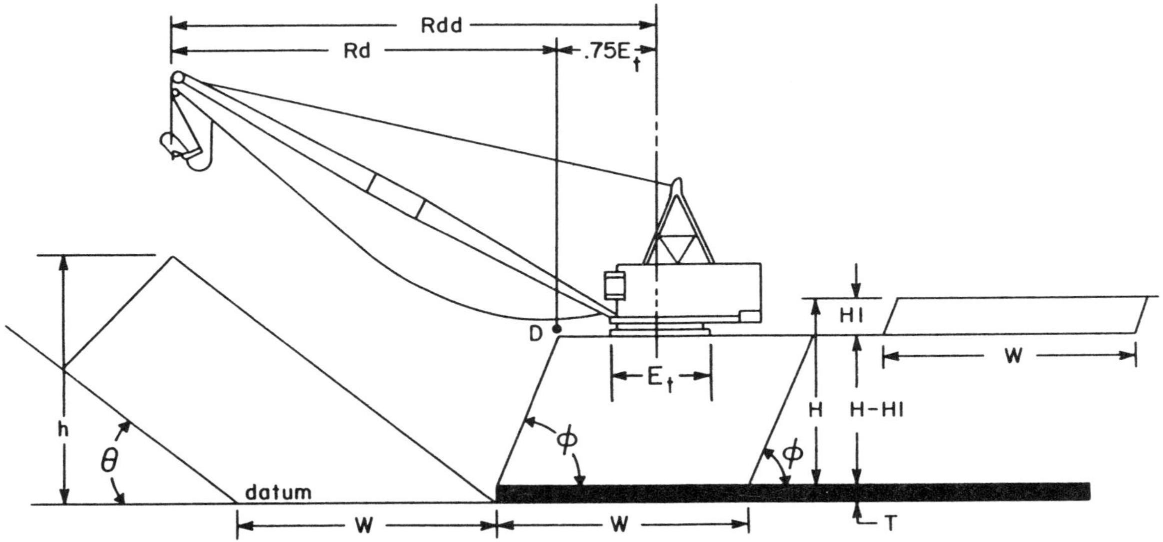

Figure 95. Dragline pit with working bench (Stefanko, et al., 1973).

produced MUF curves for shovels and draglines. The curves of MUF plotted against machine weights are straight lines for different sizes of shovels and draglines, which implies that the weight of the machine in pounds divided by its MUF was a constant. The value for a constant was found to be approximately 745 for shovels (Figure 96). For draglines, the values varied between 467 and 575 (Figure 97).

Rs = Reach of shovel in ft
Bs = Bucket size of shovel in cu yd
FACS = MUF constant for shovel in lb/MUF
SMW = Weight of shovel in lb
SMW = (Rs)(Bs)(FACS)
Rd = Reach of dragline in ft
Bd = Bucket size of dragline in cu yd
FACD = MUF constant for dragline in lb/MUF
DMW = Weight of dragline in lb
DMW = (Rd)(Bd)(FACD)

Dragline Digging Its Own Bench

The dragline sometimes digs its own bench in order to increase its range. In this type of operation, the machine chops material above its operating bench and therefore, the cycle time is no longer the normal cycle time but rather a weighted average cycle time. This can be calculated from the normal and the bench "chopping" cycle times as follows:

SMH = Scheduled monthly hours
NTCPM = Number of theoretical cycles per month
COB = Cubic yards of overburden to be removed
TBC = Theoretical bucket capacity in cu yd
NBSIZE = The actual bucket capacity in cu yd
OF = Operating Efficiency
BF = Bucket fill factor
CT = Normal cycle time in sec
CTB = Cycle time for "chopping" its bench in sec
CTW = The weighted average cycle time in sec
H1 = Depth to dragline bench in ft
H = Overburden height up to dragline bench from coal in ft
Ht = Total height of the overburden in ft
HT = H + H1

$$CTW = CT\left(\frac{H}{H1}\right) + CTB\left(\frac{H1}{HT}\right)$$

$$NTCPM = \frac{(SMH)(3600)}{CTW}$$

$$COB = \frac{(P)(TT)(HT)}{(1-A)(U)(COAW)(T)}$$

$$TBC = \frac{COB}{NTCPM}$$

$$NBSIZE = \frac{TBC}{(OF)(BF)}$$

If the dragline digs its own bench in an operation, the weighted cycle time for both the bench digging and the normal operation will be greater than the normal cycle time. The net result in the selection of the dragline is one with a larger bucket size. The reach of the dragline must be calculat-

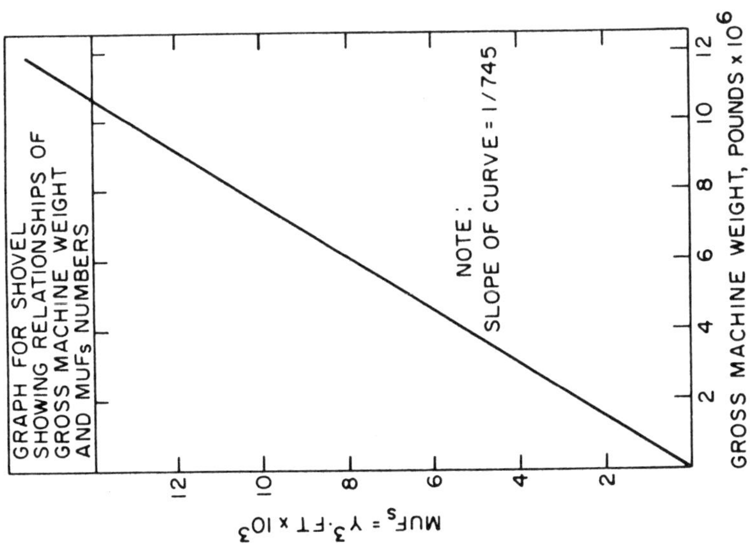

Figure 96. Shovel MUF graph (Rumfelt, 1961).

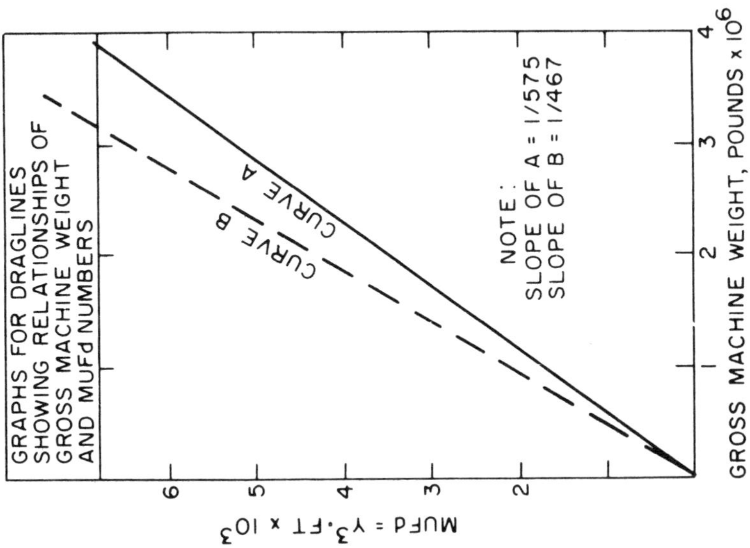

Figure 97. Dragline MUF graph (Rumfelt, 1961).

ed for the total overburden height. It is obvious that both of these will increase the maximum usefulness factor and therefore, the machine weight (Figure 95).

Multiple-Seam Mining

In multiple-seam mining, the dragline removes the overburden over the first coal seam, casts the spoil, and the coal seam is loaded out. Following this, the top of the spoil is leveled to form an operating bench for the dragline. The dragline is repositioned on this bench to strip the parting. The dragline action here is a chopping rather than a digging one. In chopping, the bucket fill factor decreases because of reduced contact angle between the bucket teeth and the overburden. Also, there is usually an increase in the bucket swing angle; i.e., when the dragline works from the leveled spoil it has to swing more than the normal swing angle in order to dump the parting material behind the original spoil. With reference to Figures 98 and 99, the weighted average cycle time can be calculated as follows:

SMH = Scheduled monthly hours
BF = Bucket fill factor
T = Coal thickness in ft
H = Height of overburden in ft
P = Production per month in tons
TT = Equivalent lb per ton of coal
COAW = Equivalent lb per cu yd of coal in solid
A = Reject fraction in run-off mine coal
U = Recovery fraction from coal in ground
COB = Cubic yards of overburden removed
CT = Normal cycle time of dragline in sec
CT2 = Cycle time for chopping overburden 2 in sec
CTM = The weighted cycle time in sec
P1 = Production from seam 1 in tons
P2 = Production from seam 2 in tons
H1 = Height of overburden 1 in ft
H2 = Height of overburden 2 in ft
T1 = Thickness of seam 1 in ft
T2 = Thickness of seam 2 in ft
BF1 = Bucket fill factor for seam 1
BF2 = Bucket fill factor for seam 2
COB1 = Cubic yards of overburden removed in 1
COB2 = Cubic yards of overburden removed in 2
T = T1 + T2
H = H1 + H2
P = P1 + P2

$$CTM = CT\left(\frac{H1}{H}\right) + CT2\left(\frac{H2}{H}\right)$$

$$NTCPM = \frac{(SMH)(3600)}{CTM}$$

$$BF = BF1 \times \frac{H1}{H} + BF2 \times \frac{H2}{H}$$

$$P1 = P\left(\frac{T1}{T}\right)$$

$$P2 = P\left(\frac{T2}{T}\right)$$

$$COB1 = \frac{(P1)(TT)(H1)}{(1 - A)(U)(COAW)(T1)}$$

$$COB2 = \frac{(P2)(TT)(H2)}{(1 - A)(U)(COAW)(T2)}$$

$$COB = COB1 + COB2$$

$$TBC = \frac{COB}{NTCPM}$$

$$NBSIZE = \frac{TBC}{(OF)(BF)}$$

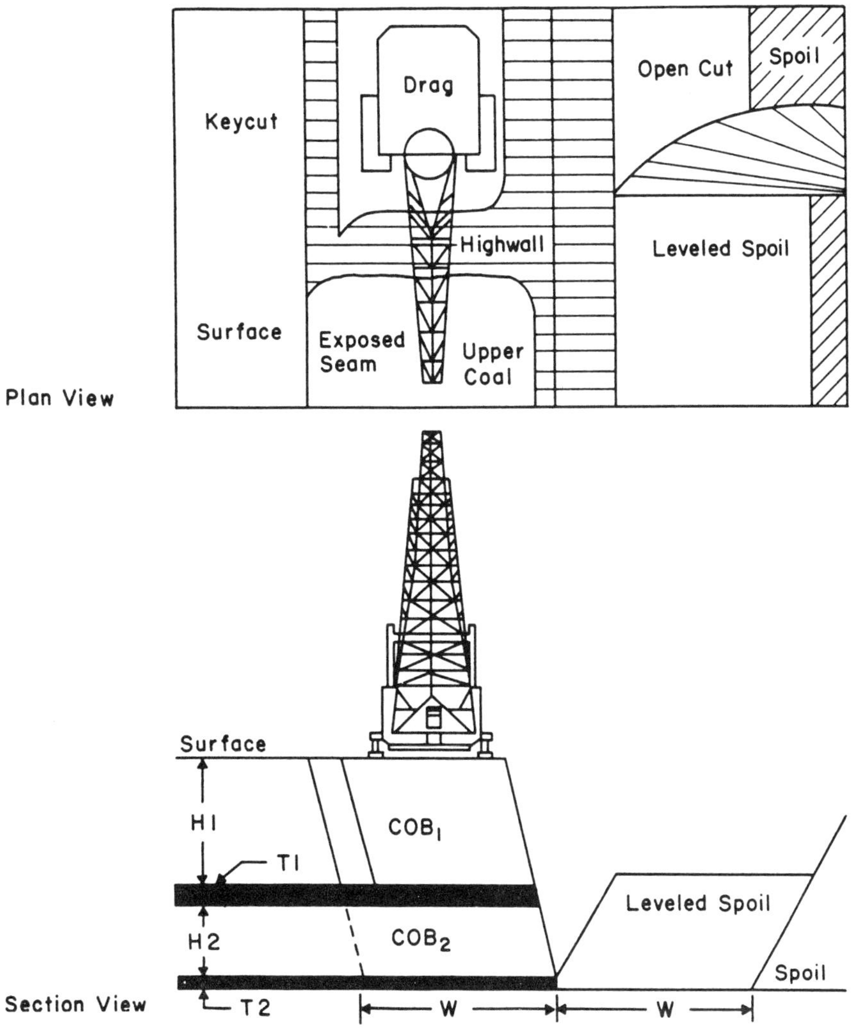

Figure 98. Multiseam dragline operation with upper seam exposed (Stefanko, et al., 1973).

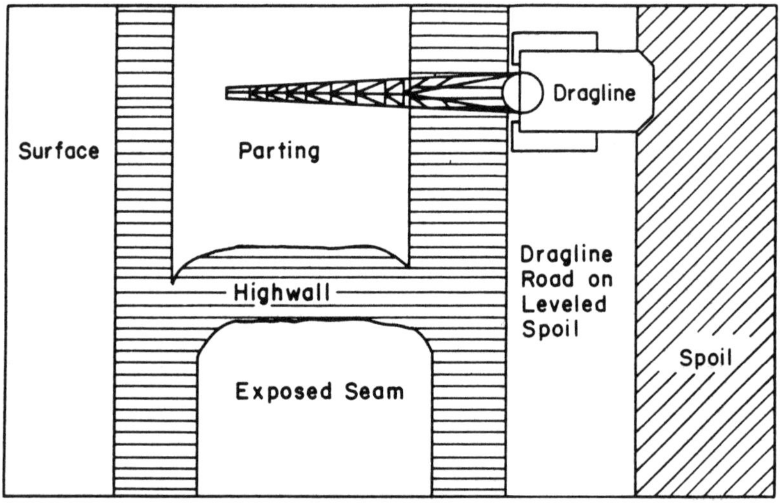

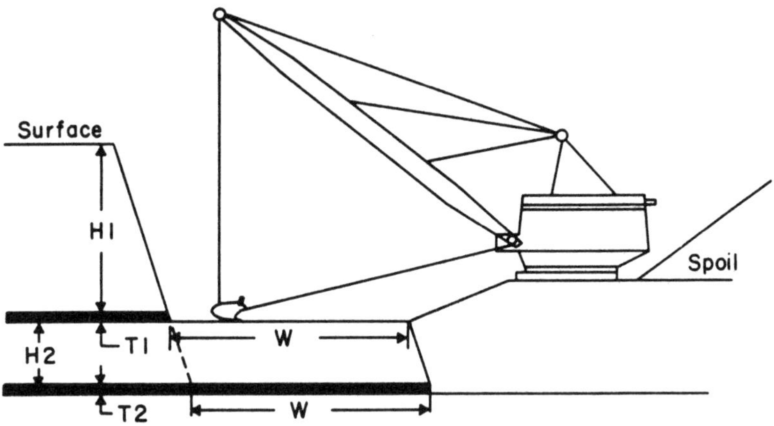

Figure 99. Dragline positioned on leveled spoil removing parting (Stefanko, et al., 1973).

The total effect of multiple-seam mining on dragline selection is that the weighted cycle time increases, the number of passes per hour decreases, the bucket-fill factor decreases, and the bucket size increases. Therefore, the maximum usefulness factor increases. Even if the reach remains the same, the machine weight will increase to meet the required production per month. The calculation of the reach in this case is involved since the dumping positions for the two modes of operation are different.

BUCKET-WHEEL EXCAVATOR SELECTION

The design of a bucket-wheel excavator (BWE) for the excavation of ore or waste is determined by the digging power required and the abrasiveness of the material to be excavated. A BWE consists of too many assembly groups and has to fulfill too many simulataneous functions to allow for the calculation of any individual component by one definite standard formula. Lengths and dimensions of a wheel excavator are highly interrelated, and their combination is always a compromise between the designed performance of the machine and its cost. In general, these machines are job-tailored and, as for every other type of excavating equipment, the cutting, tearing, loosening and splitting force required to excavate material is the deciding factor in the design of a BWE. In the United States, wheels are operated in tandem with shovels and/or draglines thereby necessitating considerations of equipment matching and working area limitations in pit design. Since bucket-wheel excavators are essentially German development, the standard units associated with BWEs are in the metric system. Therefore, in the discussion here metric units are used.

SOIL CLASSIFICATION

The BWE is an ideal machine for working in soft and medium hard strata, and it is gradually being developed for use in harder strata. Specifically, the digging resistance of the soil is the most sensitive parameter affecting the BWE operation. For excavation purposes, soils can be classified as follows:

1. The soils that can be excavated by a spade, e.g., unconsolidated strata like sand, gravel and silt.
2. The soils that can be loosened by a pickaxe before excavation, e.g., partly consolidated strata like clay beds.
3. The soils that can be loosened by blasting before excavation, e.g., hard shales, sandstone, etc.

The above classification is broad and general and cannot be of specific assistance to the BWE designer. Classification of soils according to their diggability is vital; but unfortunately, there is not any clearcut, standard method for determining this factor. Some common methods of expressing the digging resistance are:

1. Kg/cm, where the cutting resistance is calculated on the basis of load per unit length of the cutter in contact with the soil.
2. Kg/cm^2, where the load is related to the area of the cut slice.
3. Kg/cm^3, where the digging resistance is calculated as a function of the excavated volume per bucket.

The theories that are being advanced and the discussions taking place on what is considered to be the most suitable reference quantity vary widely. So far, most design calculations have been based on the cutter lengths or on the cross-sectional area of the slice cut, i.e., methods 1 and 2; but it must be recognized that the volume of material excavated per bucket has no small influence on the digging resistance. At the Fifth International Earthmoving Conference held in Prague in October, 1963, Professor N. G. Dombrowsky (1964) gave the following specific digging forces for various types of ground:

a. Light ground	1.8 to 2.5 kg/cm^2
b. Medium ground	3.0 to 3.5 kg/cm^2
c. Heavy ground	7.0 to 18. kg/cm^2

While it is not uncommon in non-German technical literature to consider the resistance to digging in relation to the cross-section of the slice, manufacturers of wheel excavators in Germany are still largely using the effective cutter length for determining the cutting forces. This is so because of the large number of test results available on various soils. Based on 150 large-scale tests of BWE operations, Himmel (1961) has calculated the specific cutting forces required to dig the types of ground encountered in open-cut mines. On an average, these were found to be:

1. About 20 kg/cm for light ground such as sand and gravel.
2. About 30 to 40 kg/cm for medium heavy ground such as sandy loam, pure loam, loess, lean clay, etc.
3. About 50 to 60 kg/cm for heavy ground such as heavy compact and plastic clays.

It is not possible to draw any conclusions on the above proposed values because the cutting power depends on the type of ground, the cross-sectional area and the shape of the slice cut, the configuration and sharpness of the cutter blades, the shape of the teeth, and the cutting speed. Hard ground requires a high specific cutting force, high cutting speeds, and additional cutting blades between the buckets. Reducing the output and the rate of swing decreases the slice cross-section and increases the specific cutting force for a given wheel drive rating. This will be clear from the following formula (Rasper, 1964):

$$N_G = \frac{K}{\eta \times c} \sqrt{Q \times S \times R}$$

where,

N_G = Power required for cutting in kw
Q = Actual digging capacity in cubic meters per hour
S = Number of bucket discharges per minute
R = Radius of the cutting wheel in meters
c = Constant depending on the bench height/wheel diameter ratio. For the cutting height = 2/3 wheel diameter, its value is 171
η = Efficiency of the wheel drive
k = Specific cutting force in kg/cm

Shown in Tables 44 and 45 are reference values for specific cutting forces applied to the more important bulk materials and virgin soils. The tables must be viewed with reservation as to the nonhomogeneity of the different soils from one geographical location to the other.

Devising methods for measuring the cutting resistance is another difficulty because the results from such theoretical considerations vary with the size of the sample tested, the angle of loading, the rate of loading and other such experimental procedures. The methods commonly employed for the determination of the cutting resistance are:

1. Simulating equipment pulled through the soil to be excavated for direct measurement of the force applied.
2. Laboratory tests with tooth and tooth impression measurements on the rock or soil samples.
3. Determination of the compactness by Proctor needles.

In a method developed by a German manufacturer (LMG), the entry of an excavator tooth into soil is copied in the laboratory on a soil specimen (Rasper and Ritter, 1961). This gives some information with respect to the soil split resistance. If numerous tests have been made with soils on which the cutting effect of BWE has been studied and analyzed, a correlation can be drawn between the theoretical laboratory values and the actual digging resistance. From observations so far, theoretical analysis can at best serve only as a guideline in the selection of the wheel because of the vast complex of influences which exist in practice that cannot be considered theoretically. Practical tests on the site of employment will confirm the suitability of the machine selected for a particular job.

TABLE 44. SPECIFIC CUTTING FORCES OF VIRGIN SOILS FOR BWE EXCAVATION (GARTNER, 1965)

SOIL TYPE	SPECIFIC CUTTING FORCES KG/CM
Earth	10 - 30
Loess	20 - 40
Sand (fine, coarse, wet or dry)	10 - 40
Clayey sand	10 - 50
Gravel fine	20 - 50
Gravel coarse	20 - 80
Sandy loam and wet loam	20 - 60
Dry loam	20 - 80
Clay wet	30 - 65
Clay dry	50 - 120
Clay schistose	35 - 120
Sandy clay	20 - 65
Clayey slate	50 - 160
Slate	70 - 100
Sandstone (easy digging)	70 - 160
Sandstone (hard digging)	160 - 280
Gypsum	50 - 130
Lime	30 - 120
Phosphate	80 - 200
Marl	60 - 140
Limestone	100 - 180
Weathered granite	50 - 100
Alluvial light consolidation	30 - 60
Alluvial heavy consolidation	70 - 150
Alluvial medium consolidation	50 - 80
Hard coal normal	50 - 100
Hard coal frozen	100 - 160
Lignite	20 - 70
Brown iron ore	190 - 210

In tests conducted on glacial till, sandstone and devonian shale, it was confirmed that earth containing a high percentage of boulders is unsuitable for wheels (Cheek, 1966). The capacity of any excavator operating in earth material with boulders will be affected adversely in proportion to the frequency of boulder occurrence. Two possible reasons can be attributed to this poor performance. Some boulders, like basalt and granite, are too hard for economical cutting. On the other hand, boulders and rocks composed of materials which the machine can cut in the solid state become loosened during excavation and cannot be rehandled.

TABLE 45. SPECIFIC CUTTING FORCES OF MATERIALS SUITABLE FOR BWE LOADING OPERATIONS (GARTNER, 1965)

MATERIAL TYPE	SIZE mm	SPECIFIC CUTTING FORCE KG/CM
Sand		10 - 20
Gravel fine	0 - 100	20 - 25
Gravel coarse	100 - 400	10 - 25
Earth		10 - 15
Phosphate		10 - 20
Bauxite		10 - 20
Iron-Mn-Cr	0 - 150	20 - 40
Ores	150 - 300	20 - 50
	300 - 450	20 - 80
Loose hard coal		10 - 30
Solidified hard coal		10 - 40
Lignite		10 - 25
Coke		10 - 30
Pellets (ore; cement)		10 - 25
Limestone		10 - 30
Slag		10 - 25

BWE OUTPUT CONSIDERATIONS

The theoretical output of a BWE is based on the bucket size and the number of bucket discharges per minute. If

I = Nominal bucket capacity in cubic meters
Z = Number of buckets in the wheel
V_1 = Cutting speed of the wheel in meters per second
D = Diameter of the wheel in meters,

then

$$S_s = V_1 \times Z/\pi D$$

where

S = Number of bucket discharges per second

$$Q_t = I \times S_s \times 3600$$

and

Q_t = Theoretical capacity of the excavator in cubic meters per hour.

As can be seen from the above equations, the number of bucket dischargers is dependent on the peripheral speed. The peripheral speed of a bucket wheel is limited by the ability of the wheel to discharge its bucket content on the chute against the counteracting centrifugal force. In theory, the maximum peripheral speed must be such that the bucket discharge will be ensured. Mathematically,

$$M \times g = M \times V_1^2 / R$$

where

M = Mass of material in the bucket, kg
R = Radius of the wheel in meters
g = Acceleration due to gravity, meters per second^2

which yields the following expression:

$$V_1 = \sqrt{g \times R} = V_{max}$$

Practical values of speed lie between 0.4 V_{max} to 0.6 V_{max}. To keep the wear on bucket's cutting knives or teeth at a minimum, speeds do not exceed 5 meters per second. The peripheral speed selected will also depend greatly on the nature of material to be excavated. In principle, however, a higher peripheral speed will be decided upon if hard material is to be cut, in which case maximum output may not be attained. Based on a constant output, the doubling of the peripheral speed will halve the amount of material excavated by each bucket; thus, cutting performance will be reduced.

Yet another factor that affects the output of a BWE is the bucket-filling capacity. It has been found that, in hard ground, the bucket-filling is around 30% to 40% of the nominal capacity. The relationship between digging resistance and the hourly capacity of the BWE is

$$Q_1/Q_2 = k_2^2 / k_1^2$$

where

Q_1 = BWE hourly capacity in cubic meters in soil with specific cutting resistance k_1
Q_2 = BWE hourly capacity in cubic meters in soil with specific cutting resistance k_2

Thus, the actual capacity of the BWE in any soil is given by:

$$Q_a = I \times B_f \times S_s \times 3600$$

where

B_f = Bucket-filling capacity in the soil expressed as a fraction of the nominal bucket capacity
S_s = Number of bucket discharges per second
Q_a = Actual capacity of the BWE in cubic meters/hour

In soils with high cutting resistance, higher cutting speeds with lower bucket-filling will result in a very small Q_a as compared to Q_t. The ratio may be as small as 0.2. One can visualize that in hard soils the excavating operation of the BWE has changed to a milling operation.

In most BWE calculations, the ratio of the bench height to the wheel excavator diameter is taken at two-thirds because the machine performance is optimum at this ratio. Obviously, with a lesser bench height full advantage of the wheel capacity is not taken. Also, increases in bench height above this limit result in undercutting which may lead to excessive spillage to be rehandled; and in soft soils sliding of the burden over the whell may result. The specific cutting resistance of the soil which can be calculated for a given digging power rating from the following formula,

$$K = \frac{N_G \times \eta \times C}{\sqrt{Q \times S \times R}}$$

where

K = Specific cutting resistance in kg/cm
S = Number of bucket discharges per minute
Q = Actual capacity in the soil in cubic meters/hour
R = Radius of the bucket wheel in meters
N_G = Digging power rating of the drive in kw
η = Efficiency of the drive
C = Constant of the bench height to wheel diameter ratio

is found to be increasing with decreasing bench height for a given wheel diameter. The change in the value of "C" with a change in the height/diameter ratio can be observed in the following table (Rasper, 1964).

Height/Diameter Ratio	0.1	0.2	0.3	0.4	0.5	0.6	0.67	0.7
Value of "C"	295	248	222	203	189	178	171	168

BUCKET-WHEEL DESIGN

At present, two types of bucket design are recognized, cellular and noncellular. In noncellular wheels, a ring chute isolates the bucket from the material flow behind it during the digging and lifting motion. Material is carried in the bucket to a point where the ring chute clears the way for the flow of material to the ladder conveyor. In the cell-type wheels, a buffer or cell is arranged behind each bucket for the excavated material. The cells side discharge this material onto a ladder conveyor. A cell-type wheel is usually stronger and more resistant to wear; but with the application of BWEs to harder formations which require higher cutting speeds, the cell-less wheels have superseded the cellular wheels. This preference for cell-less wheels arises because the discharge of excavated material from cells is governed by the rotary motion of the wheel. At high speeds, cellular wheels do not discharge completely.

The most important factor in bucket-wheel design is the diameter of the wheel. Much stress is placed on the correct size of the wheel since the service weight of the machine varies as the square of the diameter; cutting torque varies inversely as the diameter; and the cutting speed should never exceed $\sqrt{0.2gD}$, where g is the acceleration due to gravity and D is the wheel diameter. The diameter should also be sufficient to accommodate the chute and roller bar and to transfer material from the buckets to the ladder-conveyor. Since the ladder-conveyor is taken beyond the axle of the wheel, it is a controlling factor in wheel design.

OTHER CONSIDERATIONS

Wheel-boom and discharge-boom lengths have to be derived from operating conditions. The length of the wheel boom depends on the specified cutting height above and cutting depth below the grade and on the block width. The gradient limitation on the ladder-conveyor is an important factor which is decided by the type of material being handled. The length of the discharge boom depends on the amount of conveying necessary and the height, or possibly, the depth of discharge below grade. In an open-cut mine, the length of the discharge boom should be such that "spoiling" is properly affected on the dump zone and not within the pit limits.

Restrictions on the size of a modern-day BWE are imposed not by the design of the wheel but by the conveying of material from the bucket to the discharge point on the stacker boom. Also, with the design of BWEs for deep mines, a dependable steep angle conveyor is necessary. Belt speeds of 300 to 400 meters per minute and belt widths of 200 centimeters are not uncommon. In cases where the inclination angle of the ladder boom is more than 35 degrees, sandwich belt conveyors have proved to be highly satisfactory. Because of the types of material that the BWE handles, the conveyors must be liberally dimensioned and properly maintained.

MODE OF OPERATIONS

Modern BWEs generally excavate in blocks. Figure 100 shows a BWE working in an established cut. The wheel is positioned to travel on the pit floor in line with the top edge of the old highwall. As it advances, a new highwall is exposed in the direction of excavation. Digging is done by rotating the wheel, swinging it from side to side in long parallel arcs, and "crowding" into the bank, by advancing the entire machine, or by the travel of the digging boom if an automatic crowd is available (Figure 100). A second way by which the wheel can be advanced into the bank is by the falling cut method. A brief description of each of these methods is given below.

Cut with a Crowding Machine

At the end of every swing, the digging boom can be extended by the thickness of cut desired and the boom swung back in the reverse direction. Obviously, the thickness of bank excavated does not vary with the boom position; therefore, the slewing motion of the boom is fairly constant for uniform output. The thickness through which the digging boom can be advanced into the bank is theoretically calculated from the formula (Mani, 1966):

$$t = 0.133 \times \sqrt{Q_a / R \times S}$$

where

Q_a = Actual capacity of the excavator in cubic meters/hour
S = Number of bucket discharges per minute
R = Radius of the wheel in meters
t = Thickness of slice in meters

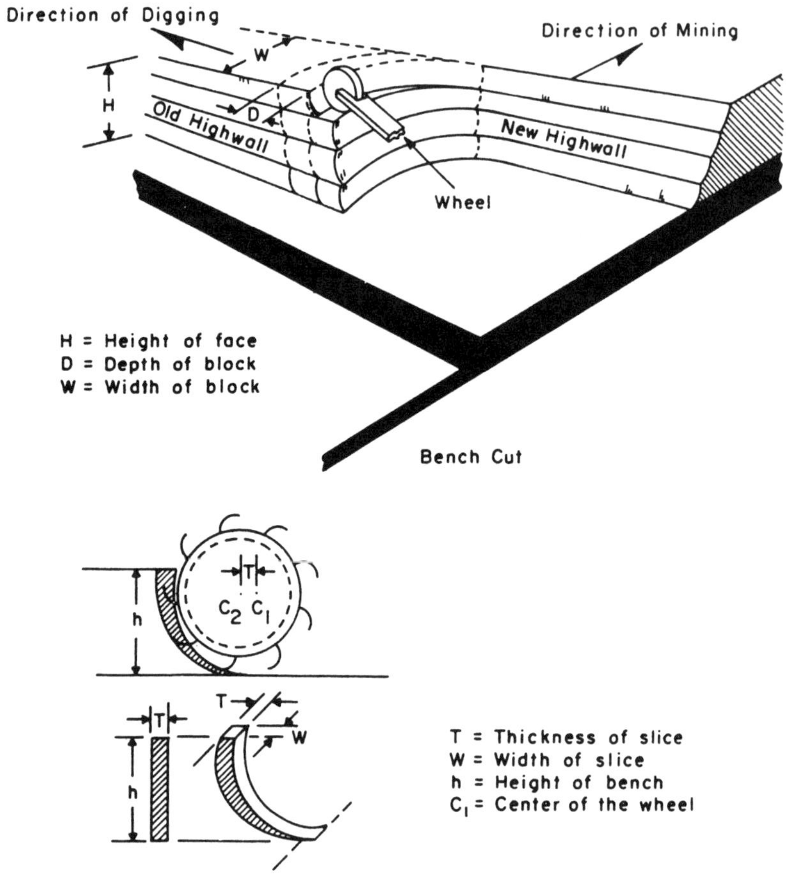

Figure 100. Block digging operation with BWE.

Cut with a Noncrowding Machine

The essential difference between this cut and the former is that here the machine is moved forward bodily in the direction of the bench by the thickness of the cut. The slew axis progresses by successive distances, t, along the line of advance; and the successive segments that are cut are not parallel but sickle-shaped. The thickness of slice varies constantly with angle Θ, both to the right and to the left of the direction of advance. A very close approximation of slice thickness at any point P is given by:

$$t_p = t\ (\mathrm{Cos}(\Theta) + \frac{\mathrm{Sin}^2\Theta}{2P_1})$$

where

$P_1 = S_r/t$
S_r = Slewing radius in meters
t = Thickness of slice in meters
Θ = Angle between the direction of advance and direction of P

Therefore, to maintain a uniform output as the boom slews away from the center, the loss in the thickness of slice must be compensated for by increasing the width of the slice cut. This is achieved by increasing the slewing speed such that $V_p = V/\cos(\Theta)$, where V_p is the slewing speed at the point P, and V is the starting slewing speed in the direction of the advance.

Two factors limit the angle of slewing. At angles more than 70 degrees, the slewing speed increases disproportionately to the increase in output; therefore, usually with noncrowding machines, the width of the block is such that the swing on either side of the direction of advance is maintained within limits. The width of the bench is given by:

$$S_r(\mathrm{Sin}\ \Theta_L + \sin\ \Theta_R)$$

where

S_r = Radius of slewing in meters
Θ_R = Angle to the right of the direction of advance $\leq 70°$
Θ_L = Angle to the left of the direction of advance $\leq 70°$

Falling Cut Method

Depending on stratification and fissuring, it is often advantageous in firm ground to operate by the drop or falling cut method. At the commencement of a block of maximum face height, a top cut about half the wheel diameter is taken by the bench cut method, the depth of the cut being great enough to leave a face formed for the top cut of the subsequent block. After this initial preparation, the excavator retreats a short distance and the rest of the block is excavated in segments which are proportioned by lowering the wheel to the desired thickness (t). The slice taken here is very similar to that of the bench cut. The geometry of the cut is explained in Figure 101 including the slice cross-section at the line of advance. If

A_h = Depth of slice normal to the working face at 0° slew in meters
A = Depth of advance in meters

the variation in the sickle depth in the falling cut method can be expressed as:

$$A_{h\Theta} = A_h\ (\mathrm{Cos}\Theta + \frac{\mathrm{Sin}^2\Theta}{2P_2})$$

where

$P_2 = \frac{S_r}{A}$
S_r = Radius from slew axis to outer edge of segment
A = Depth of advance = A_h Cosec α
α = Angle of frontal batter
Θ = Angle slewed from the direction of advance

The slew angle in the falling cut can be considerably more than that of the bench cut because $A_{h\Theta}$ does not become negligible until Θ exceeds 85°. Theoretical calculations will show that the slewing speed is to be increased in the same relation as before and that the output is identical by either of the two cuts. However, this conclusion can be erroneous if the material is not homogeneous in all the planes, which is usually the case with soils.

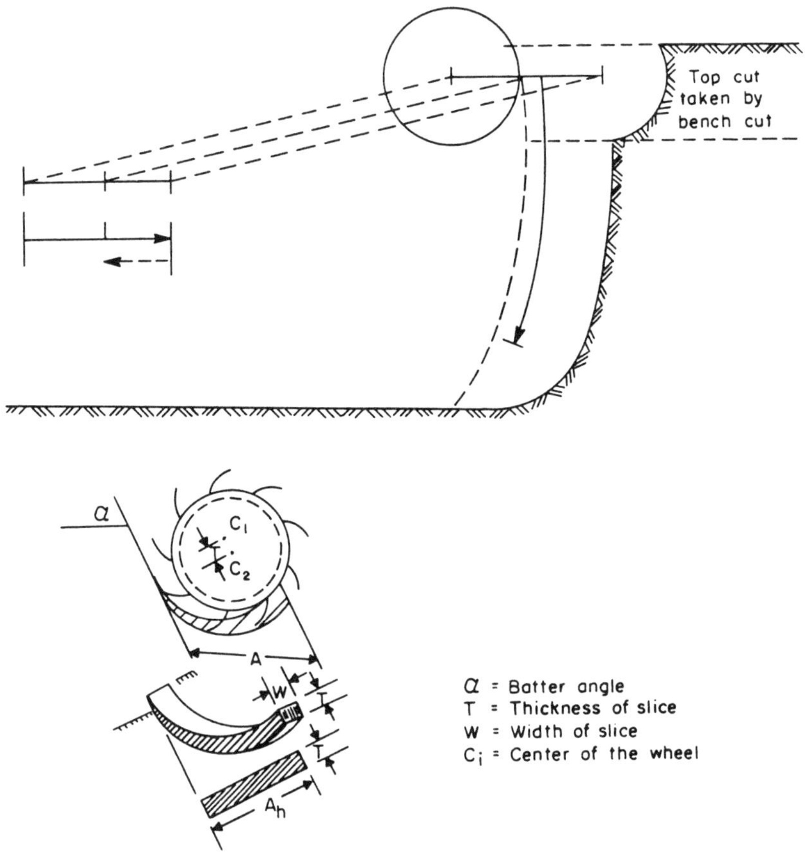

Figure 101. Fall cut block digging operation with BWE.

BWE OPERATIONS IN USA

Reference has already been made to the introduction of wheel excavators in Illinois strip mines during the 1940's. There are only a very few wheels in operation in the United States. Except for one operation each in the states of Washington and Wyoming, there are six wheel excavators in the Illinois coal fields. In American strip mining practice, bucket wheels have been used to remove loose top soil and soft beds, whereas the harder beds and the coal itself are handled by large stick shovels or draglines. In order to fully utilize the capacity of the two units in a tandem operation, the output capacity and dimensions of the wheel and shovels must be matched.

The Kolbe wheel excavator can operate only by the frontal block digging method on benches of limited width. Overburden is stripped with the excavator operating in a long trench about 30 to 32 meters wide at the bottom and transferring it directly to the spoil heap. With burdens of 30 meters, the width of the cut is only about 15 meters (Figure 31). Since there will be occasions when the wheel and the tandem shovel have to cross each other, the pit must be designed wide enough to allow this.

BWE SELECTION PARAMETERS

The required data for input are divided into four categories:

1. Mining Dimensions: These include (a) the length of the mining property; (b) the number of benches; and (c) the elevations of the benches.
2. Soil Characteristics: The data required here are (a) the various soil types; (b) their specific cutting resistances; (c) the operating speed of the wheel in the soil types; (d) the bucket filling capacity; and (e) the swell factor.

In soils with higher specific cutting resistances, the wheel is speeded up and the successive slices that are taken are thinner. This seriously affects bucket filling capacities and in some cases less than 25 percent of the theoretical output is mined. Swell factors for sand and similar loose, broken material is not very significant, whereas a compacted material may increase in volume by 30 percent to 40 percent on fragmentation.

3. Wheel Specifications: The data input here include (a) the diameter of the wheel; (b) the number of buckets in the wheel; (c) the length of the slew axis; (d) the overall weight of the machine, and (e) the width of the machine.
4. Miscellaneous: Realistic efficiency figures are necessary for the mechanical and electrical components of the BWE to arrive at proper production figures.

Much research and trial-and-error experience have gone into the most recent designs of the digging wheel, its bucket and other appurtenances. The result is a device that will satisfactorily handle the variety of materials encountered, especially sticky clays, soils containing some floating boulders, and shale of varying hardness and composition. BWEs are machines designed for continuous digging and discharge. The continuous steady action requires less specific power and greatly reduces load fluctuations, characteristic of intermittent methods like that of shovels and draglines.

The range of application of BWE is not limited to mining alone. The classic method of stacking out and reclaiming bulk materials by conventional machines has been replaced today by far more economical use of conveyors, stackers, and bucket wheels.

A BWE is an asset to land reclamation. Spoil piles are much more regular in size and contour than those produced by shovels or draglines. The hard rock is spoiled at the bottom of the mined area by the shovel or the dragline, and the original loose top soil is returned to relatively the same position, making replanting easier. In many cases, the crop and grazing value of the reclaimed spoil areas is greater than prior to excavation.

DRILLING AND BLASTING

Since this subject has been covered in detail in an earlier section, here the important considerations will be briefly outlined. Selection of drilling equipment depends on the types of materials and their strengths. Thus, penetration resistance to drilling should be studied before selection of a drill and a method of drilling. Soils and other loose materials do not require fragmentation, but are drilled in the process of drilling underlying consolidated materials, when these loose materials overlie the rocks. Although there are several methods of drilling, only mechanical methods which utilize percussion drills or roller bit rotaries, or a hybrid of both, are considered.

Selecting a particular drill involves the evaluation of energy and power consumed, bit wear and replacement cost, drilling rate, and cost of drilling.

This evaluation depends on several factors, namely: 1) the design and operating parameters comprised of drill, rod, and bit which should be selected to match the rock type; 2) lubrication and cleaning variables, i.e., circulation fluids usually either water, air, or mud; 3) the hole geometry, i.e., size and depth, which are exogenously determined; 4) environmental and geologic factors such as the rock properties, structural and petrological properties of the rock, as well as the overburden stress, all of which determine the drillability of the overburden; and 5) factors external to drilling such as cost of labor, job site, scale of operation, type and availability of power to the site, climatic conditions, and management and supervisory efficiency.

Selection of Drilling Equipment

Even after good mathematical analysis, the pit engineer has to use judgement to arrive at the selection of drilling equipment. The following procedure would aid in decision making:

1. Determination and specification of job conditions, i.e., labor, weather, and site location.
2. Determination or goal setting for the unit operations, e.g., excavation, haulage, crushing capacity, production required, pit geometry, fragmentation, and throw required.
3. Determination of number of drill holes, size, design, and area of influence expected.
4. Determination of rock strength and drillability and selection of feasible drill methods.
5. Specification of operating criteria for each system to be considered, i.e., rock, bit, and circulation fluid.
6. Estimation of the performance characteristics including costs for depreciation, maintenance, power, and fluids.
7. Selection of the drilling system and equipment which best meet the above requirements as well as has the lowest overall cost (if possible). Table 46 gives information for selection of drill type based on rock strength and type.

TABLE 46. APPLICATION OF DRILLING AND PENETRATION METHODS TO TYPES OF ROCK (PFLEIDER, 1968)

	ROCK TYPE			
RATING →	1	2	3	4
CATEGORY →	Soft	Medium Hard	Hard	Very Hard
EXAMPLES →	(shale, weath. limestone)	(limestone, weath. sandst.)	(granite, chert)	(quartzite, taconite)
DRILLING METHOD ↓				
rotary, drag	x	x		
rotary, roller	x	x	x	x
percussion		x	x	x
jet piercing			x	x

Number of Drills

In surface coal mining, drilling and blasting is done for removal of the overburden. Amount of drilling and blasting depends on the types of strata present in the area. Hard rocks require more drilling and blasting, and often the soil is not fragmented. The number of drills needed for drilling can be calculated as follows:

SYE = Square yards of coal to be exposed
COB = Cubic yards of overburden removed
DS = Spacing of the drill holes, ft
DB = Burden of the drill holes, ft
DP = Drilling pattern $(ft)^2$
LDPM = Length of drilling required per month (ft/month)
NH = Number of drill holes
DR = Drilling rate (ft/hour)
HPS = Hours scheduled per shift
SPD = Shifts per day
DPW = Drilling days per week
SPM = Weeks worked per month
SMH = Schedule monthly hours
DPM = Number of drills required
DP = DS x DB

$$NH = \frac{(SYE)(9)}{DP}$$

$$LDPM = \left(\frac{COB}{DP}\right)(27)$$

$$SMH = HPS \times SPD \times DPW \times WPM$$

$$DPM = \frac{LDPM}{(SMH)(DR)}$$

EXPLOSIVES REQUIRED PER MONTH

The explosive commonly used in stripping is ANFO. Ammonium nitrate is not readily explosive until the addition of the fuel oil, which acts like a catalyst to enhance its explosive strength. It is much safer, less expensive, easier to prepare and more effective for rock blasting than most comparable explosives. ANFO is available in bulk premixed, on-spot mix and bagged forms.

Fundamental to effective blasting is the use of correct amount of explosive for the volume of overburden to be fragmented. This relationship is usually expressed as the "powder factor," which is defined as the ratio of cubic yards of overburden fragmented per pound of explosives. It varies from 2.5 to 9.0 (from hard rocks to soft soils).

To determine the amount of explosive needed per foot of hole, the area of influence of a drill hole (DP) is divided by 27, and then divided by the powder factor "PK".

$$\left(\frac{DP}{27}\right) \cdot \left(\frac{1}{PK}\right)$$

Therefore, knowing the number of holes drilled per month, the drilling pattern and the overburden depth, the amount of explosive required per month (AANFO) can be calculated:

DC = Length of explosive charge in hole in feet
AANFO = Amount of explosive required per month

$$AANFO = (\frac{DP}{27}) \cdot (\frac{1}{PK}) \cdot (NH) \cdot (DC) \text{ lb}$$

Alternatively, if the entire overburden is to be blasted, then

$$AANFO = \frac{COB}{PK}$$

where

COB = Cubic yards of overburden removed

But it is common knowledge in stripping, that not all the overburden requires fragmentation before removal. This is particularly true for topsoil and soft soil. Therefore, actual consumption of explosive must be calculated on the basis of experience with the overburden, i.e., for individual application, the fraction of overburden that should be fragmented must be known.

COAL LOADING

Although there are several methods of coal loading, the method considered in this report is a shovel-truck combination.

Here, shovels are selected to match the advance of the stripping equipment and the capacity of the trucks. Even when matching is achieved, the general trend is to select a loader with a slightly higher loading ability than calculations show for normal operations. The calculations for selection of a loader are as follows:

CTL = Loader cycle time in seconds
BL = Bucket size of loading shovel in cu yd
TPB = Tonnage per bucket
F = Factor for converting 1 cu yd of loose coal to tons
SMH = Scheduled monthly hours
SMH = HPS x SPD x DPW x WPM
NPH = Number of cycles

$$NPH = SMH \times \frac{3600}{CTL}$$

$$TPB = (\frac{P}{1 - A}) \div NPH$$

$$BL = \frac{TPB}{F}$$

A loader with this capacity could easily handle the monthly production.

Truck Fleet Selection

The truck capacity and loader capacity should be matched so that the truck waiting time on loader and loader waiting time for trucks are minimized. The calculations for matching the loader with the truck and for selecting the required number of trucks are as follows:

HD = Hauling distance in ft

FT = Fixed time in min
DT = Delay time in min
MT = Maneuver time in min
TLK = Truck loading time in min
CTK = Truck cycle time in min
TKC = Truck capacity in tons
TPB = Tonnage per bucket
NT = Number of trucks required

$$TLK = [\frac{TKC}{TPB}] [\frac{CTL}{60}]$$

$$FT = DT + MT + TLK$$

$$CTK = Ft + [\frac{HD}{(S1)(88)} + \frac{HD}{(S2)(88)}]$$

$$NT = \frac{CTK}{TLK} + 1$$

The travel time and return time of the truck used above should be calculated using average velocities for loads (S1) and empties (S2). Figure 102 will aid in the sizing of trucks with regard to shovel bucket capacity.

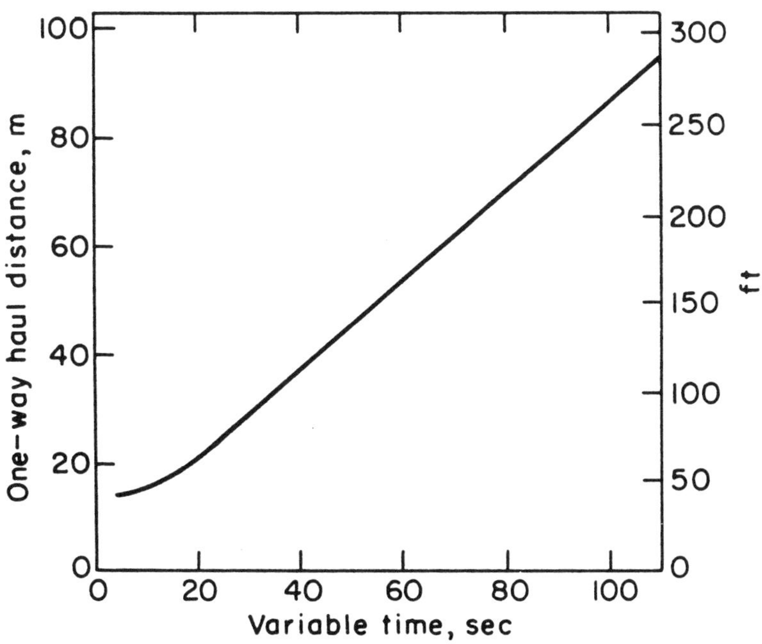

Figure 102. Front-end loaders variable time (Atkinson, 1971).

SECONDARY EQUIPMENT SELECTION

The secondary or support equipment in surface mining ranges from highly sophisticated construction equipment to small farm machines for reclamation and revegetation purposes. In this section, the support equipment considered are bulldozers, scrapers, and front-end loaders. These machines are used for initial land clearing, mine road construction, rehandling of spoil, loading, and haulage and reclamation jobs. They are virtually indispensible in tight spots where larger machines cannot easily maneuver.

As with shovels and draglines, the secondary equipment are associated with fixed and variable components in their cycle times. For example, Atkinson (1971) gives the following values for the fixed and variable components of the cycle time for front-end loaders (Table 47 and Figure 102). According to Caterpillar Tractor Company (1975), a fixed time of 0.40 minutes (24 seconds) for front-end loaders working in loose granular material on a hard-smooth operating surface is recommended. This fixed time includes load, dump, four reversals of direction, full cycle of hydraulics and minimum travel.

TABLE 47. FIXED TIMES FOR FRONT-END LOADERS (ATKINSON, 1971)

BUCKET CAPACITY	FIXED TIME (SEC)		
yd^3	Easy Digging	Medium Digging	Medium-Hard Digging
5	32	33	41
6	33	34	42
7	33	35	44
10	37	39	51
12	39	42	56
15	41	44	60

Note: The application of front-end loaders in hard digging conditions is marginal and requires comprehensive investigation.

Since material type affects the fixed time of the loader, Caterpillar Tractor Company (1975) recommends the corrections given in Table 48 for other material characteristics and operating conditions.

For scrapers, Caterpillar (1975) recommends the use of 0.7 minutes (42 seconds) each for loading, maneuvering, and spreading.

There are several methods by which the secondary equipment can be selected. They should match the performance of the primary stripping equipment. Two alternative selection methods are outlined here:

A. Procedure I. Briefly, this procedure consists of the following steps:

1. Select a suitable bucket (or blade) size. The capacity is usually expressed in cubic yards.
2. Calculate the variable part of the cycle time.
3. Determine the fixed times from time study. Sometimes, this can be obtained from manufacturers' handbooks.
4. Calculate the total cycle time by adding the variable cycle time to the fixed cycle time.
5. Calculate the number of passes per hour.
6. Determine the hourly production by multiplying the number of passes (cycles per hour) by load per cycle (capacity).

TABLE 48. MISCELLANEOUS CORRECTION FACTORS (CATERPILLAR, 1976)

MATERIALS	MINUTES TO BE ADDED TO 0.4 MINUTES
Mixed	+ 0.02
Up to 1/8 inch	+ 0.02
1/8 inch to 3/4 inch	- 0.02
6 inches and over	+ 0.03 and up
Bank or Broken	+ 0.04 and up
Conveyor or Dozer piled 10 feet and up	0.00
Conveyor or Dozer piled 10 feet or less	+ 0.01
Dumped by truck	+ 0.02
Management Conditions	
Common ownership	up to - 0.04
Independently owned trucks	up to + 0.04
Constant operation	up to - 0.04
Inconsistent operation	up to + 0.04
Small target	up to + 0.04
Fragile target	up to + 0.05

To arrive at the actual production, correction factors from the manufacturer's handbook or from experience are applied.

B. Procedure II. Here, one can use the maximum production per hour given by the manufacturer for some standard condition. To determine the actual production, job condition correction factors must be applied to the maximum production.

Correction Factors. Most of the correction factors discussed earlier for draglines and shovels equally apply to the secondary equipment. Additional factors that apply to secondary equipment are discussed here. Some of these factors are presented in Tables 49 and 50.

Material Factor. The performance of bulldozers, scrapers and front-end loaders are affected by material constituents and characteristics. Since these equipment have varied applications, from loading and grading in loose material to excavating hard frozen grounds, their capacity for production as given by manufacturers must be corrected by a factor known as material factor. Table 50 gives such factors for bulldozers.

Grade Factor. Since secondary equipment are mobile, their production capability is affected by grade of the roadway and the work area conditions. Factors are applied to production capabilities as given by manufacturers to correct for the grade effect.

TABLE 49. BULLDOZER CORRECTION FACTORS (CATERPILLAR, 1976)

Job Condition Corrections:	Track-Type Tractor	Wheel-Type Tractor
OPERATOR - Excellent	1.00	1.00
Average	0.75	0.60
Poor	0.0.60	0.0.50
MATERIAL -		
Type-		
Loose stockpile......................	1.20	1.20
Hard to cut; frozen-		
with tilt cylinder...............	0.80	0.75
without tilt cylinder............	0.70	----
cable controlled blade...........	0.60	----
Hard to drift; "dead" (dry, non-cohesive material or very sticky material)............	0.80	0.80
Rock, ripped or blasted.............	0.60-0.80	----
SLOT DOZING...........................	1.20	1.20
SIDE BY SIDE DOZONG...................	1.15-1.25	1.15-1.25
VISIBILITY - Dust, rain, snow fog or darkness.....................	0.80	0.70
JOB EFFICIENCY - 50 min/hr.	0.84	0.84
40 min/hr.	0.67	0.67
DIRECT DRIVE TRANSMISSION............. (0.1 min. fixed time)...............	0.80	----
*BULLDOZER - Angling (A) blade.........	0.50-0.75	----
Cushioned (C) blade..................	0.50-0.75	0.50-0.75
D5 narrow gauge...........	0.90	----
Light material U-blade (coal)..........	1.20	1.20
Blade bowl (stockpiles)...........	1.30	1.30

*Note: Angling blades and cushion blades are not considered production dozing tools. Depending on job conditions, the A-blade and C-blade will average 50-75% of straight blade production.

TABLE 50. SWELL FACTOR - BULLDOZED MATERIAL (ATKINSON, 1971)

MATERIAL	SWELL FACTOR
Broken rock	1.65
Heavy clay (wet)	1.45
Earth with boulders	1.35
Earth	1.25
Sand and small gravel	1.1

Visibility Factor. Impairment of operator's visibility due to atmospheric conditions such as snow, fog, rain, etc., affect the performance of the secondary equipment. Visibility factors are applied particularly to bulldozers and scrapers.

Transmission Factors. This factor accounts for the transmission mode of the equipment: direct or automatic drive. The effect of the transmission train mode on the equipment's productivity can be corrected by applying transmission factors to the production capability given by manufacturers.

Dozing Factor. This factor accounts for the mode of operation of a bulldozer, whether slot dozing or side-by-side dozing. It applies to bulldozers only and relevant values are given in Table 49. Factors for operating efficiency and job efficiency, as applied to primary equipment, are also applied to secondary equipment. Values for these factors are given in Table 49.

1. Bulldozers. Bulldozers are used extensively in surface mining. They are extremely useful for dozing material through short distances under severe conditions. Selection of a suitable bulldozer can be accomplished by calculating,

Bb = Bucket or blade size (machine capacity) in cu yd

The choice of blade type depends on the kind of job application and soil conditions. Table 51 shows bulldozer blade selection chart according to job application.

TABLE 51. BULLDOZER SELECTION CHART FOR RIPPING AND DOZING APPLICATION (SKELLY AND LOY, 1975)

Job	Ground Condition	Rubber Tired Dozers			Track Dozers			Remarks
		Small	Medium	Large	Small	Medium	Large	
Clean up Around Small Shovel	Friable	x			x			No Ripping
			x					Medium Ripping
	Hard Friable					x		Hard Ripping
		x			x			No Ripping
			x			x		Medium Ripping
				x		x		Hard Ripping
	Hard Jagged				x			No Ripping
						x		Medium Ripping
							x	Hard Ripping
Clean up Around Large Shovel	Friable		x					No Ripping
			x					Minimum Ripping
			x			x		Medium Ripping
	Hard Friable		x					No Ripping
			x			x		Medium Ripping
				x		x		Hard Ripping
	Hard Jagged					x		No Ripping
						x		Medium Ripping
							x	Hard Ripping
Waste Dump Dozing Capabilities	Friable and Hard Friable	x						9 (26 Ton Trucks)
			x					32 (59 Ton Trucks)
				x				>32 (50 Ton Trucks)
	Hard Jagged				x			9 (26 Ton Trucks)
						x		32 (59 Ton Trucks)
	Wet Muddy				x			90 to 180 Tons/Hour
						x		270 to 720 Tons/Hour
							x	>720 Tons/Hour

FT = Fixed time in min
TF = Travel loaded time in min
TE = Travel empty time in min
VT = Variable time in min
MT = Maneuver time in min
CT = Cycle time in min
LT = Loading time in min
NPH = Number of passes per hour
PPH = Production per hour (theoretical) in cu yd
MAXPH = Actual Production per hour in cu yd
HD = Haul distance in ft
S1 = Loaded speed in mph
S2 = Empty speed in mph
Bb = Bucket size of bulldozer in cu yd
LF = Load factor
BF = Bucket fill factor
AO = Operating factor

The selection procedures are outlined here.

$$FT = MT + LT$$

$$TF = \frac{HD}{(S1)(88)}$$

$$TE = \frac{HD}{(S2)(88)}$$

$$VT = TF + TE$$
$$CT = TF + FT \text{ (minutes)}$$

$$NPH = \frac{60}{CT} \text{ (cycles per hour)}$$

$$PPH = (NPH)(Bb)$$

$$MAXPH = (PPH)(LF)(BF)(AO)\left(\prod_{i=1}^{k} CF_i\right)$$

where k is the number of other applicable correction factors from the manufacturer's handbook, and CF_i is the value for the ith correction factor. Alternatively, the actual production can be calculated as,

$$MAXPH = (MAXPHG)(LF)(BF)(AO)\left(\prod_{i=1}^{k} CF_i\right)$$

where MAXPHG is the maximum production obtained from catalog or handbook under specified conditions.

2. Scrapers. There are several types of scrapers, namely, elevating and conventional scrapers, and single- and double-engine powered scrapers to mention a few. The major distinction is between wheel-tractor scrapers and crawler-tractor scrapers. Wheel scrapers, because of their mobility, are used in surface coal mining for topsoil removal, road building, etc.

Shown in Figure 103 are the scraper cycle times as functions of distance and grade.

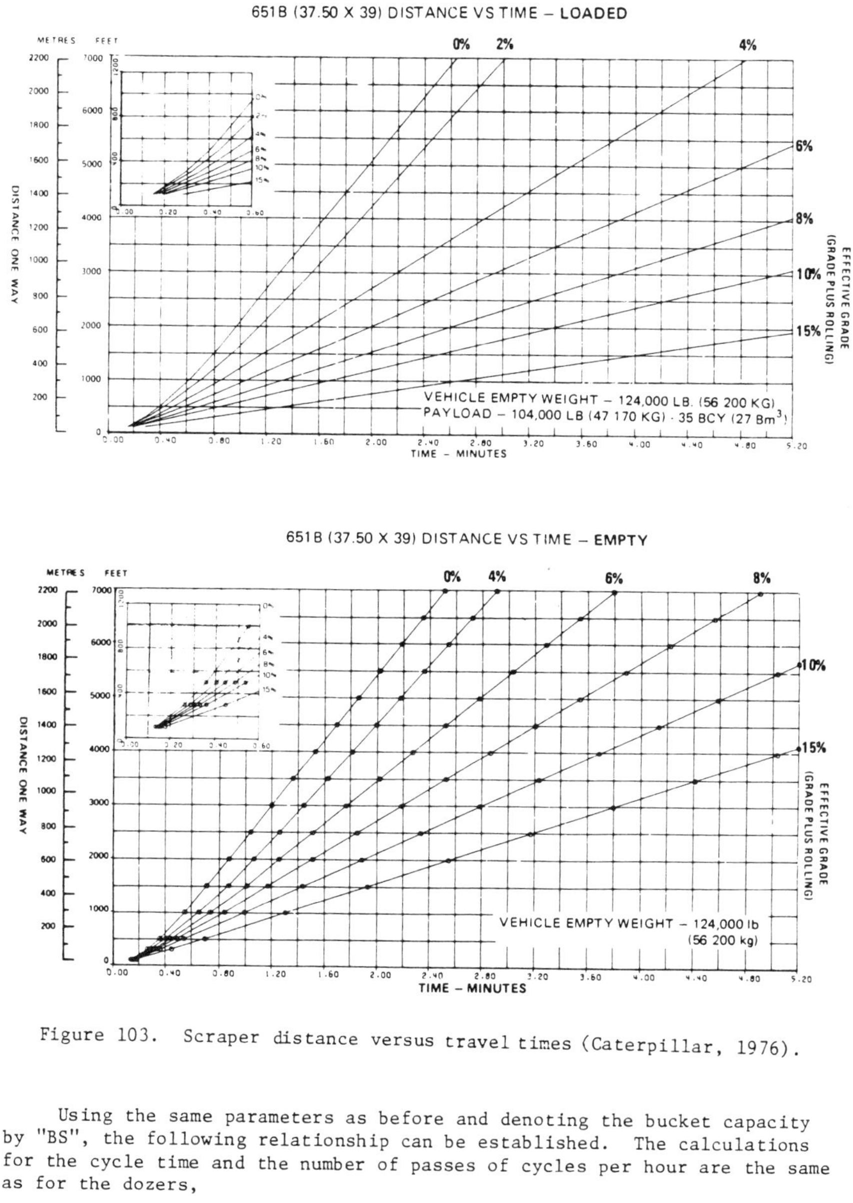

Figure 103. Scraper distance versus travel times (Caterpillar, 1976).

Using the same parameters as before and denoting the bucket capacity by "BS", the following relationship can be established. The calculations for the cycle time and the number of passes of cycles per hour are the same as for the dozers,

$$BS = \text{Bucket size of scraper in cu yd}$$

$$PPH = (NPH)(BS)$$

$$MAXPH = (PPH)(LF)(BF)(AO)\left(\prod_{i=1}^{k} CF_i\right)$$

Alternatively,

$$MAXPH = (MAXPHG)(LF)(BF)(AO)\left(\prod_{i=1}^{k} CF_i\right)$$

where k is the number of other applicable correction factors from the manufacturer's handbook and CF_i is the value if the ith correction factor. For calculating the number of pusher tractors, the pusher cycle time must be known. The maximum number of scrapers that can be served by one pusher tractor is given by rounding the following ratio to the next lower integer,

$$\text{Number of scrapers per pusher tractor} = \frac{\text{Scraper cycle time}}{\text{Pusher cycle time}}$$

$$\text{Number of pusher tractors} = \frac{\text{Total number of scrapers}}{\text{Number of scrapers per pusher}}$$

3. Front-end Loaders. In selecting front-end loaders, one must avoid overestimating the capacity of the bucket. Improper bucket-sizing may lead to instability of the equipment. There are three Society of Automotive Engineers (SAE) ratings for front-end loader bucket capacities: a) struck capacity; b) heaped or rated capacity; and c) static tipping load.

BL = Bucket capacity in cu yd
DP = Dumping time in min
MS = Maneuvering time in min
Ft = Total fixed time in min
LT = Loading time in min
CT = Cycle time in min
MAXPH = Actual production per hour in cu yd
LF = Load factor
BF = Bucket fill factor
AO = Operating factor
NPH = 60/Ct
PPH = (BL)(NPH)

$$MAXPH = (PPH)(LF)(BF)(AO)\left(\prod_{i=1}^{k} CF_i\right)$$

where k is the number of other applicable correction factors from the manufacturer's handbook and CF_i is the ith correction factor.

Alternatively, Table 52 provides manufacturer production capacities for front-end loaders, from which actual production can be calculated by application of the correction factors,

$$MAXPH = (MAXPHG)(LF)(BF)(AO)\left(\prod_{i=1}^{k} CF_i\right)$$

In using the correction factors average conditions should be assumed where performance data is not available. Some correction factors are a combination of two or more other factors. For example, "AO," the operating factor, is a product of operator efficiency factor and job management factor. Where accurate data are available, results would reflect the actual performance.

SUMMARY

This chapter has briefly reviewed the engineering considerations in equipment selection and pit design. Equipment selection formulas have been provided to select the bucket size, reach of machines, width of pit, number of drills, loading shovels, trucks, and bulldozers to achieve a production target. Formulas have also been provided to select bucket-wheel excavators, scrapers, and front-end loaders. Important job correction factors were discussed. In almost all cases, the selection of other equipment is dictated by major stripping equipment; because of this, great care must be exercised in their selection.

TABLE 52. WHEEL-LOADER PRODUCTION ESTIMATING TABLE (CATERPILLAR, 1976)

Estimated Cycle Times		CUBIC YARDS/60 MIN. HR. ESTIMATED BUCKET PAYLOAD - BANK CUBIC YARDS Bucket Sizes																		
Hundredths	Cycles Per Hr.	1.0	1.5	2.0	2.5	3.0	3.5	4.0	4.5	5.0	5.5	6.0	6.5	7.0	7.5	8.0	8.5	9.0	9.5	100
.35	171																			
.40	150	150	225	300	375	450	525													
.45	133	133	200	268	332	400	466	530	600	665	730	800	865							
.50	120	120	180	240	300	360	420	480	540	600	660	720	780	840	900	960	1003	1080	1140	1200
.55	109	109	164	218	272	328	382	436	490	545	600	655	705	765	820	870	925	980	1008	1090
.60	100	100	150	200	250	300	350	400	450	500	550	600	650	700	750	800	850	900	950	1000
.65	92	92	138	184	230	276	322	368	416	460	505	555	600	645	690	735	780	830	875	920
.70	86							342	386	430	474	515	560	600	645	690	730	775	815	860
.75	80													560	600	640	680	720	760	800

JOB EFFICIENCY	EFFICIENCY FACTOR	BUCKET LOAD FACTOR
Worktime/Hr.		
60 Min. Hr.	100%	Bucket Size X 1.00
55	91%	.95
50	83%	.90
45	75%	.85
40	69%	.80
--	---	.75

Reclamation

INTRODUCTION

Competition for land to be used for agriculture, living space, recreation, and industry is becoming more intense. Mineral exploration, development, extraction and processing all have different impacts and conflict with these essential uses. The very nature of the non-renewability of a mineral deposit necessarily makes mining a temporary activity, to be completed sooner or later. The importance of mining, particularly surface mining, however, can hardly be overemphasized as it is doubtful if there is any aspect of human life which does not require the products of mining. It is known that over 90% of the mined products are obtained through the surface method. In fact, more than 75% of the land affected by surface mining is the result of the extraction of coal, crushed stone, and sand and gravel. Therefore, it is necessary to be concerned with the concurrent and sequential land use for optimum land utilization. The allocation of land for agriculture, forest, wildlife, urban growth, recreation, transportation, and so forth, is compounded by political, economical, sociological, and ecological considerations. When some of these lands hold mineral values, the big question is: How can one mine these riches without impairing the utility value for satisfying land demands during and after mining? While this is too frequently interpreted as a problem, it should also be recognized as an opportunity to remold the surface for a desirable end-use.

LAND EFFECTS

Even though surface mining of coal constituted only 0.3% in 1914 and 7% in 1937 of the total U.S. coal production, the use of surface mined lands for other productive purposes was demonstrated and pursued by some coal companies voluntarily as early as 1920. In fact, prior to 1965, there were only seven states which had specific laws requiring land reclamation with West Virginia enacting the first legislation in 1939. The post-1965 period has been one of rapid growth in laws and regulations with regard to environmental control in general, and surface mining in particular, both at the federal and state levels. This discussion of land effects is provided only to serve as an overview of the genesis of the environmental problems associated with surface mining.

Land disturbances from surface mining, in terms of area affected, have grown to enormous proportions in recent years, for the following technological reasons: as easily mined and richer deposits are fast being extracted, and in some cases, completely exhausted, attention has been directed towards mining larger tracts of lower-grade, higher cost deposits; in surface mining of coal, the ratio of overburden thickness to that of coal has been increasing; and current operations are involved with handling more material, and often from greater depths, than heretofore. The need to exploit vast reserves with thick cover and heavy relief has resulted in the development of larger and more powerful equipment which can displace and remove enormous amounts of materials at a much faster rate.

Land damages that are cited against surface mining are mainly due to the destruction of surface topographies and of soil conditions that existed before mining commenced. Often, the potential productivity of the soil for plant growth is greatly reduced after mining. Soils that are disrupted by these operations are often chemically active and toxic, thereby becoming a source for water pollution. Also, if the overburden is a massive rock formation, huge blocks of rock occur in the graded spoil which make it difficult for the smooth passage of farm machinery. Much larger areas are also affected by the unconsolidated spoil heaps and voids because these conditions affect drainage patterns. Here, the natural processes of erosion and sedimentation are accelerated, moving huge volumes of soil into receiving streams. When not properly graded and maintained against dumping waste and refuse, these sites become potential health hazards. Finally, the steep walls created as a result of thick burdens and heavy relief, coupled with the possibility of ground movement with lapse of time, present a safety hazard in inhabited neighborhoods and are aesthetically unattractive. Additional areas may also be affected by the associated haulroads and preparation and processing plants.

A summary of the major environmental effects due to the various surface mining unit operations is presented in Table 53. A detailed report on this subject has been prepared by Grim and Hill (1974). While this section is concerned with surface mining effects on land and land reclamation, the effects on water, air, wildlife, and other resources must also be considered. In fact, all these effects are not mutually exclusive. For example, air pollution may eventually lead to water pollution and both may degrade the land. Surface mining without control may result in serious blights and hazards to public health, contaminate air and water resources, adversely affect land values, create public nuisances, and generally interfere with community development. In fact, the controversy with regard to surface mining of coal is so intense that some have sought a complete ban on surface mining of coal and a greater emphasis on underground mining.

RECLAMATION PRACTICES

Rehabilitation of surface-mined lands is, however, a tangible program. In fact, strip-mined land, abandoned open pits, and spoil and waste piles are being shaped and graded, stabilized and seeded, and returned to productive use with proper care and management. Mined-area spoil banks that can be revegetated present minor problems and have great potential for development. Various reclamation programs that are being actively pursued include returning the ground for agricultural and livestock farming, for reforestation, for recreation, and for housing and industrial sites (Figures 104, 105, 106, 107, 108, 109). The possibilities for development under these conditions are limited only by cost benefit and innovation. There are, however, marginal and problem spoils (acid, toxic, etc.) which require special attention and even better planning. Concern for good mining practices and effective land reclamation programs is witnessed by the adoption of new federal regulations and an increased research and development effort.

On August 3, 1977, the U.S. Congress enacted Public Law 95-87 titled The Surface Mining Control and Reclamation Act of 1977. It represents a significant effort on the part of the Federal Government to establish a nationwide program to protect society and the environment from the adverse effects of surface mining operations. Among the stated purposes of the 1977 Act, the following are particularly relevant to mine planning and operation:

TABLE 53. POTENTIAL MAJOR EFFECTS OF SURFACE MINING UNIT OPERATIONS ON LAND, WATER AND AIR RESOURCES

Land and Soil

Surface Mining Unit Operations	Soil Erosion	Overburden Swelling	Toxic Strata	Soil Inversion	Soil Stability	Landslides	Spoil Piles	Oil Spills	Coal Spills
1. Exploration	x								
2. Area Dewatering Diversion, Etc.	x								
3. Drilling									
4. Blasting		x	x			x			
5. Stripping (Over-burden Removal)			x	x	x		x	x	
6. Haulage	x							x	x
7. Top Soil or Other Soil Storage	x		x				x		
8. Maintenance								x	
9. Beneficiation									

Water

Surface Mining Unit Operations	Aquifer Effects	Runoff Alteration	Sediments	Toxic Substances	Groundwater	Contamination	Industrial Effluents	Sludge
1. Exploration	x							
2. Area Dewatering Diversion, Etc.		x	x	x	x			
3. Drilling	x	x						
4. Blasting	x	x						
5. Stripping (Over-burden Removal)								
6. Haulage			x					
7. Top Soil or Other Soil Storage			x		x			
8. Maintenance								
9. Beneficiation				x	x	x	x	x

Air

Surface Mining Unit Operations	Exhaust Emissions	Dust	Noise	Other (Welding, etc.)	Blasting Fumes
1. Exploration	x	x	x		
2. Area Dewatering Diversion, Etc.					
3. Drilling	x	x	x		
4. Blasting		x	x		x
5. Stripping (Over-burden Removal)	x	x	x	x	
6. Haulage	x	x	x		
7. Top Soil or Other Soil Storage		x			
8. Maintenance	x	x	x	x	
9. Beneficiation		x	x		

Wild Life

Surface Mining Unit Operations	Habitat Altered	Species Diversity	Aquatic Life	Animal Essentials	Accident/Deaths	Soil Organisms	Vegetation Potential	Wildlife Disturbed
1. Exploration								x
2. Area Dewatering Diversion, Etc.	x	x	x	x		x		x
3. Drilling								x
4. Blasting	x				x			x
5. Stripping (Over-burden Removal)	x					x		
6. Haulage	x				x	x		x
7. Top Soil or Other Soil Storage	x					x	x	x
8. Maintenance								
9. Beneficiation								

Other

Surface Mining Unit Operations	Aesthetic	Dangerous Material
1. Exploration		
2. Area Dewatering Diversion, Etc.		
3. Drilling		
4. Blasting	x	x
5. Stripping (Over-burden Removal)	x	
6. Haulage		
7. Top Soil or Other Soil Storage	x	
8. Maintenance	x	x
9. Beneficiation	x	

Figure 104. On land reclaimed with grass and trees after strip mining, responsible coal companies have created public recreation areas like this roadside park and campground in southeastern Ohio.

Figure 105. Charolais cattle and Giant Canadian Geese on mined lands.

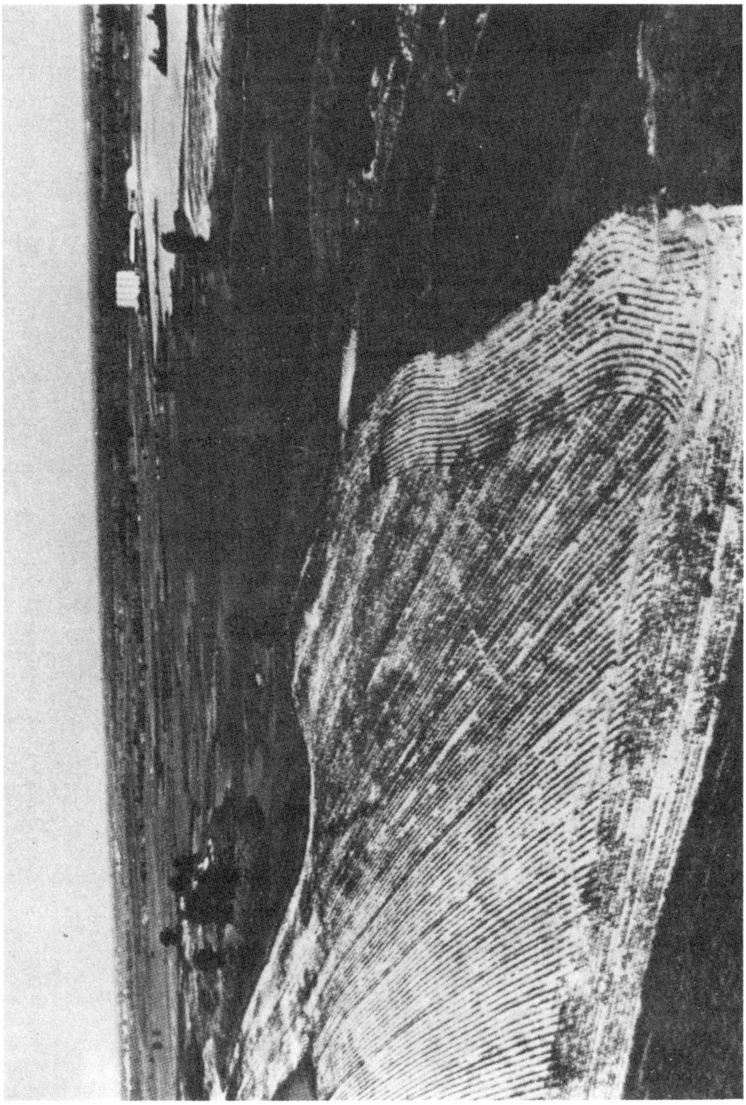

Figure 106. Corn, as well as forage, can grow on former coal land. The ponds in the background (former pits) provide watering places for cattle.

Figure 107. Charolais calves fatten in a feedlot on strip-mined land in Indiana. The farm machinery shed in the background was a repair shop for mining equipment in the days when the area was strip mined.

Figure 108. A golf course on strip mined land.

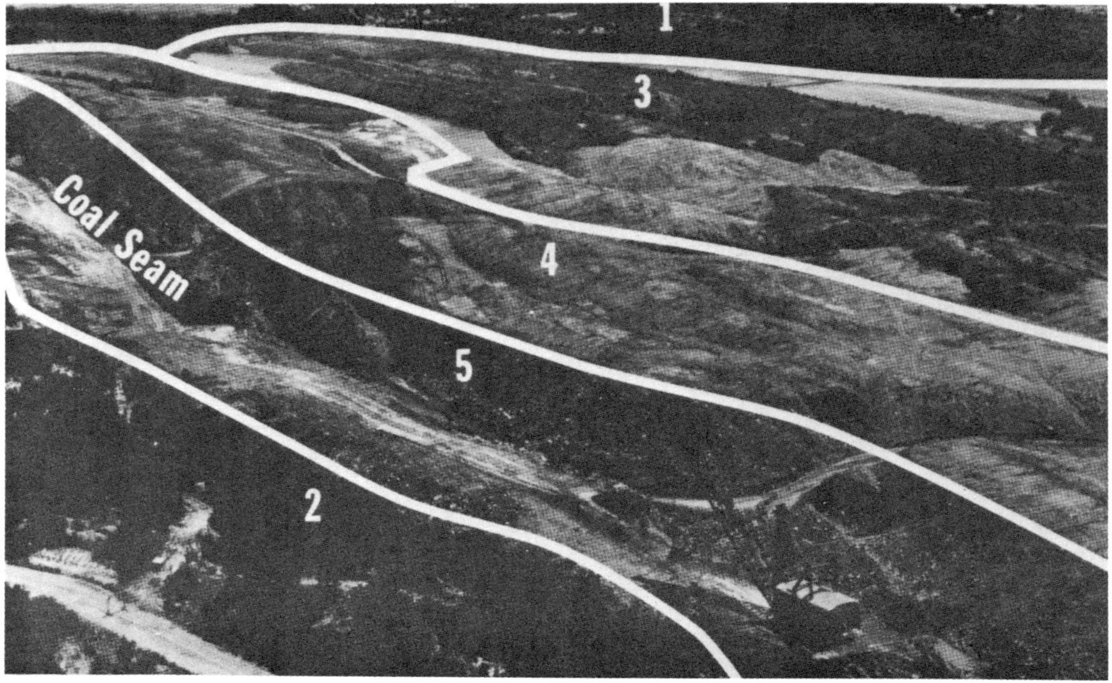

Figure 109. Various phases of surface mining and reclamation in a Kentucky coal mine: (1) virgin farm lands, (2) reforested area mined in 1947-1951, (3) reclaimed area mined 1954-1968, (4) reclaimed areas mined 1970-1971, partially contoured but not yet seeded, and (5) current mining operation with exposed coal seam.

1. To assure that surface coal mining operations are so conducted as to protect the environment.
2. To assure that adequate procedures are undertaken to reclaim surface areas as contemporaneously as possible with the surface coal mining operations.
3. To assure that the coal supply essential to the Nation's energy requirements and to its economic and social well-being is provided, and strike a balance between protection of the environment and agricultural productivity and the Nation's need for coal as an essential source of energy.

Even without the federal law, new or amended surface mine regulations were being enforced in 38 states. At the present time, several states are revising their statutes to be in accordance with the federal law.

Currently 99% of the nation's surface mined coal is being produced under land reclamation programs (Daley, 1974). In the history of the U.S., surface mining has disturbed 0.177 percent of the land and about one third of that has been judged as reclaimed by the U.S. Department of Interior.

In 1972, there were over two million acres of land requiring reclamation, and about two million not requiring reclamation for a total of 4,004,900 acres of disturbed land.

In the seven-year period between 1965 and 1971 private landowners reclaimed over a third of a million acres. More than 90 percent of the four million acres disturbed up to 1972 is privately owned. The USDA Soil Conservation Service (SCS) began participating in land restoration in the 1930's. Soil Conservation Service is active in developing new plants that can survive under the difficult slopes and acid conditions found on most surface-mined land. Most research now underway in government and industry is in six categories: revegetation, chemistry of overburden and spoils, hydrology, earth movement and placement, haulroads, and land-use potentials.

RECLAMATION PLANNING

The specific planning problem with regard to reclamation is one of maximizing production and maintaining environmental quality to be consistent with the constraints imposed by the regulatory agencies, and the need to satisfy the profit objectives of the companies. Mining actions, on one hand, use some resources to produce coal, and at the other extreme, produce waste products, both of which affect the environment. There are direct costs and revenues to the company and public, as well as indirect costs and benefits which must be coordinated in the planning. The necessary planning steps are (1) to make an inventory of the premining conditions, (2) to evaluate and decide on the postmining requirements of the region to accomodate the needs and desires of the affected groups, (3) to analyze alternative mining and reclamation schemes to achieve the requirements optimally, and (4) to develop an acceptable integrated mining and reclamation scheme most suitable under the technical, social, and economic conditions.

For planning reclamation, information is needed in several areas. The initial concern from the environmental standpoint was the ability to reclaim the land in a minimal manner; today, however the environmental studies prior to launching a strip mine are much more detailed. The ecosystem factors that need to be considered today are set forth in Table 54. The listing is not complete, but points to the recognition of the fact that any attempt to extract mineral resources involves a change in the physical, biological, social, physiosocial, biosocial, and psychosocial factors. From the initial

TABLE 54. INFORMATION NEEDS ON ECOSYSTEM FACTORS FOR SURFACE MINE PLANNING AND LAND USE

I. SOCIOECONOMIC INVENTORIES

(a) Land demands for mining and associated activities, and availability, encroachment into critical areas, other resources affected and foregone benefits.

(b) Past, current and future land use plans for the area-industry, agriculture, wilderness, etc.

(c) Transportation network

(d) Recreation networks

(e) Archeological and historical interests

(f) Demographic and population characteristics
- (i) employment patterns
- (ii) tax base
- (iii) land value alterations and their impacts on local government
- (iv) attitudes of public, political, and educational institutions in the community/region

II. GEOLOGIC ANALYSIS

(a) Topography

(b) Overburden Characteristics
- (i) stratigraphy
- (ii) physiography
- (iii) geomorphology
- (iv) chemical nature

(c) Soil and Toplayer Characteristics
- (i) physical
- (ii) chemical
- (iii) limitations for vegetation and future land uses
- (iv) limitations for engineering

III. COAL CHARACTERIZATION

(a) Coal seam attitude

(b) Coal seam thickness

(c) Coal seam analyses
- (i) proximate analyses
- (ii) ultimate analyses
- (iii) washability analyses

IV. TERRESTRIAL ECOLOGY

(a) Natural vegetation, characterization, identification of survival needs

(b) Crops

(continued)

TABLE 54. (continued)

IV. TERRESTRIAL ECOLOGY (continued)

(c) Game animals

(d) Resident and migratory birds

(e) Rare and endangered species

V. AQUATIC ECOLOGY

(a) Aquatic animals - fish, waterbirds, resident and migratory

(b) Aquatic plants

(c) Characterization, use and survival needs of aquatic life system

VI. WATER RESOURCES

(a) Surface hydrology
- (i) watershed considerations
- (ii) flood plain delineation
- (iii) surface drainage patterns
- (iv) amount and quality of surface run-offs

(b) Ground water hydrology
- (i) ground water table
- (ii) aquifers
- (iii) amount and quality of ground water flows
- (iv) recharge potential

VII. CLIMATOLOGY

(a) Precipitation

(b) Wind - airflow patterns, turbulence

(c) Humidity

(d) Temperature

(e) Climate type

(f) Growing season

VIII. OTHER

(a) Noise

(b) Particulates in the air

(c) Aesthetics

construction of access facilities to prospect to the final completion of the mining and beneficiation operations, the ecosystem balance is continually being altered. A discussion of the major factors as they affect the reclamation decision follows.

Physical and chemical attributes of the premining environment are factors in several areas of design and planning, particularly in the engineering design of excavation, materials handling, and ground support systems. They are also necessary in assessing water quality control of the operation and determining the potential future land uses.

Geological aspects of the coal occurrence (e.g., attitude, depth, thickness, stratigraphy, etc.) is the basis for establishing alternative mining and reclamation schemes including slope stability. The nature of the overburden, especially those components as sulfur, having potential environmental impacts must be studied. The structural and stratigraphic evaluation which effect the hydrological conditions are critical for effective mining and reclamation. Figure 110 and Table 55 show the type of analyses that were done in an exposed highwall in southern Ohio.

OHIO POWER HIGHWALL STUDY

East Pit Highwall

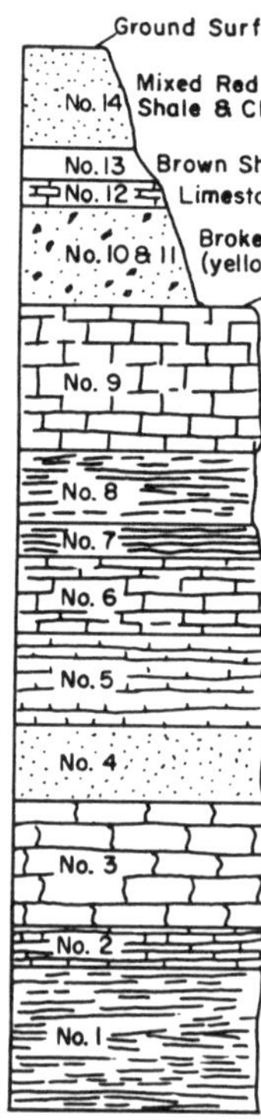

Layer	Thickness (feet)	pH	Fe (ferric)	Sulfates	Soluble Salts	Toxic Aluminum
			POUNDS PER ACRE			
No. 14 Mixed Red Shale & Clay	4	5.3	.4	40	320	.42
No. 13 Brown Shale	1.5	7.7	.2	22	420	0
No. 12 Limestone	1	8.1	.3	30	100	0
No. 10 & 11 Broken Shale (yellow—brown)	4	8.2	.1	25	360	0
No. 9 Sweet Water Limestone	6					
No. 8 Green Shale	3	8.1	.3	91	800	0
No. 7 Red Shale	1.5	7.8	.5	82	940	0
No. 6 Yellow Material (rock-like)	3	8.0	.5	61	520	0
No. 5 Blue Shale Rock	4	8.2	.3	61	660	0
No. 4 Yellow Material	3	3.8	22.5	106	3,200	.72
No. 3 Massive Sandstone	5	3.4	23.5	289	5,000	.66
No. 2 Fibrous Coal-Like	1.66	3.1	25.0	310	5,500	.54
No. 1 Black Shale	6	8.4	1.3	61	1,200	0

Source: Farm Clinic of U.S.

Figure 110. Analysis of the east pit highwall, a problem wall 5 years old, Ohio Power Highwall Study (Krause, 1973).

TABLE 55. CHEMICAL ANALYSES OF THE EAST PIT HIGHWALL OHIO POWER HIGHWALL STUDY, 1967 (SOURCE: THE FARM CLINIC, WEST LAFAYETTE, IND.) (KRAUSE, 1973)

Sample No.	Organic Matter		Total Sol. Salts	Nitrogen						Fer-ric	Fer-rous						Toxic	
				NO_3	NH_3	P_2O_5	K_2O	Ca	Mg	Fe	Fe	Mn	B	Cu	Zn	SO_4	Al	Mo
	%	pH	lb/acre															ppb
4	0.9	3.8	3200	10.8	42	1.1	91	1365	270+	22.5	0.2	3.0	0.34	0.8	0.65	106	0.72	12
6	0.8	8.0	520	9.2	46	4.4	112	3500	270+	0.5	0.1	4.5	0.36	1.0	0.13	61	0	62
10	0.9	8.2	360	12.7	43	4.2	40	3450	270+	0.3	0.1	0.6	0.40	0.8	0.11	25	0	75
12	0.8	8.1	180	8.5	47	2.6	44	3640	270+	0.4	0.3	0.5	0.39	0.5	0.09	30	0	50
13	1.0	7.7	420	9.7	45	6.1	82	3250	270+	0.3	0.2	1.4	0.42	0.8	0.16	22	0	58
14	0.8	5.3	320	8.1	49	1.5	52	3080	114	0.5	0.4	4.5	0.37	2.1	0.50	40	0.42	25
Interpretation																		
Low	---	---	---	<20	<30	<5	<80	<500	<100	---	---	<1	<0.32	<1	<1	---	---	<25
Medium	---	---	---	20-40	30-50	5-20	80-150	500 to 1000	100-180	---	---	1-4	0.32 to 0.37	1-4	1-4	---	---	25-50
Satis-factory	---	---	---	>40	>50	>20	>150	>1000	>180	---	---	>4	>0.38	>4	>4	---	---	>50
Excessive or toxic	---	---	---	---	---	---	---	---	---	---	---	>20	>0.45	--	--	---	>1	--

Soils must be considered to ensure their proper removal, placement, and postmining disposition. While the definition and identification of the material to be classified as topsoil is fraught with difficulties, the characteristics of the material to serve as soil after mining must be known. The characteristics of topsoil should be such that it should ensure an effective growth of vegetative cover than other materials. Topsoil and subsoil, though not always the most ideal materials, do have certain advantages over other materials, including higher organic matter, better moisture-holding capacity, natural seed source, and soil microorganisms. The removal and storage of topsoil and replacement of it on top of the graded soil, in many instances leads to better revegetation potential, which in turn leads to better erosion control. In many older mining districts, spoil banks are only sparsely revegetated. This is due in part to the lack of nutrients in the spoil and the water flow characteristics of the restored strata. In cases where bedrock material, shale, sandstone, siltstone, and fireclay have been mixed with the original topsoil covering, inadequate fine-grained matrix material and organic matter may be available to provide nutrients or moisture-holding capacity. Too frequently, the fine grained material sinks through the interstices in the coarse matrix. High infiltration capacity, high porosity and high permeability are too commonly encountered when no attempt is made at toplayer restoration and compaction. Color of the spoil has an effect on the surface temperatures of the spoil banks. Temperatures in excess of 150°F have been reported on spoil surface composed of dark shales.

Regrading of spoils alone may be inadequate to provide a growth medium. In fact, the mining operation provides an opportunity to modify the soil strata to include burial of toxic soil strata and complete rebuilding of soil layers at the top. A few of the soil modification and reconditioning methods include treatment with lime or limestone, fertilizers, and other additions such as mulch, fly ash or sewage sludge to adjust pH, nutrient and physical conditions. Two important factors to be determined here are the amount and depth of application of these modifiers.

Topography influences the selection of the type of mining, and mining equipment. These topics were discussed in an earlier section. Severe terrains also influence the methods of revegetation, and limit alternatives for future land use.

Extremes in climate affect the selection of equipment. Seasonal fluctuations determine desirable planting periods. In addition to the problem of permafrost in severe cold climates, operation of mobile equipment becomes difficult. Heavy rains can cause erosion complicating water quality control, sediment formation, and the establishment of vegetation.

Hydrology, both surface and underground, together with the physical and chemical characteristics of overburden and coal, is important for any reclamation program planning. A knowledge of water movement and quality in the mine area and modification during mining is essential. In all cases, erosion and sediment production are immediate concerns in the critical postmining, pre-revegetation time period, although their control during storms is a long-term problem that requires detailed planning. The Surface Mining Control and Reclamation Act of 1977 requires that the surface mining operation be so conducted as to minimize the disturbance (1) to the prevailing hydrologic balance at the mine site and in associated offsite areas, and (2) to the quality and quantity of water in surface and ground water systems both during and after surface coal mining operations. Among the specific provisions of the law for the avoidance of acid or toxic mine drainage, included are such measures as:

1. preventing or removing water from contact with toxic producing deposits.

2. treating drainage to reduce toxic content which adversely affects downstream water upon being released to water courses.
3. casing, sealing, or otherwise managing boreholes, shafts, and wells to keep acid or other toxic drainage from entering ground and surface waters.

With regard to hydrologic balance, other provisions of the Act deal with factors such as suspended solids and recharge capacity of the mined area. The negative impacts of surface mining are to be minimized by:

1. conducting surface coal mining operations so as to prevent, to the extent possible using the best technology currently available, additional contributions of suspended solids to streamflow or runoff outside the permit area. In no event shall contributions be in excess of requirements set by applicable State or Federal law.
2. constructing any siltation structures, prior to commencement of surface coal mining operations. Such structures are to be certified by a qualified registered engineer and constructed as designed and as approved in the reclamation plan.
3. cleaning out and removing temporary or large settling ponds or other siltation structures from drainways after disturbed areas are revegetated and stabilized; and depositing the silt and debris at a site and in a manner approved by the regulatory authority.
4. restoring recharge capacity of the mined area to approximate premining conditions.
5. avoiding channel deepening or enlargement in operations requiring the discharge of water from mines.
6. preserving throughout the mining and reclamation process the essential hydrologic functions of alluvial valley floors in the arid and semiarid areas of the country.
7. other such actions as the regulatory authority may prescribe.

While the scope of the present study does not include a detailed discussion of the biological characterization of the mining area, such typification is important in the selection of a reclamation plan. In many cases, the law provides for the protection of unique or endangered species of plants and animals which require specific consideration. Such provisions may even prevent the mining of coal.

Sociohumanistic factors, though very important, are another study area which is outside the scope of the present report. Nonetheless, a reclamation plan must consider attitudes of the public and current and future uses of the land. Aesthetic or historic value of the land prior to mining may also be important enough to preclude mining.

LAND-USE ANALYSIS

The current land-use patterns may provide the best guideline for reclamation. For example, if there is a need for pasture lands, forested hillside land adjoining a pasture may be reclaimed for pasture after mining. Similarly, orphaned mined land in a wilderness area may be more worthwhile if restored as a forest. Future land-use plans contrary to premining use are unusual and not the common practice. Acceptance of mountaintop removal and valley fill mining methods which create new flat areas in regions where such areas are in short supply reflect the permanent, positive potentials of the temporary, drastic disturbance of lands due to mining. Identification of the current and probable future land uses can be summarized into the following five categories:

1. "Wilderness" or unimproved use
2. Limited agriculture or recreation, with little development, such as grazing, hunting or fishing or as timber land
3. Developed agriculture or recreation, such as crop land, water sports, vacation resorts
4. Suburban housing or light commercial and industry
5. Urban housing or heavy commercial and industry

Analysis of alternative long-term uses requires compliance with socioeconomic and public constraints as well as the economic goals of the mining company. Discussion on this subject is necessarily limited here since another report on land-use planning has dealt with this subject in greater detail. Irrespective of the choice of short-term and long-term land use, these plans must be linked to the mining plan in space, time and method. Definition of the long-range plans do achieve a purpose by minimizing disturbance unnecessary for the determined future use and delimiting those reclamation activities that are not a part of the normal mining or land-use plans. Reference has already been made to the enactment of legislation at the federal and state levels pertaining to surface mining and reclamation.

Several provisions of the Surface Mining Control and Reclamation Act of 1977 and the interim regulations deal with land-use planning. Specifically, with regard to obtaining a mining permit, a reclamation plan must be submitted which shall be of sufficient detail to demonstrate that reclamation can be achieved as required by Federal or State programs. Such a plan must cover, among other items, the following important points with regard to the uses of land:

1. the uses existing at the time of the application, and if the land has a history of previous mining, the uses which preceded any mining.
2. the capability of the land prior to any mining to support a variety of uses giving consideration to soil and foundation characteristics, topography, and vegetative cover.
3. the use which is proposed to be made of the land following reclamation, including a discussion of the utility and capacity of the reclaimed land to support a variety of alternative uses and the relationship of such use to existing land use policies and plans, and the comments of any owner of the surface, State, and local governments or agencies thereof which would have to initiate, implement, approve, or authorize the proposed use of the land following reclamation.
4. a detailed description of how the proposed post-mining land use is to be achieved and the necessary support activities which may be needed to achieve the proposed land use.
5. the consideration which has been given to making the surface mining and reclamation operations consistent with surface owner plans and applicable State and local land-use plans and programs.
6. the consideration which has been given to developing the reclamation plan in a manner consistent with local physical environmental and climatological conditions.

Even otherwise, mining companies today have to meet with nongovernmental, historical, wildlife, and other environmental groups and allay their fears. The integrated mining, reclamation and land-use planning diagram (Figure 111) sets out an iterative procedure for the evaluation of mining and reclamation plans. The feedback loops in the diagram serve to indicate the procedure which, when the plans are optimal will lead to the mining and reclamation of the coal seam. Shown in the Figure is also another set of interactions that takes place before a mining plan is put into practice. The iterative loop through which the mining and reclamation plans have to pass includes not only the federal, state and local agencies but the special interest groups and the general public. Finally, the ecosystem factors to be considered in the development of a mine today are varied, many, and increasing. The planning of surface mines, thus, is very involved and thorough, and the objective of planning is the long-term use of the land as a prime natural resource.

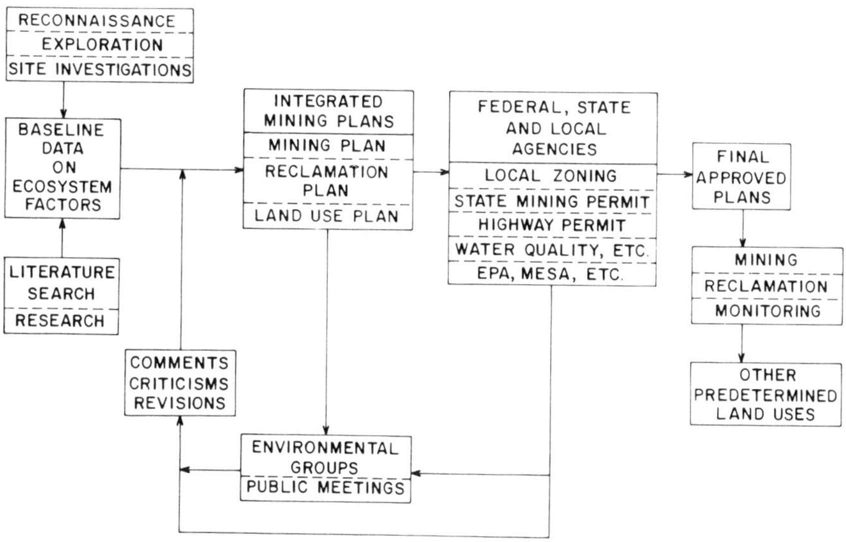

Figure 111. Integrated and interactive surface mining, reclamation and land-use planning.

ECONOMIC FEASIBILITY STUDIES

In a free enterprise society, it is difficult to continue in business unless there is an adequate rate of return in monetary terms on the investment. The goals of environmental protection and mineral resource extraction must therefore be analyzed in terms of both the economic return and the constraints and incentives of the public and the government. At the mine level, it is necessary to establish the direct cash cost of reclamation. The consideration of internal, direct costs, and income from mineral sales, by more easily quantifiable rate of return analysis, is traditional.

Grim and Hill (1974) have summarized the results of several studies on reclamation costs. Mathematica Incorporated (1973) has estimated that reclamation costs will be about 8% of the total mining cost (Table 56). Other cost figures in literature vary from $1,236 to $19,768 per hectares ($500 to $8,000 per acre) depending on the severity of the land effects and the reclamation requirements. Doyle et al. (1974), have prepared a useful report on the analysis of pollution control costs. Their report addresses the costs involved in revegetation, acid mine drainage, stream diversion, strip mine spoil and waste, backfilling and regrading, etc. Table 57 is extracted from that report to indicate the various factors that affect backfilling and grading costs.

Since reclamation costs are site specific, the requirements of labor, equipment and supplies must be estimated. A convenient basis for estimation purposes may be to identify the acreage disturbed for the recovery of coal from underneath an acre of ground. The costs of reclamation are calculated for the disturbed land and then assigned to the tons of coal recovered. Reclamation costs are preferably indicated both in $/acre and $/ton.

To minimize environmental disturbances and to maximize land use, it is necessary to include the sociohumanistic benefits and costs. These are not as easily quantified, defy traditional engineering approaches, and raise conflicts between the various organizations' objectives. The consideration of these benefits and costs are outside the scope of this report. The future of surface mining hinges partially upon the ability to reclaim land in a satisfactory manner. Where the environmental impacts cannot be clearly ascertained, it may be that the mineral resource should not be mined. However, in a majority of the instances, knowledge and technology are available to achieve acceptable impacts. Results of several studies reveal that no major breakthroughs are indicated in the near future. Only through concerted efforts will evolutionary processes result that adapt present equipment with mining cycles which reflect in greater social concern for environment.

TABLE 56. ESTIMATED AVERAGE PRODUCTION COSTS @ AVERAGE STRIPPING RATIO = 8:1 and 0.5 ACRES DISTURBED PER THOUSAND TONS OF COAL PRODUCED* (MATHEMATICA, 1973)

COST ELEMENT	COST ($/TON)	COST (% OF TOTAL)
Sediment structure	0.12(1)	2.9
Acreage fees	0.01(1)	0.2
Bonding	0.00	0.0
Scalping	0.08(1)	1.9
Stripping	2.40	57.5
Overburden haulage	0.05	1.2
Coal loading	0.10	2.4
Coal haulage	0.50	12.0
Backfilling and grading	0.08(1)	1.9
Revegetation	0.03(1)	0.7
Royalties	0.50	12.0
Severence tax	0.30	7.3
Total	$4.17	100.0

* Acre = 0.40 hectares; short tons = 0.907 metric tons

(1) Reclamation costs. These costs are equivalent to $0.32 per ton or 7.6% of the total costs

Note The primary assumption underlying these estimates is that the stripping ratio is 8:1; the cost balance would change somewhat if the assumed stripping ratio were to change.

TABLE 57. STRIP MINE RECLAMATION PROJECT VARIABLES AFFECTING BACKFILLING AND GRADING COSTS (DOYLE, 1974)

1. Geographic location

2. Topographic setting (original, pre-reclamation and final ground slopes)

3. Type of strip mine:
a) area, b) contour, c) area-contour, d) other

4. Coal seams mined and thickness

5. Inclination of coal seams in back of highwall:
a) dip, b) rise, c) horizontal

6. Condition of coal seams in back of highwall:
a) not mined, b) auger mined, c) drift mined (entries opened or caved, d) mine workings exposed by stripping operation

7. The probable hydraulic head that could develop if coal in back of highwall was mined

8. Strip mine area information:

   A. Length, width, and area covered by spoil before reclamation
   B. Highwall height (maximum and average height)
   C. Highwall length
   D. Number of cuts
   E. Total area affected during reclamation in acres (including area above highwall and outside of slopes)
   F. Volume of spoil to be moved
   G. Average haul distance for backfilling and grading
   H. Texture of spoil
   I. Amount of large rock and material requiring special handling of mining timbers, machinery, and debris, junked cars, and other solid waste
   J. Amount and reactivity of pyritic material (minerology and mode of occurrence. For example finely dispersed; single crystals or crystal aggregates; coatings on joint surfaces; in form of lenses, layers or modules; "sulfur balls"; pyritic shales, etc.)
   K. Clearing and grubbing requirements

9. Type of backfill:
a) contour, b) pasture-reverse slope, c) swallowtail, d) head of hollow, 3) submergence, f) other

10. Physical sealants for covering toxic material
a) none, b) clay, c) bituminous material, d) plastic material, e) other

11. Compaction desired:
a) none, b) only toxic materials, c) all spoil material with exception of upper layer up to 0.91 meter (1 to 3 feet)

12. Accessibility factors:

   A. Right-of-way problems
   B. Ingress and egress construction (include clearing and grubbing for access and post-construction revegetation)
   C. Other factors affecting access

(continued)

TABLE 57. (continued)

13. Surface and subsurface ownership of strip-mined area. Also, ownership of properties for ingress and egress: a) public, b) private, c) in process of being acquired or line placed on property, d) abandoned, 3) temporary easement, f) other

14. Time of year reclamation performed

15. Weather conditions during reclamation period(s)

TECHNICAL FEASIBILITY STUDIES

Technical feasibility is a prime requisite for the adoption of any reclamation plan. It is preferable to key-in the reclamation plans with mining plans since nonconcurrent reclamation methods tend to be environmentally less optimal and more expensive. Reclamation efforts initiated long after the mining activity has ceased generally result in reclamation that serves a limited land use. Concurrent reclamation with positive controls aids in the development of well-conceived future land uses with a long-term program to finish and maintain the reclamation plans. Concurrent methods allow total spoil manipulation and/or complete spoil profile construction.

Technical requirements of reclamation will vary depending upon the degree of restoration. In the standard surface mining approaches of today, selective replacement of material is relatively simple. Contour mining approaches such as block cut, box cut, haul back, mountaintop removal and modified area mining are generally flexible to allow segregation and selective replacement. Two such plans are illustrated in Figures 112 and 113 for contour mining. Larger scale area methods employing traditional overcasting are less flexible. It is common practice to handle overburden by direct overcasting preferably with no rehandle. The problem with simple overcasting with a large machine is that the spoil can be cast only at a limited distance from the cut, and then only at a limited range of angles to the line of travel of the equipment. However, selective placement of spoils in these cases can also be achieved by a combination of equipment and methods. For example, the requirement to save topsoil and to spread it over the regraded land has led the mining companies to the use of scrapers that remove the topsoil, run around the pit and dump over formerly graded spoil. Selection of shovel-truck systems, as opposed to that of conventional overcasting, also provides the needed flexibility in soil placement. In fact, the overburden conditions can be closely monitored, and the placement of the overburden so achieved to ensure 1) burial of toxic spoils and less desirable material such as boulders, or impermeable clays, 2) topsoil replacement, 3) desired stratigraphic sequence in the spoil, and 4) flatter or other desired valley configurations. Additionally, truck shovel systems permit blending in the pit for coal quality and greater recovery of coal. The goal for premining planning and scheduling for resource extraction and reclamation is to develop such steps as an integral part of the mining method.

The question of restoration versus adaption of contour is one which is decided by the general mining approach. In strict contour mining, in which the seam is followed around a hill, the restoration to approximate original contour with excess spoil properly disposed in the valleys is generally most appropriate. Where there are flat lands, and the need is for small hills, terrace and shallow hollows, the indicated approach may be to restore some degree of relief by leaving the mined-out areas and spoils in such a reclaimed state that the future potential may be for grazing, wildlife, or recreation. In areas where the lands are used for agriculture prior to mining, the plan may be to re-establish all mined areas to a more tillable topography.

In backfilling, three major areas of long-term concern are 1) slope stability, 2) ground water control, and 3) water quality maintenance. The primary source of slope failure is the prolonged action of water, through erosion, and mass wasting. The primary result of water's actions on slopes is erosion and sediment production. Long uninterrupted slopes are undesirable as compared to ones with terraces and diversions which decrease slope lengths and direct runoff water to safe outlets. Additionally, while backfilling, the permeability of the fill for water perculation can be modified. Such modifications are achieved by sealing or leaving the coal seam open and by selective placement of the spoil. One of the concerns of water quality control, in addition to the acid potential, is the sediment. The interconnection between slope stability and water quality (sediment) is enhanced by apparently conflicting approaches to these problems. Gross stability is enhanced by establishing grade and cover to minimize water infiltration and permeation. Conversely, erosion and sediment control are achieved by encouraging infiltration and permeation to minimize runoff. The proper approach would entail, first, the establishment of as heavy a cover crop after backfilling as possible, to encourage vegetative interception and evapotranspiration as opposed to surface or subsurface runoff. However, not all of the water can be handled in this manner. Provision must be made for storms by designing clean-running interceptions and diversions for excessive rainfall.

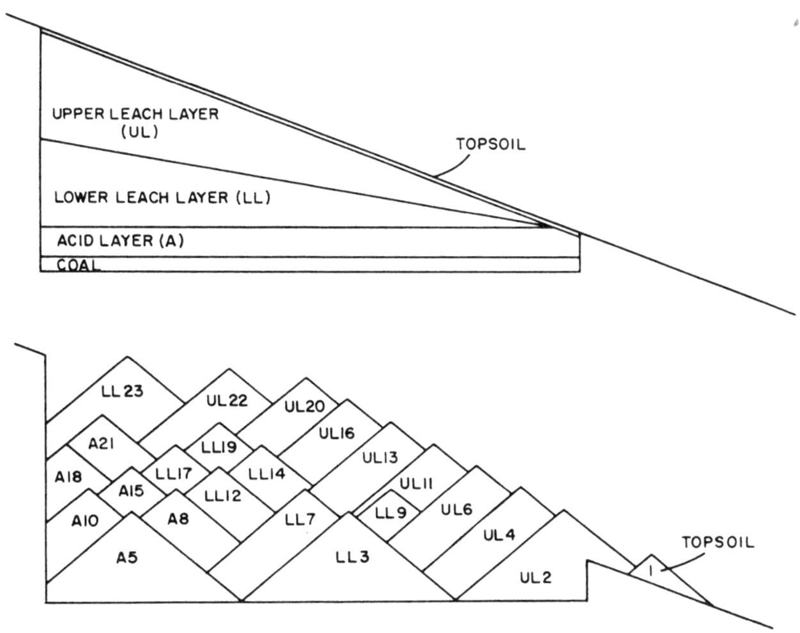

Figure 112. Movement of spoil within a block to insure segregation and burial of acid-bearing material (Saperstein and Secor, 1973).

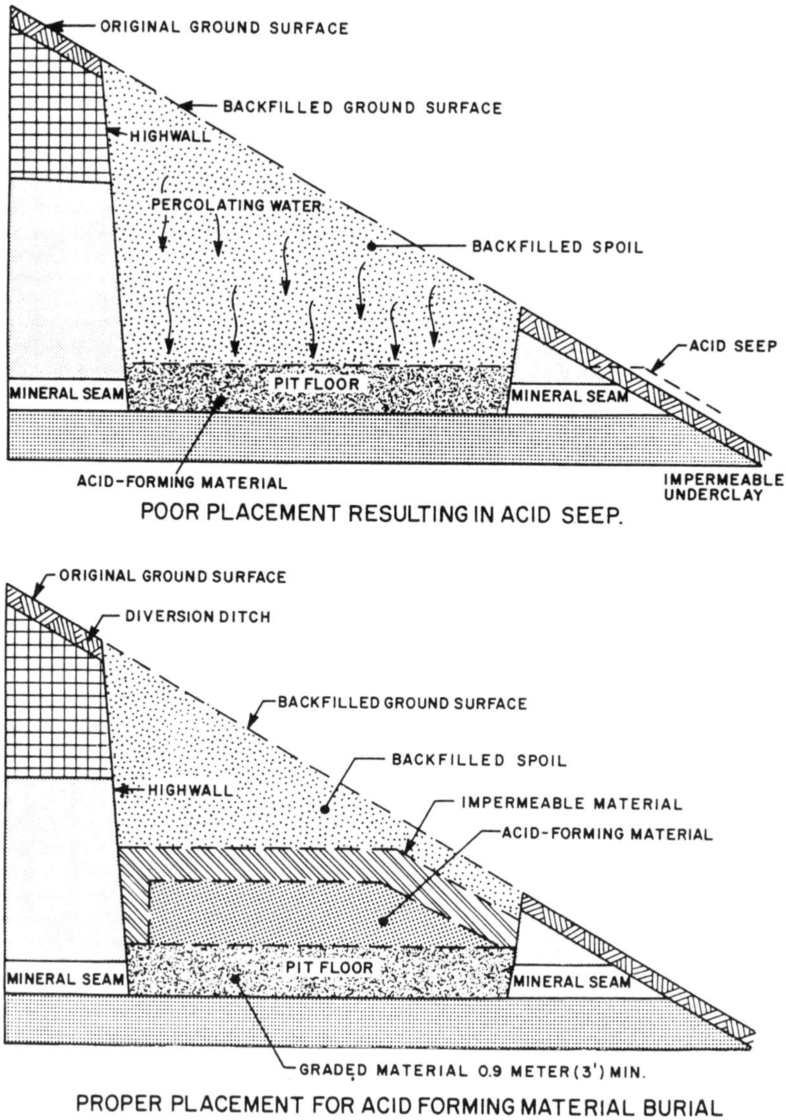

Figure 113. Spoil placement for better environmental impact (Hill and Grim, 1975).

Stability and water quality are inherent in the design of fills and backfills. If a suitable, large-rock drainage base and contact zone is established at the base contact of the fill, slippage can be minimized while rapid drainage of groundwater is enhanced. Such approaches to water quality and the control of stability are exemplified by the head-of-hollow or valley-fill concept, explained in the earlier section on mining methods.

Replacement of topsoil is practiced in mining where required by regulations. The dictum of topsoil replacement could be viewed as a requirement to replace or establish a layer of material of specific thickness, having certain physical and chemical characteristics, and primarily, capable of supporting some specific vegetation. In pre-mining planning, attention should be directed at identifying all strata potentially able to adapt as a surface covering. If topsoil is not available, it becomes necessary to consider soil amendments to establish vegetation. Common approaches to soil amendments involve the addition of lime, mulch, and fertilizer. Many mulches have been used to assist in establishing vegetative cover. Straw or hay can be manually applied, but they, as well as wood chips, are often spread mechanically or hydro-pneumatically. Mulch should be tagged down with tar, etc. Rapid growing annual plants and small grains can be used to establish ground cover, to leave organic matter in the soil, and to serve as a mulch for perennial establishment. The common practice today is to establish such quick-cover mulch crops prior to topsoiling to assure vegetative stabilization and to enhance the organic content and water-holding capability of the surface layer. Specification of the graded surface, furrowing or harrowing, is also done to act as seed and water traps. Care must be taken in selecting species to avoid competition.

Fly ash, the residue of coal-fired boilers, can be applied to increase nutrients, moisture and alkalinity and to ameliorate toxic materials present. The trace elements contained in fly ash are often important micronutrients, while fly ash is also rich in the common macronutrients except nitrogen. Nitrogen deficiency can be handled by planting inoculated clover, alfalfa, vetch, or other legumes. The use of fly ash may be particularly attractive when the power plant is located at or near the mine and the ash must be disposed of in some fashion (Adams, 1971).

Stabilized sewage sludge is another recyclable amendment to reclaimed-mined land. Sludge is rich in nutrients and may reduce acidity, as well as provide stable organic matter to form a humus. A related type of amendment is composting with municipal waste, which has been practiced in other countries but has seen limited application in the United States. Care must be taken, however, to avoid poisoning the soil with too high a level of heavy metals and salt from the sludge (Grim and Hill, 1974).

In surface coal mining, positive control of environmental impacts is achieved by proper mining and reclamation techniques which include concurrent reclamation, overburden segregation, toxic material burial, and topsoiling. Grading and backfilling should immediately follow mining as the freshly spoiled material is easier to move, backfill, and grade. Revegetation should follow grading as soon as possible. The primary purpose of vegetation is to stabilize the spoil to prevent erosion and to ensure water quality and quantity control. In most cases, therefore, revegetation is best accomplished by the establishment of grasses and legumes, as soon after mining as possible, though the long range plan may call for reforestation or other land uses. It is possible to develop a time sequence for the various reclamation activities. Table 58 presents such a sequence. For operational management, this sequence must have specific dates for a particular operation.

MONITORING AND MAINTENANCE

Even with proper reclamation, rededication of the land for productive purposes may require more time. Grazing, for example, should not be allowed immediately on a new cover crop. In any case, use of the land is not definitely accomplished at the end of mining. A careful plan of monitoring and control, and maintenance and development are required.

TABLE 58. TIME SEQUENCE FOR RECLAMATION ACTIVITIES

I. DURING SITE PREPARATION:

1. Install control measures (diversion, sediment traps and basins, etc.).
2. Clear and grub, market lumber if possible; stockpile brush for use as filters; run brush through woodchipper and use chips for mulch.
3. Stabilize areas around temporary facilities such as maintenance yard, power station, and supply area.

II. DURING OVERBURDEN REMOVAL:

1. Divert water away from and around active mining areas.
2. Remove topsoil and store it if possible and/or necessary (Figure 114).
3. Selectively mine and place overburden strata if possible and/or necessary.

III. DURING COAL REMOVAL:

1. Remove all coal insofar as possible.
2. Break or prevent damage to the strata immediately below the coal seam as desired.

IV. IMMEDIATELY AFTER COAL REMOVAL:

1. Seal the high wall if necessary.
2. Seal the low wall if necessary.
3. Backfill--bury toxic materials and boulders, dispose of waste, ensure compaction.

V. SHORTLY AFTER COAL REMOVAL:

1. Rough grade and contour, taking these factors into consideration (Figures 115, 116 and 117):

    a. Time of grading, specific time limit; tied to advance of mining; seasonal considerations.
    b. Slope steepness.
    c. Length of uninterrupted slope.
    d. Compaction.
    e. Reconstruction of underground and surface drainage patterns.

2. If necessary, make mine spoil amendments (root zone), taking these factors into consideration:

    a. Type of amendment, fertilizers, limestone, fly ash, sewage sludge, or others.
    b. Depth of application.
    c. Top layer considerations--temperature (color), water retention (size consist, organics), mulching and tacking.

VI. IMMEDIATELY PRIOR TO FIRST PLANTING SEASON:

1. Fine-grade and spread topsoil, taking seasonal fluctuations into consideration (Figure 118).
2. If necessary, manipulate the soil mechanically by furrowing, deep-chiseling or harrowing, or constructing dozer basins.
3. Mulch and tack.

(continued)

TABLE 58. (continued)

VII. DURING THE FIRST PLANTING SEASON:

Seed and revegetate, considering--time and methods of seeding; choice of grasses and legumes (Figures 119 and 120).

VIII. AT REGULAR, FREQUENT INTERVALS:

Monitor and control: slope stability; water quality, both chemical (ph, etc.) and physical (sediment); vegetation growth.

Monitoring slopes to assure stability and control to maintain water quality are two important aspects. To minimize these problems initially and to prevent them from occurring in the future, at the time of grading it is necessary to keep all surfaces as minimally sloped as possible and under the angle of repose. Additionally, through regular inspections any failure or runoff is corrected by refilling, compacting, and rapid establishment of cover. To control the erosion affects of running water, it is necessary to place interceptor and diversion channels to reduce downslope flows.

Where the coal mine drainage phenomenon cannot be prevented, a long-term program to neutralize the discharge is necessary. In most surface-mining situations, much of the necessary layout is present in sediment-control ponds. Addition of limestone, lime, anhydrous ammonia (where no discharge to surface drainage is encountered), soda ash, sodium hydroxide, or other suitable neutralizing reagents can minimize acid formation. However, isolation of toxic material and concurrent backfilling is probably the most desirable approach to minimization of acid formation problems.

One other aspect of maintenance is the sediment and erosion control installations. With appropriate planning and no unusual storms, such maintenance of these features may be minimal. However, control structures may become clogged or filled and should be cleared and cleaned through regular inspections.

SUMMARY

This chapter has briefly reviewed the premining planning for reclamation. A time sequence for reclamation activities was provided, along with examples of good reclamation practices.

Figure 114. A pusher-scraper combination for topsoil and overburden removal.

Figure 115. Large crawler tractor in the process of rough grading coal mine spoil banks in hilly terrain.

Figure 116. A giant bulldozer with an experimental 40-foot blade levels mined land in southeastern Kansas.

Figure 117. Completed rough grading of coal mine spoil banks in hilly terrain.

Figure 118. Bulldozer spreading topsoil on graded spoil.

Figure 119. Coal mine spoil banks reclaimed to farmland.

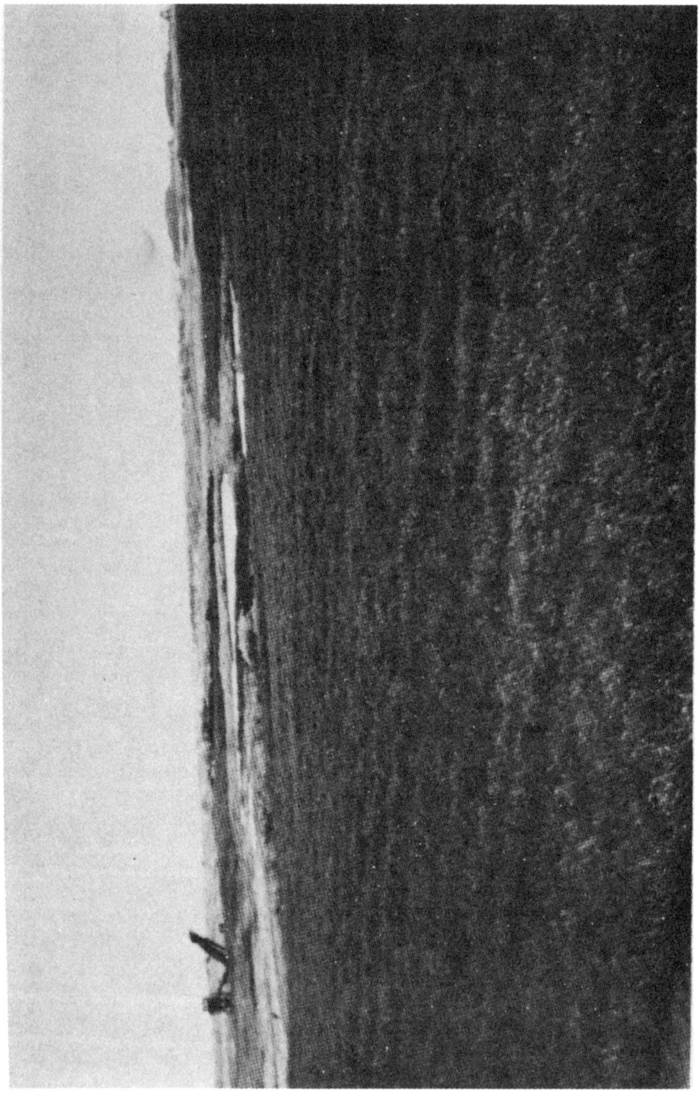

Figure 120. Area which was mined and reclaimed in 1975. The stripping equipment can be seen in the background (Photo taken in July 1976).

Surface Mining Costs

INTRODUCTION

Economic evaluation of various mining plans is necessary to make a comparison of the alternatives. The key to preliminary analyses is the stripping ratio since, for a given situation, it usually dictates the initial decision to mine or not to mine by surface methods. Capital cost considerations can hardly be over-emphasized. However, equipment costs vary widely and are a function of the amount of steel and the fabrication in the design and construction of the equipment. Cost involved for other surface facilities (e.g., storage, office space, buildings, etc.) are also subject to great regional variances. Also, the financing and accounting procedures of companies differ, thereby making it difficult to arrive at meaningful strip mining costs.

Estimation of the mining costs must be, by necessity, based on the company's experience. The labor and material cost must be estimated for drilling, explosives, overburden removal, reclamation, pit cleaning, coal loading, haulage, road building, fuel, oil, grease, maintenance, supervision, depreciation, etc. Additionally, costs for transporting, erecting, dismantling, and moving the primary stripping and other equipment must be considered. Since the viability of a project must be determined over the mine life, these have to be projected into the future, taking into account the inflationary and productivity trends.

The selling price of coal is an uncertain factor. It is a complex function of the demand and availability of other energy resources and their prices. The correlation between the selling price and the mining and preparation cost, on one hand, and the attractiveness of investment in stripping, on the other, is strong.

STRIPPING RATIO

The stripping ratio is the proportion of overburden to coal removed. Though there are several units in which this proportion can be defined, the most common in the industry is in the <u>in situ</u> cubic yards of overburden removed per ton of clean coal (1 ton = 2000 lb). The equivalent unit in the metric system will be cubic meters of overburden removed per tonne (1000 kg) of clean coal. Simple arithmetic leads to the conversion factor between these two units:

$$\left(\frac{m^3 \text{ of overburden}/}{\text{tonne of coal}}\right) = \left(\frac{yd^3 \text{ of overburden}/}{\text{ton of coal}}\right) \times \left(\frac{1}{1.2075}\right)$$

The stripping ratio can also relate the selling price of coal with the costs of mining the coal and stripping the overburden. In technical literature, the calculations are sometimes based on average overburden depths, though in reality, the break-even stripping ratio is a point value beyond which the coal seam cannot be economically stripped. As the overburden depth increases, more money is spent on exposing the coal seam until a limit is reached when the value of the recovered material (clean coal) is just enough to pay for all the cost involved in mining, preparing, and selling the material.

The removal of overburden, or stripping, is generally regarded as the most significant component of surface coal mining costs. Variations in stripping ratio affect the scale of the equipment and the efficiency of its operation as far as overburden handling is concerned, while procedures for handling, preparing and marketing coal, and costs associated with these three steps are, in comparison, fixed from mine to mine.

Distinction is often made between several different ratio calculations, such as actual or overall stripping ratio, ultimate or break-even stripping ratio, daily or working stripping ratio, and the stripping ratio developed on the basis of underground mining costs vs. surface mining costs excluding stripping costs. The considerations of various ratios are also dependent on the scale of the operation, the capabilities of equipment and the general economic structure of the company.

Usually the first analysis is the calculation of a stripping ratio at which recovery by surface mining is cost equivalent to recovery by underground methods. Once a surface vs. underground ratio is established, the amount of surface mineable reserves is established. In addition, whatever resources can be recovered at a lower stripping ratio, it is being mined more economically than is possible by an underground approach.

For example, assume that the underground mining costs for a 1.0-m (3.3-ft) seam are \$8.50 per tonne (\$7.74 per ton). For the same seam, suppose the removal costs by surface methods are \$0.35 per tonne (\$0.32 per ton), and that the overburden can be removed by large equipment for \$0.20 per m (\$0.18 per yd^3). Costs for regrading and reclaiming the mined area are assumed to be \$700 per ha (\$0.07 per m^2 or \$283 per acre). The underground vs. surface-stripping ratio (R_{su}) may be calculated as follows, realizing that there are 16145 tonnes/ha-m (1800 tons/acre-ft.):

$$R_{su} = \frac{\begin{array}{c}\text{Underground mining cost}\\ \text{per tonne coal}\end{array} - \begin{array}{c}\text{Surface mining and reclamation}\\ \text{cost per tonne coal}\end{array}}{\text{Surface stripping cost per m}^3\text{ overburden}}$$

$$R_{su} = \frac{\$8.50 - ((\$0.35) + (\frac{\$700}{16145}))}{\$0.20}$$

$$R_{su} = \frac{\$8.50 - (\$0.35 + \$0.04)}{\$0.20}$$

$$R_{su} = 40.55 \text{ m}^3 \text{ overburden/tonne coal}$$

Coal should therefore be mined only in that portion of the deposit where the ratio does not exceed 40.55 m^3 overburden per tonne coal. In a deposit to be area mined, these calculations would show that the deposit could be exploited by surface mining as long as the overburden covering was less than 64.88m (197 ft). These conditions would be encountered in the Ohio eastern regions which incorporate the largest stripping equipment to such depths of overburden.

Total design of the mine requires that an ultimate break-even stripping ratio be calculated. This ratio is that which establishes the ultimate limits of mining, and is not the overall stripping ratio. The break-even ratio is that ratio at which the total cost of marketing one tonne of coal is equal to the price of that tonne of coal. The overall ratio must be less than the break-even ratio or no profit would be realized. The break-even ratio (R_{BE}) is calculated as:

$$R_{BE} = \frac{\text{Recoverable value per unit of coal} - \text{Production cost per unit of coal}}{\text{Stripping cost per unit waste}}$$

Working in common units of tonnes of coal and m^3 of overburden, a family of analyses and curves may be prepared based on various costs and prices which can be estimated.

Table 59 shows an analysis accounting for variations in percent reject and market price of coal.

TABLE 59. ANALYSIS OF STRIPPING RATIOS

PER CENT REJECT	5	10	15	20
Recovered Coal per mined tonne	0.950 tonne	0.900 tonne	0.85 tonne	0.80 tonne
Costs: Mining	$2.00 per tonne			
Washing	2.00 per tonne			
Reclamation	0.05 per tonne			
Marketing	6.50 per tonne			
Total Cost, excluding stripping	$10.55 per tonne			
Given stripping cost =	$0.20 per m^3			
For market value of				
$15 per tonne -				
Value =	$14.25	$13.50	$12.75	$12.00
Net =	3.70	2.95	2.20	1.45
Ratio =	18.5:1	11.75:1	11.0:1	7.25:1
$30 per tonne -				
Value =	$28.50	$27.00	$25.50	$24.00
Net =	17.95	16.45	14.95	13.45
Ratio =	89.75:1	82.25:1	74.75:1	67.25:1
$45 per tonne -				
Value =	$42.75	$40.50	$38.25	$36.00
Net =	32.20	29.95	27.70	25.45
Ratio =	161:1	149.75:1	138.5:1	127.25
$60 per tonne -				
Value =	$57.00	$54.00	$51.00	$48.00
Net =	46.45	43.45	40.45	37.45
Ratio =	232.25:1	217.25:1	202.25:1	187.25

In reality, R_{BE} (or any stripping ratio) is most commonly a linear polynomial function in several variables, including those as summarized in Table 60.

The break-even stripping ratio function may be expressed as:

$$R_{BE} = \frac{(P_{CC} - O_{HD}) \times (1.0 - P_{RE}) - (C_{MI} + C_{CL} + C_{RE})}{C_{ST}}$$

or,
$$R_{BE} = \frac{P_{CC}}{C_{ST}} - \frac{(P_{RE})(P_{CC})}{C_{ST}} + \frac{(O_{HD})(P_{RE})}{C_{ST}} - \frac{C_{MI}}{C_{ST}} - \frac{C_{CL}}{C_{ST}} - \frac{C_{RE}}{C_{ST}} - \frac{O_{HD}}{C_{ST}}$$

In this second form, it is apparent that the calculation of a break-even stripping ratio may be performed by determining each term separately and accumulating the result.

TABLE 60. FACTORS IN STRIPPING RATIO CALCULATIONS

FACTOR	SYMBOL	TYPICAL RANGE[1] LOW	HIGH
Percent reject, as decimal	P_{RE}	0	60
Mining cost per tonne mined	C_{MI}	\$ 0.20	\$ 2.00[2]
Cleaning cost per tonne	C_{CL}	\$ 0.00	\$ 5.00
Site preparation and revegetation cost per tonne mined	C_{RE}	\$ 0.003	\$ 1.00
Stripping cost per cubic meter overburden	C_{ST}	\$ 0.04	\$ 1.00
Market price of coal, per clean tonne	P_{CC}	\$10.00	\$100.00[3]
Transportation, sales and overhead per clean tonne	O_{HD}	\$ 0.00[4]	\$ 15.00

Notes 1. Although some values may appear high at the present time, allowances are made for trends in the future.

2. Some mines do encounter long (>10 km) truck haulage, which is considered as secondary transportation in this analysis and included in O_{HD}.

3. Market price for metallurgical coal could conceivably inflate to this level in the future.

4. Mine-mouth plants have essentially no O_{HD} costs.

A sample calculation is displayed as follows:

Given: P_{RE} = 0.10 (i.e., 10% reject)

C_{MI} = \$ 1.00 per tonne coal

C_{CL} = \$ 3.00 per tonne coal

C_{RE} = \$ 0.35 per tonne coal

O_{HD} = \$ 8.00 per clean tonne coal

P_{CC} = \$22.50 per clean tonne coal

C_{ST} = \$ 0.25 per cu m overburden coal

then,

$P_{CC} \times (1. - P_{RE}) \div C_{ST}$	= \$22.50 x 0.9 ÷ 0.25 =	+ 81.0 m^3/t
$-O_{HD} \times (1. - P_{RE}) \div C_{ST}$	= \$ 8.00 x 0.9 ÷ 0.25 =	28.8 m^3/t
$-C_{MI} \div C_{ST}$	= \$ 1.00 ÷ 0.25 =	4.0 m^3/t
$-C_{CL} \div C_{ST}$	= \$ 3.00 ÷ 0.25 =	12.0 m^3/t
$-C_{RE} \div C_{ST}$	= \$ 0.35 ÷ 0.25 =	1.4 m^3/t
R_{BE}	= Total	+ 34.8 m^3/t

It is possible to analyze variations in stripping ratio and sensitivity with regard to variations in any single factor or combination of factors. Rather than attempt such an involved analysis, the following figure presents the variation in stripping ratio with respect to the total net value per tonne of coal, excluding of stripping cost, and stripping cost per cubic meter of overburden. Figure 121 shows the stripping ratio in cubic meters of overburden per tonne of coal. It is possible to obtain the stripping ratio in meters of overburden per meter of coal by multiplying this ratio by 1.34. Of course, not all of these situations are probable in today's conditions. However, even the unlikely costs of \$100 for coal and \$0.05 for stripping could be possible if high-grade metallurgical coal continues to rise in price, while earth-moving technology continues to develop along its current trends.

STRIPPING RATIO IN MULTIPLE-SEAM MINING

The analysis of stripping ratio may be complicated by considering the case where several seams are mined from one cut. Multiple-seam planning requires that both total stripping ratio and the incremental stripping ratios be analyzed. A necessary condition is that the cost of stripping all overburden, innerburden, and partings be less than or equal to the total net value of coal extracted from the seams. However, the total analysis is not, of its own, sufficient to justify multiple-seam mining. Even if it is feasible to remove two seams in terms of the overall stripping ratio, the lower seam will not be mined unless the incremental stripping ratio for the lower seam is also acceptable.

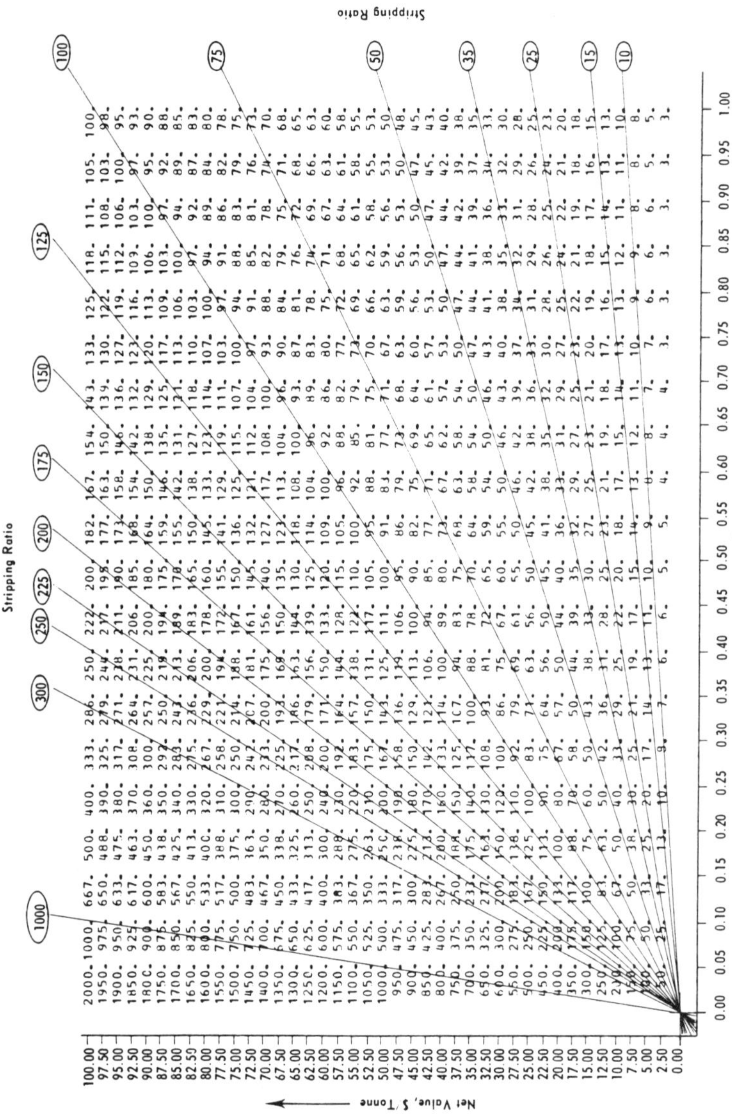

Figure 121. Stripping ratio in m^3/tonne for various coal values and stripping costs.

Consider, for example, a situation where the deposit consists of 10 m of overburden, 1.0 m of coal, 20 m interburden, and 0.5 m coal. The overall stripping ratio is calculated as:

$$\frac{10 + 20 \text{ m overburden}}{1.0 + 0.5 \text{ m coal}} = \frac{30}{1.5} = 20{:}1,\ \text{m/m}$$

If the break-even ratio for the company were 25:1, m/m, then the total ratio would be acceptable, and seemingly both seams should be mined. However, the incremental stripping ratios are:

Upper seam, $\frac{10 \text{ m}}{1\text{m}} = 10{:}1$ m overburden to m coal

Lower seam, $\frac{20 \text{ m}}{0.5 \text{ m}} = 40{:}1$ m interburden to m coal

While the upper seam is quite favorable for mining, the incremental analysis of the lower seam fails to justify mining both seams.

A complication might exist when considering more than two seams: a situation where mining the upper two seams is not feasible while mining three or more seams is feasible. Consider a situation as above, but in which 10m of interburden lie over a third, lowest seam, 2.5 m thick. The incremental analysis would become:

Overall, $\frac{10 + 20 + 10}{1 + 0.5 + 2.5} = 10{:}1$ m/m

Upper seam, 10:1, as above

Middle seam, 40:1, as above

Lowest seam, without considering middle seam

$$\frac{20 + 0.5 + 10}{2.5} = \frac{30.5}{2.5} = 12.2{:}1,\ \text{m/m acceptable}$$

Lowest seam, including middle seam as byproduct

$$\frac{20 + 10}{3.0} = 10{:}1$$

It would then be advisable to mine to the lowest seams. Of course, since the lowest seam is to be mined, and since the stripping ratio considering the second seam as "byproduct" is better than mining the lowest alone, the second seam will also be recovered. Unique cases may exist when the particular economics of coal handling and cleaning may indicate that the middle seam should be treated as interburden. With stringent requirements for the burial of refuse and conservation of natural resources, the practice may not be justified.

STRIPPING RATIO AND COSTS

The determination of stripping ratio has been shown to be quite dependent upon and sensitive to a variety of costs. Such a dependence on costs means that the break-even stripping ratio is related to the scale of the operation. The scale and type of operation, economic nature of the company, and reclamation approach will affect the development of the break-even stripping ratio and the daily or working stripping ratio of the operation.

The working stripping ratio is generally much less than the break-even ratio. It varies with the depth of cover at the time and provides an overall ratio less than or equal to the break-even stripping ratio. In general, lower stripping ratios imply lower production costs. An objective in planning is often to mine with the contour. Mining with the contour means starting where the depth of overburden is least or at the outcrop, if the seam is outcropping. The working stripping ratio then increases with the progress of mining. Such an approach allows higher unit profits (i.e. low production costs) in the first years of life, when the time value of money is highest, and defers high unit production costs to a later point in time. Additionally, if improvements in technology or a higher price for the product is attained at a future date, the break-even stripping ratio can be increased. However, where the topography is severe (steep hill), the break-even point is reached in a very small width. Therefore, the entire width is mined to the break-even point and then reclaimed. However, in these operations, scale and bonding are minimized.

In area mining and mountaintop mining there is usually some coal recovered under working stripping ratios in excess of the break-even ratio. To be justified, the incremental gain in total profits attributable to mining the coal so burdened must still balance to an overall ratio less than the break-even ratio. This practice is required in the area directly under a mountaintop or under a small knoll in an area mine where leaving the mountaintop or knoll would require that different equipment of smaller scale be used in the entire operation. The extraction of a small tonnage of coal which does not, strictly speaking, pay for itself may be justified on the basis of economy of scale that could not be otherwise realized.

COST ANALYSIS

Determination of mining acceptability is not completely restricted to stripping ratio. There can also be established a minimal mineable thickness, a minimal value of coal, and technological limits to exploitation.

Where sufficient tonnage and market are indicated, a more detailed analysis is performed. In this analysis, specific alternative mining methods or groups of subsystems will be specified as indicated by the geometry of the deposit and the stripping ratio. The design problem can be one of selection of primary earthmoving methods and equipment from several alternatives. For each alternative method proposed, capital and operating expenses are detailed and more exact stripping limits and recoverable tonnages can be determined. There are at least three stages of analysis which are carried out:

1. Mine vs. no-mine analysis, based on stripping ratio, gross cost, and gross return estimates.
2. Selection and analysis of alternative methods, based on gross geological aspects.
3. Final design analysis, based on detailed comparison of specific equipment, production requirements, and legal and social constraints.

Costs incurred by mining companies are broken down in two ways: 1.) capital costs are large outlays of money for land, equipment, and buildings. For tax purposes, these costs are not regarded as directly deductible expenses, but must be either amortized, depreciated or depleted. 2.) operating costs (or expenses) are for such items as labor and supplies, and are deductible, in total, in the year of occurrence. Costs are also treated as direct or indirect, with fixed or variable costs, depending on their relationship to coal production.

Some capital expenditures are usually expected by the government to be treated as amortizable items. These include organization costs of the corporation under certain conditions, acquisitions, or nonmineral leases, research and development expenses not treated as expenses, amounts other than purchase connected with acquisition of trademarks, copyrights and the like, and the cost of certified pollution control facilities placed in service in the years 1969 through 1974, inclusive. Amortization is calculated by a straight line method. For example, for an amortizable item of five-year life, one fifth of the amount is deducted from the gross income in each year of the project life.

Depreciation is an allowable deduction prior to taxes for the wear and tear of equipment during their normal use. It is in effect a measure of the decrease in value of capital equipment items, and book value decreases as equipment depreciates with time. In any instance, depreciation is regarded as the yearly decrease in real value of property which is used or employed in the course of business. Depreciation may be calculated strictly on the basis of time or may be calculated strictly on the basis of units of production. By either method, the first step is to determine the depreciable amount for the item. This is generally the purchase price less the anticipated salvage value at the end of a useful life. While mining companies must establish meaningful and useful lives for buildings, machinery, and other depreciable items, sometimes guidelines and acceptable lives are established by the Internal Revenue Service (IRS). The purchase price, salvage value, and equipment life must be consistent with the IRS guidelines and acceptable to the IRS.

A company exploiting a wasting asset like coal is allowed to deduct a certain sum from the income it receives from mining coal to reflect the loss of the capital asset, which is the coal in the ground. This sum is the depletion allowance. Two alternative methods are available for calculating this depletion allowance, cost depletion and percentage depletion. The company can select whichever method gives the greater deduction and is also allowed to switch from one to another, if desired, from year to year. Methods of depreciation and depletion are discussed in several engineering economy texts.

The operating costs can be broken down into direct and indirect, and fixed and variable costs. Certain items of the mining work can be directly related to the production of a given number of units of coal: the amount of work the men in the pit perform with the utilization of a shovel or dragline. Other costs, such as taxes, sewer, mine plant maintenance, dry-change facilities, etc., do not relate directly to the production of specific units of coal but accrue indirectly to the operation as a whole. The former may be regarded as direct costs while the latter are regarded as indirect costs. In a similar analysis, certain costs are fixed without regard to the quantity of coal being mined, such as the maintenance of facilities, engineering staff pay, supervisory pay, etc. Other costs, such as direct labor pay, operation expenses for drills, shovels, trucks, blasting, etc., are incurred directly by and can be accounted directly to the production of given quantities of coal. In most cases, these fixed costs are prorated for accounting purposes to a unit of production basis.

Where appropriate accounting and industrial engineering records have been kept, the estimation of costs for new methods and new approaches is greatly facilitated, as the increments of cost involved in the new approach may well be historically documented at present in previous, traditional approaches.

COST ESTIMATION

It is not the purpose here to go into details of cost analysis models. From the discussion thus far, it is obvious that there is no one model or method that will give correct answers. Even if the models or analysis procedures are analytically sound, the input data in most cases is proprietary to the company. The U.S. Bureau of Mines has published several models that address the cost of surface mining of coal.

Boyce (1975) reports on a computerized model to analyze the technical and financial aspects of large-scale surface mining. The model is intended to be used for first-order planning of large scale surface coal mines and for cost evaluation under various conditions. It was developed for the Energy Research and Development Administration (ERDA) by Flour Utah, Incorporated.

Otte and Boehlje (1975) have developed a model for calculating the cost of mining and reclamation with various materials handling techniques and procedures in different pit configuration. This computer-oriented model calculates production and costs for each individual task performed to uncover and remove coal and to reclaim the mined area. Total costs per ton of coal removed and the capital outlay required to obtain a specified level of annual production are also determined by aggregating the cost of each task.

Manula et al (1976) have developed a cost-model for anthracite open-pit mining which determines a rate of return for a specified realization (i.e., selling price) or vice-versa through a discounted cash-flow analysis. The model divides costs into four groups; ownership, labor, supplies, and miscellaneous costs. The last category includes the charges for welfare, royalties, etc.

NUS Corporation has developed a cost model for the Electric Power Research Institute (EPRI, 1977). The model incorporates the essential features of surface coal mine equipment selection and cost engineering principles to analyze the costs of producing coal from a new surface mine. It is intended to provide an engineering approach to determination of production costs to be used in developing coal price-availability relationships.

A computer model has been developed by Ramani, Murray and Manula (1977) to generate the selling price of coal based on a combination of geologic, engineering and economic variables, as these are related to area, contour and open-pit mining. Much of the background information used in model development was based on the report by the NUS Corporation to Electric Power Research Institute (EPRI, 1977).

Surface Mining Cost Models

Katell and Hemingway (1972, 1976) have estimated typical capital investments, operation costs, and selling prices for coal strip mines that will fuel a 250-million-standard-cubic foot-per-day, high-BTU gas plant in Eastern, Interior, and Northern Great Plains states of the United States. These studies are regularly updated; the latest study can be found in the Bureau of Mines Information Circular 8703 (Katell, Hemingway and Berkshire, 1976).

Toth and Stinnett (1976) present a somewhat simple model of the NUS/EPRI Model (EPRI, 1977) for a preliminary estimate of surface mining costs. The model is presented here in its entirety because the principles on which the model is based can easily be extended to any surface mining operation with modifications. The accuracy of the model results may be in the range of $\pm$ 30%, typical of preliminary estimates. Since some values used here are outside the range of the graphs, instead of using the graphs, calculations are performed to arrive at the parameter values. In general, the approach presented here parallels that which an engineer would follow in performing a preliminary new mine cost estimate. The "bottom line" of the analysis is the minimum, acceptable selling price for the coal (from the seller's viewpoint), but certain other important figures are generated as well. The model centers around a set of nine graphs which allow for gross variances that are usually encountered among coal deposits, and some input, therefore, is required from the user.

The first graph (Figure 122) illustrates the relationship between the designed mine capacity and the volume of overburden which must be removed.

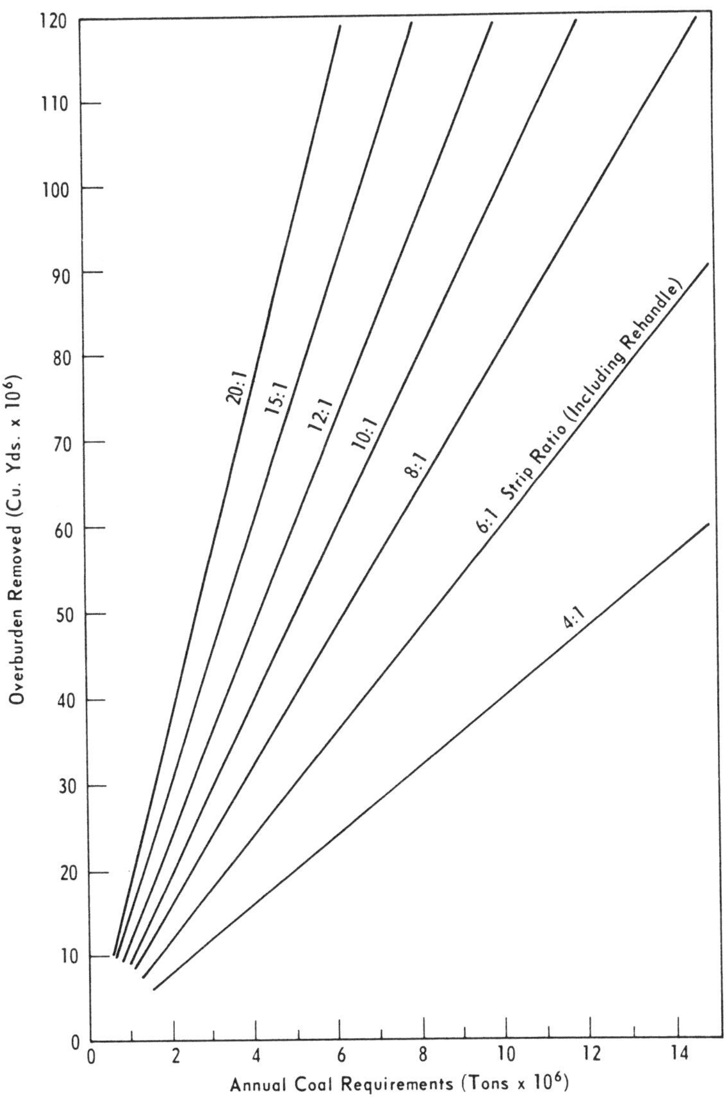

Figure 122. Production versus overburden removal.

The variable relating the two is the stripping ratio. In actual practice, a mine's stripping ratio will vary from year to year and likely will approach some upper limit late in life, beyond which the mine can no longer be operated profitably. The strip ratio suggested for this analysis is the average effective strip ratio (including rehandle), although a rough sensitivity analysis can be performed by simply varying the ratio selected and noting the effect on coal selling price. A 1,696,720 tonne/year (1,670,000 ton/year) mine located in Illinois and characterized by a strip ratio of twelve cubic yards of overburden per ton of coal is selected for illustrative purposes. The resulting overburden figure, 15,290,000 cu m (20,000,000 cubic yards), is used as input to the next graph (Figure 123).

In Figure 123, the annual overburden to be removed is translated into total machine capacity needed to affect removal. The basic overburden removing equipment in area stripping operations is assumed to be either a shovel or dragline. The variable in Figure 123 is machine "efficiency" which is dependent upon a number of operating parameters including cycle time, operating hours, operator efficiency, maintenance reliability and so forth. Additionally, if the machine is used for road building or encounters excessive rehandling or chopping down as a result of multiple-seam mining, the efficiency will be lowered considerably.

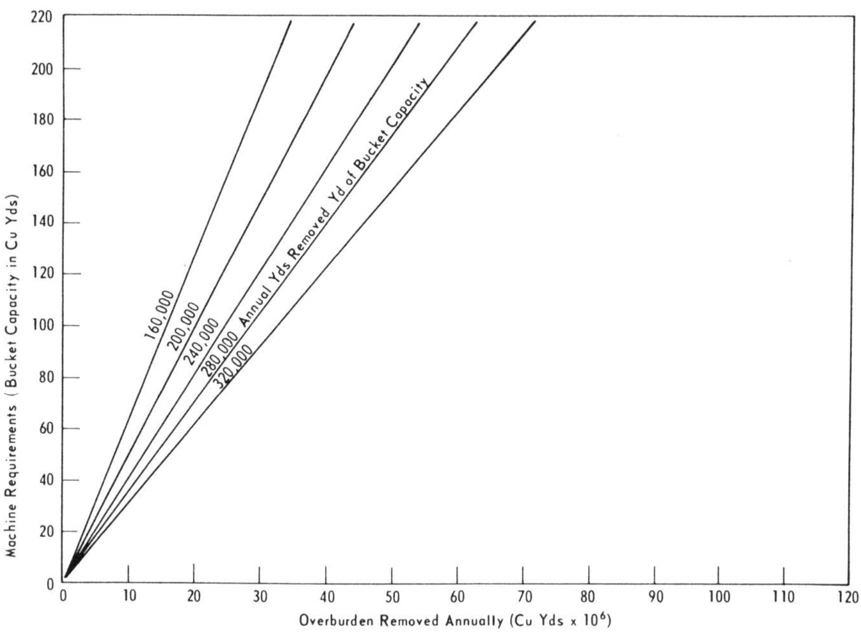

Figure 123. Overburden removal versus machine capacity.

In the example, overburden digging conditions are presumed to be good--blasting will not likely be required--and a single seam will be extracted. Thus selecting a moderately high efficiency of 366,252.45 cu m/cu m (280,000 yards/yard) of bucket capacity, 53.51 cu m (70 cubic yards) of machine capacity will be needed.

Figure 124 illustrates the relatively large component of capital costs attributable to the overburden stripping equipment, and thus suggests the rationale for focusing on the strip ratio as a primary factor in ultimately affecting the selling price of coal. There is reason to believe that machine costs may not exhibit the straight line function with size as is illustrated; however, since an operator may have the option of selecting multiple pieces of smaller capacity to meet production needs, the straight line assumption should remain with the accuracy of this analysis.

The three other lines on the graph show typical capital outlay for physical investment (the proven reserves, initial equipment, and plant needs), the start-up mine investment (including interest during construction, preproduction development, and working capital), and the total investment throughout the mine operating life (including equipment replacements).

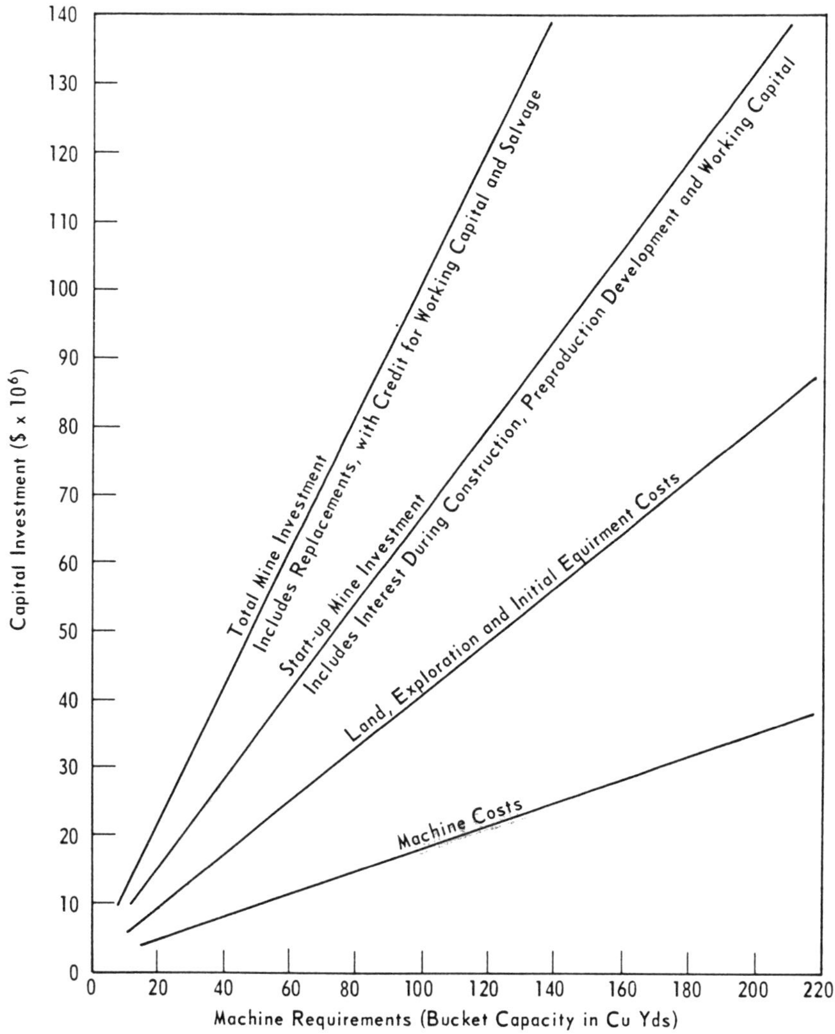

Figure 124. Capital requirements for a surface coal mine.

Referring to the example, the start-up and total mine investment may approach $48,000,000 and $73,000,000, respectively.

Using the total capital investment from the preceding graph, the straight line depreciation and amortization charges per ton of coal produced can be determined in Figure 125. All mines are presumed to have a 30-year operating life, although a similar chart could be constructed if this simplification is invalid. The number derived from Figure 125 represents the noncash portion of operating costs and will be summarized with others to arrive at the minimum acceptable selling price.

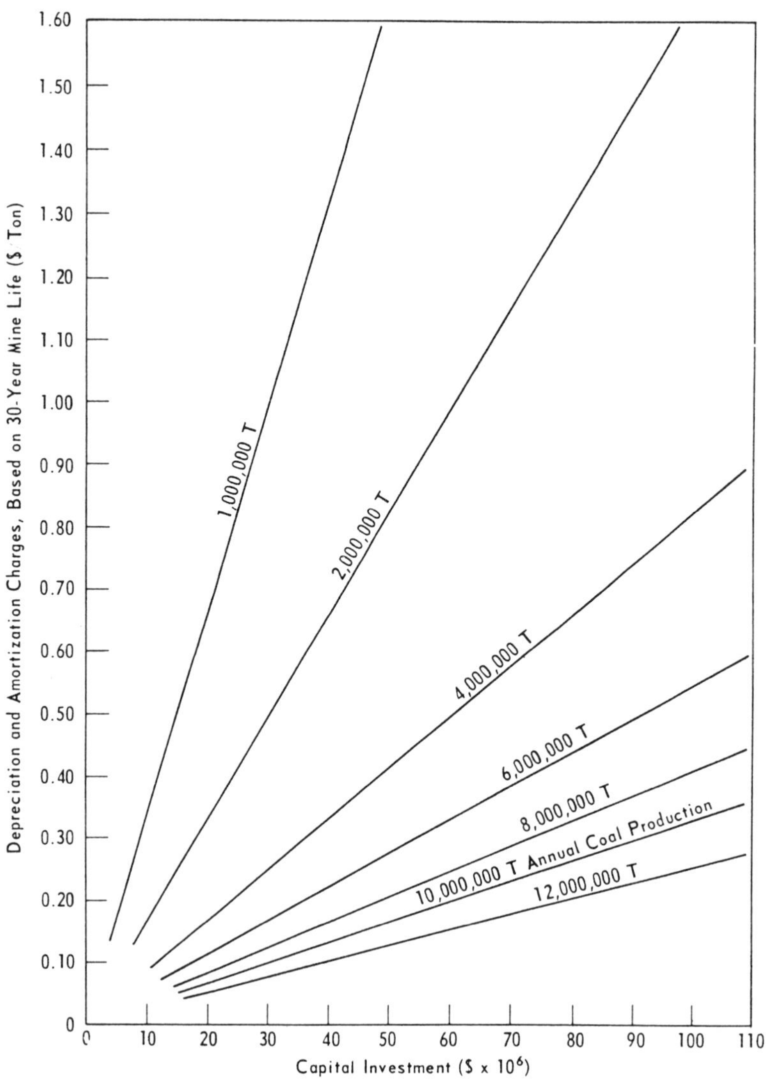

Figure 125. Graph for the computation of depreciation charges.

Since the data for the mine is outside the limits of the graph in the example for this 1,696,720 tonne (1,670,000 ton/year) operation, per ton depreciation and amortization charges can be calculated as:

$$\frac{\$73,000,000}{30 \times 1,696,720} = \$1.43/\text{tonne} \quad (\frac{\$73,000,000}{30 \times 1,670,000} = \$1.46/\text{ton})$$

This calculation also illustrates the simplicity of the use of the model.

Figure 126 is a display of the range and average productivity of surface mining operations in the United States, measured in tons/man-day.

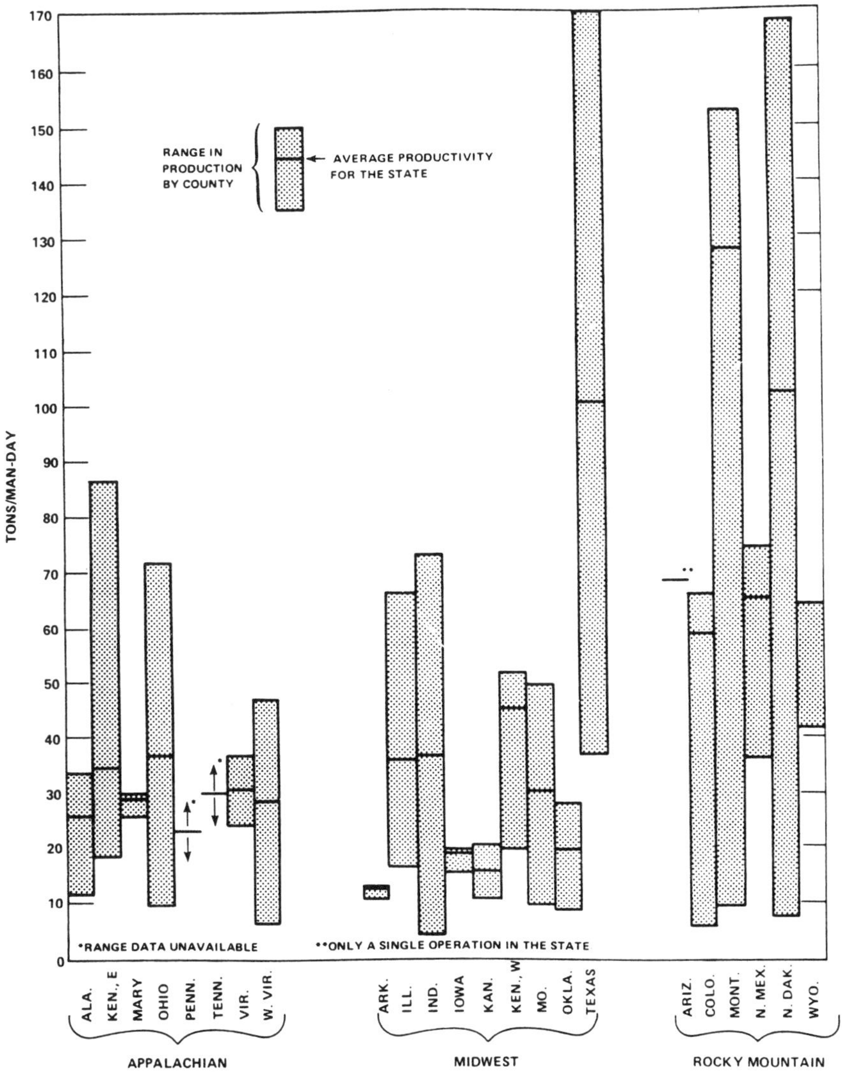

Figure 126. Productivities for surface coal mine operations in the United States, 1973.

The range in production is by county, rather than individual mine, so presumably a range chart based on the latter category would show even greater variation.

The relatively low productivity of the entire United States of 36.57 tonne/man-day (36 tons/man-day) stems from the large proportion of coal coming from the Appalachian and Interior Regions. It is reasonable to assume in the example, however, that at a large new well-engineered mine in Illinois the labor productivity will be 30.48 tonne/man-day (30 tons/man-day).

Figure 127 relates the labor cost per ton of coal produced to the average labor rate and the productivity of the mine personnel. Mine personnel in this analysis includes supervisory staff in addition to all hourly workers.

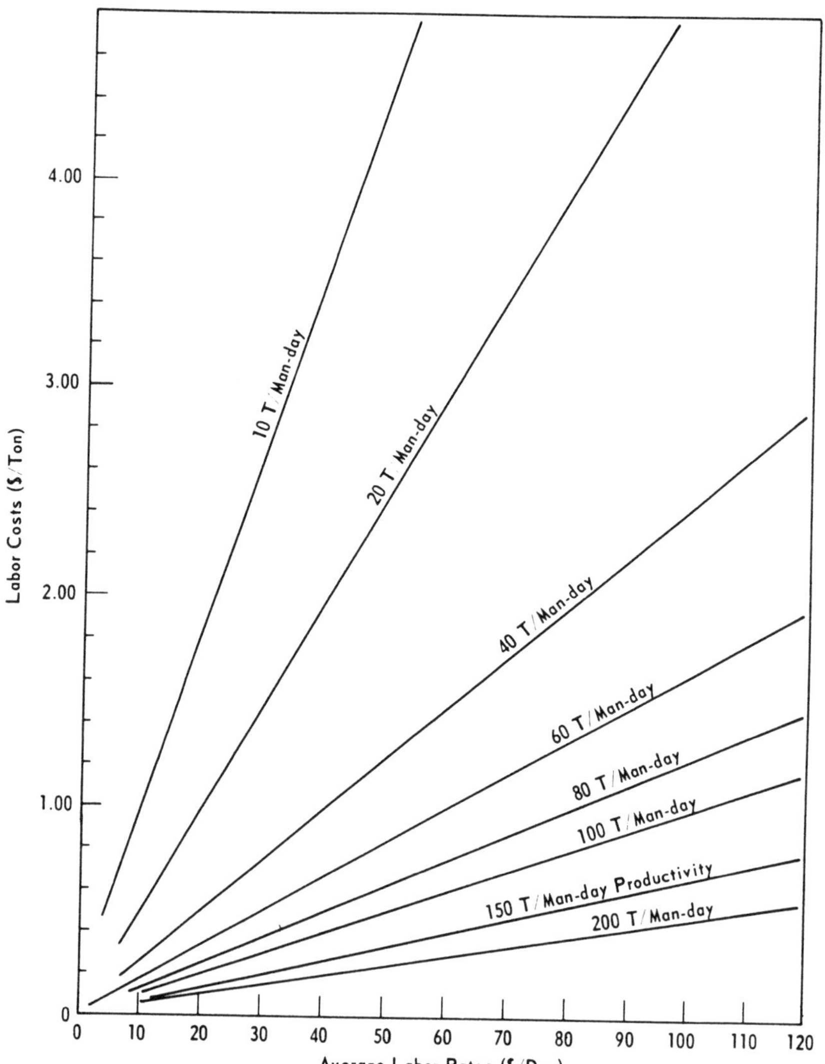

Figure 127. Graph for the calculation of labor costs per ton.

Assuming a labor rate of \$60/day and a productivity of 30.48 tonne/man-day (30 tons/man-day), for the example, labor costs are determined to be \$2.26/tonne (\$2.3/ton). It is quite possible to vary both productivity and labor rate, and arrive at different labor costs.

Assuming a productivity of 30.48 tonne/man-day (30 tons/man-day) and approximately 250 working days/year, the minimum number of men not including provisions for absenteeism, vacations and turnover is calculated below.

$$\text{No. of men required} = \frac{\text{Annual Production}}{(\text{Productivity})(\text{No. of working days/year})}$$

$$\frac{1,696,720}{30.48 \times 250} \simeq 223 \text{ employees or } \left(\frac{1,670,000}{30 \times 250} \approx 223 \text{ employees}\right)$$

Figure 128 provides the relationship between labor costs per ton of coal produced and direct operating costs for the operation.

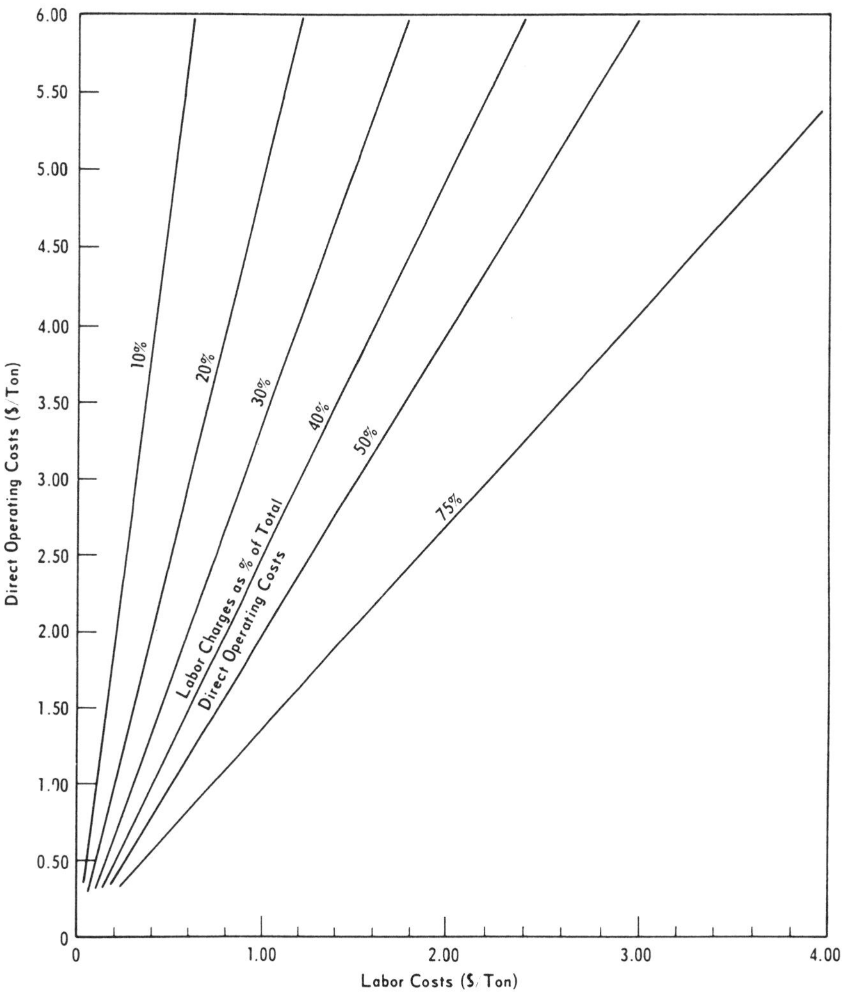

Figure 128. Relationship of labor costs and direct operating costs.

As defined here, direct operating costs include not only labor charges but also supplies, power, fringe benefits, welfare payments, insurances, royalties, property and payroll taxes, and so forth. Because of the catch-all nature of this grouping, and because of the initial premise that new, highly productive mines are to be considered for analysis, the proportion of labor costs to direct operating charges is predicated to be somewhat lower than historical records usually indicate.

The data for the mine is outside the limits of the graph. Therefore, with labor charges estimated at 30% of total direct charges, the direct operating costs are calculated to be \$7.54 per tonne (\$7.67 per ton) of coal (i.e., \$2.3/0.3). Here again, different percent values can be used for labor charges as a component of the total cost.

The method of mine financing has an important bearing on total mining costs as does the schedule of depreciation. Figure 129 shows a range of typical debt-equity ratios commonly employed by the utility and mining industries. The lines below the 50/50 line are simply the inverse of those above and are added essentially to allow full graphical application in determining mining costs.

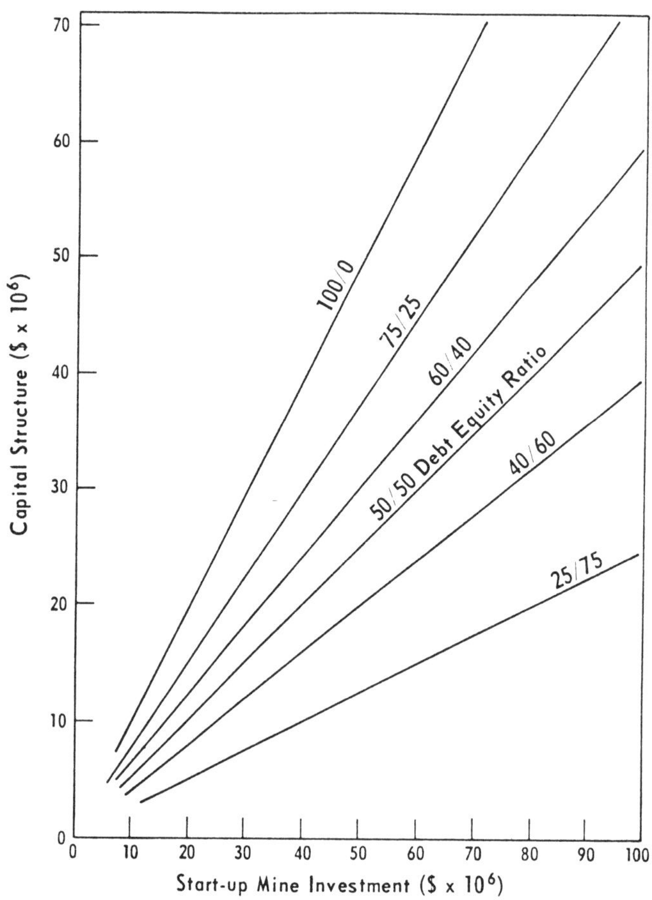

Figure 129. Capital structure for start-up mine investment.

The start-up mine investment number derived in Figure 124 provides the basis for financing as it is assumed that equipment replacements and the like will be purchased from internally generated funds. This approach results in a steadily decreasing debt/equity ratio throughout the property life and would tend to increase ultimate mining costs as the risk-taking investor will require a somewhat greater return on his exposed funds. A mitigating circumstance, though, is the fact that the remaining book value of the mine plant will approach zero over the operating life and so the equity base, on which the return is calculated, will be shrinking. The assumption in this analysis, then, is that the per ton financial charges during the life of the property will approximate those established by the capital structure at the time of mine start up.

The example shows a 75/25 debt/equity ratio which may be occasioned through utility ownership and development of the reserves. The debt and equity components of the initial mine investment are \$36,000,000 and \$12,000,000, respectively.

From Figure 130, one can determine the financing charges per ton of coal depending on a choice of interest rates. Dividing the resulting numbers from Figure 130 by the annual productive capacity will allow direct determination of these charges. The calculations are illustrated here because the values for the example mine fall outside the limits of the graph.

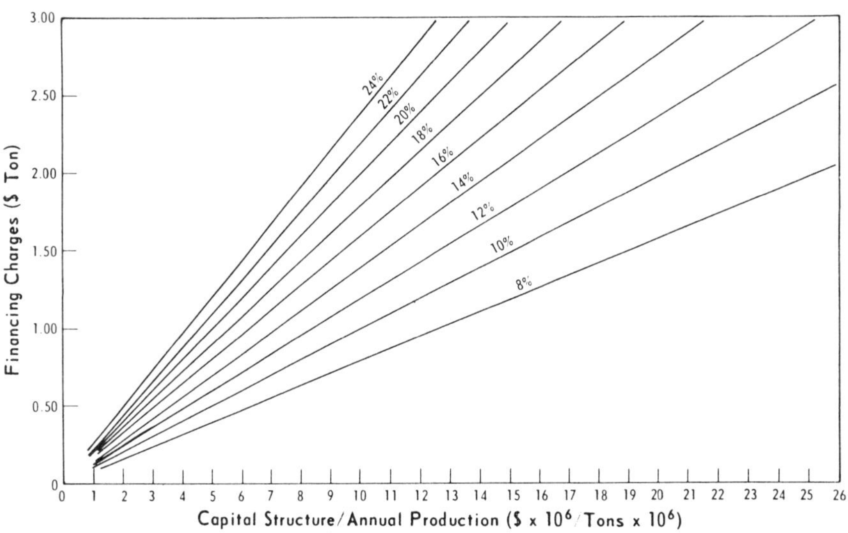

Figure 130. Graph for the calculation of financing charges and rate of return.

Capital structure/annual ton is calculated as shown below.

Total capital investment = \$73 million
Start up investment = \$48 million
Debt component of start up investment = \$36 million
Equity component of start up investment = \$12 million

Capital structure (Debt component) per annual ton $= \dfrac{\$36 \text{ million}}{\$1.67 \text{ m tons}} = \$21.56$

Capital structure (Equity component) per annual ton	= $\frac{\$12 \text{ million}}{\$1.67 \text{ m tons}}$	= \$ 7.19
Financing charges debt at 10%	= \$21.56 x 0.10=	\$ 2.16
Return on equity at 18%	= \$ 7.19 x 0.18=	\$ 1.29
Income taxes = 0.33 x \$1.29	= \$ 0.43	

Thus assuming in the example a debt servicing rate of 10% and a return to equity of 18%, the burden will come to \$2.12/tonne and \$1.27/tonne (\$2.16/ton \$1.29/ton), respectively. One additional value can be calculated from Figure 130, the approximate federal and state income taxes. If the combined tax rate can be taken as 50%, and if one assumes a depletion allowance will be limited by the 50% of net income ruling, income taxes will total 33% of the return to equity figure. As illustrated in the example, these are \$0.42/tonne (\$0.43/ton).

Table 61 is the culmination of the graphical exercise wherein the previously determined charges from Figures 125, 128 and 130 are totalled to arrive at the minimum acceptable selling price per ton necessary to allow the mine operator an appropriate return on investment.

Referring once more to the example, the selling price for the particular operating and deposit characteristics should approximate \$12.80/tonne (\$13.01/ton).

TABLE 61. CALCULATION OF SELLING PRICE

Depreciation cost/ton	(from Figure 125 or calculation)		\$ 1.46
Direct operating cost/ton	(from Figure 128 or calculation)		\$ 7.67
Debt financing cost/ton	(from Figure 130 or calculation)		\$ 2.16
Return on investment	(from Figure 130 or calculation)		\$ 1.29
Taxes @33% ROI	(from calculation)		\$ 0.43
Minimum acceptable selling price			\$13.01/ton
Start-up mine investment	(from Figure 124)	\$48,000,000	
Total mine investment	(from Figure 124)	\$73,000,000	
Mine capacity	(assumption)	1,670,000 tons	
Productivity	(assumption)	30 tons/man-day	
Personnel requirements: labor and supervisory (not including provisions for absenteeism, vacations, and turnover)	(calculation)	223 employees	

SUMMARY

There are, of course, other approaches to costing an operation. For example, one major mining company has identified optimum mine sizes and mining methods on the basis of geographical and market conditions. It has economically optimized these selections through analysis of past practice. When the marketing department notifies engineering that it has received a contract, the engineering department selects an appropriate property which is held by the company in the geographical vicinity of the contract or at least as close to the contract delivery point as is possible to meet the type of product required. For the contract size specified, e.g., a five-million ton per year contract, the engineering department has, on file, the plans for mining development at the particular geographical location. Such an approach may be desired by large companies which can exercise economies of scale by utilizing the same, or closely related, approaches and equipment throughout their operations and are willing to sacrifice some potentially better approaches. For the average coal mining company, the sizes and types of deposits will vary over a large scale. Therefore, the design plans and economic analyses to determine the optimum, ultimate mine design will proceed in the more traditional fashion. As already indicated and referenced, computer-oriented cost analysis models are available for costing total surface mine operations or even individual unit operations.

References

Adams, L. M., Capp, J. P., Eisentrout, E., Reclamation of Acidic Coal Mine Spoil with Fly Ash, U.S. Department of the Interior, Bureau of Mines Report of Investigation Number 7504, April 1971.

Andrews, A. B., "Air Blast and Ground Vibrations in Open-Pit Mining," Mining Congress Journal, Vol. 61, No. 5, 1975.

Ash, R. L., "The Mechanics of Rock Breakage," Parts I through IV, Pit and Quarry, Vol. 56, Nos. 2, 3, 4, and 5, 1963.

Ash, R. L., "Cratering and Its Application in Blasting," Unpublished report, Department of Mining and Civil Engineering, University of Minnesota, 1970.

Atkinson, T., "Selection of Open-pit Excavating and Loading Equipment," Institution of Mining and Metallurgy, Vol. 80, Bulletin No. 776, Sec. A, 1971.

Bailey, P. A., "Exploration Methods and Equipment," Chapter 2.1, Surface Mining, Ed. E. P. Pfleider, SME/AIME, New York, 1968.

Bergmann, O. R., Wu, F. C., and Edl, J. W., "Model Rock Blasting Measures Effect of Delays and Hole Patterns on Rock Fragmentation," Engineering and Mining Journal, Vol. 175, No. 6, 1974.

Boddorff, D., "That's the Way Rock Breaks," The Explosives Engineer, Spring, 1974.

Boyce, T. A., "Economic System Analysis of Large Scale Surface Mining," Proceedings of the 3rd Symposium on Surface Mining and Reclamation, Vol. II, NCA/BCR, Washington, D.C., 1975.

Burton, A. K., "Off-Highway Trucks in the Mining Industry," Mining Engineering, Vol. 27, No. 8, 1975; Vol. 27, No. 10, 1975; Vol. 27, No. 11, 1975; Vol. 27, No. 12, 1975; Vol. 28, No. 1, 1976.

Caterpillar Tractor Co., Fundamentals of Earthmoving, Peoria, Illinois, July 1975.

Caterpillar Tractor, Co., Performance Handbook, 6th Edition, Peoria, Illinois, 1976.

Cheek, E. E., "Application of a Standard Wheel Excavator in Coal," Mining Congress Journal, Vol. 52, No. 11, November 1966, pp. 33-35, 38-39.

Civil Engineering Handbook, 4th Edition, McGraw Hill Co., New York, 1962.

Condon, J. L., and Snodgrass, J. J., "Effects of Primer Type and Borehole Diameter on AN-FO Detonation Velocities," Mining Congress Journal, Vol. 60, No. 6, 1974.

Cummins, A. B., Chairman of the Editorial Board, SME Mining Engineering Handbook, AIME, New York, 1973.

Daley, D. D., "Emphasizing Environment - Natural Beauty Considered," Mining Engineering, Vol. 26, No. 2, 1974.

Dick, R. A., "Explosives and Borehole Loading," SME Mining Engineers Handbook, SME/AIME, New York, 1973.

Dombrowsky, N. G., "The New Bucket Wheel Excavator and Its Construction Problems," Bergbautechnik, Vol. 14, 1964, p. 277.

Doyle, F. J., "Variables Affecting Grading Costs for Strip Mine and Refuse Bank Reclamation," Proceedings of the First Symposium on Mine and Preparation Plant Refuse Dispoal, National Coal Association, 1974, pp. 212-227.

Drevdahl, E. R., Profitable Use of Excavation Equipment, Technical Publications, Desert Laboratories, Inc., Tuscon, Arizona, 1961.

DuPont Co., Blaster's Handbook, Wilmington, Deleware, 1967.

Ellison, R. D., and Thurman, A. G., "Geotechnology: An Integral Part of Mine Planning," A Chapter in Coal Exploration, Ed. W. G. Muir, Miller Freeman Publications, Inc., San Francisco, 1976.

EPRI, Coal Mining Cost Models - Surface Mines, Electric Power Research Institute, EPRI EA-437, Research Project 435-1, 1977.

Felix, G. L., "Development of the 350 Ton Haulage Truck," Mining Congress Journal, Vol. 62, No. 3, 1976.

Gartner, K., "Bucket Wheel Excavator on Building Sites," Forden und Heben, Vol. 3, No. 6, November 1965.

Gordon, R. L., Historical Trends in Coal Utilization and Supply, U.S. Bureau of Mines Report, Grant No. G0155116, Washington, D.C., 1976.

Grim, E. C., and Hill, R. D., Environmental Protection in Surface Mining of Coal, Report No. EPA-670/2-74-093, U.S. Environmental Protection Agency, Cincinnati, Ohio, 1974.

Haley, W. A., "Changing Methods and Equipment Use in Appalachian Coal Mining," Second Research and Applied Technology Symposium on Mined Land Reclamation, NCA/BCR, 1974.

Highway Engineering Handbook, 1st Edition, McGraw Hill Co., New York, 1960.

Hill, R. D., and Grim, E. C., "Environmental Factors in Surface Mine Recovery," Symposium on Restoration and Recovery of Damaged Ecosystems, VPI & SU, Blacksburg, Virginia, March 1975.

Himmel, W., "The Specific Digging Resistance in Dependency on the Shaving Area and Shaving Form in Different Soils," Freiberger Forschungshefte, A 265, 1961, pp. 1-40.

International Harvester, "An Analysis of Equipment Selection for Heavy Construction," Payline Division, 1975.

Katell, S., and Hemingway, E. L., "Cost Analysis of Model Mines for Strip Mining of Coal in the United States," U.S. Bureau of Mines, Information Circular 8535, 1972.

Katell, S., Hemingway, E. L., and Berkshire, L. H., "Basic Estimated Capital and Operating Costs for Coal Strip Mines," Revision of IC 8661, U.S. Bureau of Mines, Information Circular 8703, 1976.

Kimball, L. Robert (Consulting Engineers), Slope Stability, Volume 1 - Report and Field Book, Report ARC-71-66-T3 prepared for the Appalachian Regional Commission, p. 81, undated.

Krause, R. R., "Predicting Mined Land Soil," Chapter in Ecology and Reclamation of Devastated Land, Vol. 1, Ed., R. J. Hutnik and Grant Davis, Gordon Breach, New York, 1973.

Learmont, T., "Productivity in Large Stripping Machines," SME/AIME Transactions, Vol. 258, No. 3, 1975.

Maneval, D. R., "Coal Mining vs. Environment: A Reconciliation in Pennsylvania," Appalachia, A Journal of the Appalachian Regional Commission, Vol. 5, No. 2, February-March 1972.

Mani, M. S., "Design Features of Bucket Wheel Excavators," Journal of Mines, Metals and Fuels, Vol. 14, No. 3, March 1966, pp. 71-79.

Manula, C. B., Albert, E. K., and Burgos, V., A Report on Anthracite Open-pit Mining, Part V. Engineering and Mine Cost Analysis, Special Research Report SR-104, Coal Research Section, The Pennsylvania State University, 1976.

Mathematica, Inc., The Design of Surface Mining Systems in Eastern Kentucky, Department of Natural Resources and Environmental Protection, Division of Reclamation, April 30, 1973.

McGraw Hill Inc., Coal Age, Vol. 75, No. 7, 1970, pp. 173-179.

MESA, Engineering and Design Manual: Coal Refuse Disposal Facilities, Prepared by E. D'appolonia Consulting Engineer for U.S. Department of the Interior, Mining Enforcement and Safety Administration.

Meyer, C. F., Route Surveying, International Textbook Co., Scranton, PA, 1969.

Morgenstern, N. R., and Price, V. E., "The Analysis of the Stability of General Slip Surfaces, " Geotechnique, Vol. 15, No. 1, pp. 79-93, 1965.

Morrell, R. J., and Unger, H. F., "Drilling Machines, Surface," SME Mining Engineering Handbook, SME/AIME, New York, 1973.

Murray, C. W., "A Computer Model for Optimizing Surface Mine Haul Road Profile," M. Eng. Report, Mineral Engineering Management, The Pennsylvania State University, 1978.

Nicholls, H. R., Johnson, C. J., and Duvall, W. I., Blasting Vibrations and Their Effects on Structures, Bulletin 856, U.S. Department of the Interior, Bureau of Mines, 1971.

Nordness, H. J., "Horizontal Drills for Surface Mines," Proceedings 3rd Conference on Mine Productivity, The Pennsylvania State University, 1976.

Obert, L. A., and Duvall, W. I., Rock Mechanics and the Design of Structures in Rock, John Wiley, New York, 1967.

Otte, J. A., and Boehlje, M., "A Model to Analyze the Cost of Strip Mining and Reclamation," Proceedings of the 3rd Symposium on Surface Mining and Reclamation, Vol. II, NCA/BCR, Washington, D.C., 1975.

Peurifoy, R. L., Construction Planning, Equipment and Methods, McGraw Hill, New York, 1970.

Pfleider, E. P., Ed., Surface Mining, American Institute of Mining, Metallurgical and Petroleum Engineers, Inc., New York, 1968.

Porter, W. E., "Multiple Seam Strip Mining: A Survey of Economic Feasibility Model," Unpublished M. Eng. Report, The Pennsylvania State University, 1971.

Price, G. C., Manula, C. B., and Ramani, R. V., Material Handling Research, The Bucket Wheel Excavator, U.S. Bureau of Mines, Information Circular 8580, 1973.

Pugliese, J. M., Designing Blast Patterns Using Empirical Formulas, U.S. Bureau of Mines, Information Circular 8550, Washington, D.C., 1972.

Ramani, R. V., "Computer Simulation of a Bucket Wheel Excavator," Unpublished M.S. Thesis, The Pennsylvania State University, 1968.

Ramani, R. V., and Grim, E. C., "Surface Mining-A Review of Practices and Progress in Land Disturbance Control," Chapter 14 in the book Reclamation of Drastically Disturbed Lands, Ed., F. W. Schaller and P. Sutton, American Society of Agronomy, Madison, Wisconsin, 1978.

Ramani, R. V., Murray, C. W., and Manula, C. B., Application of a Total System Surface Mine Simulator to Coal Stripping, Volume IV, Surface Mine Cost Model, National Technical Information Service (NTIS), Publication No. 294634/AS, 1979.

Rasper, L., and Ritter, H., "Opening Up of the Lignite Mine of the Neyveli Lignite Corporation and Experience with Bucket Wheel Excavators in Hard Overburden," Braunkohle, Warme und Energie, Vol. 13, No. 10, October 1961, pp. 390-400.

Rasper, L., "Present Stage of the Development of Bucket Wheel Excavators in the German Federal Republic," Braunkohle, Warme und Energie, Vol. 16, No. 11, November 1964, pp. 457-471.

Rumfelt, H., "Computer Methods for Estimating Proper Machine Mass for Stripping Overburden," Mining Engineering, Vol. 13, No. 5, 1961.

Rumfelt, H., "Application and Performance of Wheel Excavators," Mining Congress Journal, Vol. 47, No. 6, June 1961, pp. 46-49.

Saperstein, L. W., and Secor, E. S., "Improved Reclamation Potential with the Block Method of Contour Stripping," Proceedings of the Research and Applied Technology Symposium on Mined-Land Reclamation, Pittsburgh, Pennsylvania, Bituminous Coal Research, Inc., NCA/BCR, March 1973.

Scott, R. F., Principles of Soil Mechanics, Addison-Wesley, Reading, Massachusetts, 1963.

Secor, E. S., "An Economic Comparison Between the Block and Conventional Methods of Contour Strip Mining," Unpublished M. Eng. report in Mineral Engineering Management, The Pennsylvania State University, 85 pp., 1977.

Skelly and Loy Engineering Consultants, Economic Engineering Analysis of U.S. Surface Coal Mines and Effective Land Reclamation, U.S. Bureau of Mines Report, Contract No. S0241049, Washington, D.C., 1975.

Starr, C., "Energy and Power," Scientific American, Vol. 225, No. 3, September 1971, pp. 37-49.

Stefanko, R., Ramani, R. V., and Ferko, M. R., An Analysis of Strip Mining Methods and Equipment Selection, Research and Development Report No. 61, Interim Report No. 7, Prepared by Coal Research Section, The Pennsylvania State University, for Office of Coal Research, U.S. Department of the Interior, Washington, D.C., May 1973.

Terzaghi, K., and R. B. Peck, Soil Mechanics in Engineering Practice, 2nd Edition, John Wiley, New York, 1967.

Thomas, L. J., Introduction to Mining, Hicks and Sons, Sydney, Australia, 1973.

Toth, G. W., and Stinnett, L. A., "Techniques of Rapid Preliminary Estimates of Coal Mining Costs," paper presented at the 1976 Mining Convention, American Mining Congress, Denver, Colorado, September 26-29, 1976.

U.S. Department of the Interior, Surface Mining and Our Environment, A Special Report to the Nation, 1967.

Wang, F. D., et al., Computer Program for Pit Slope Stability Analysis by the Finite Element Stress Analysis and Limiting Equilibrium Methods, U.S. Bureau of Mines, Report of Investigations 7685, p. 53, 1972.

Wendertz, R. L., "Coal Loading Shovels," (a review of factors to consider in loading equipment selection), Mining Congress Journal, Vol. 56, No. 4, 1970.

Bibliography

Adams, L. M., Capp, J. P., Gillmore, D. W., "Coal Mine Spoil and Refuse Bank Reclamation with Powerplant Fly Ash," Proceedings of the Third Mineral Waste Utilization Symposium, Chicago, Illinois, March 1972.

Alder, L., and Nauman, H. E., Analyzing Excavation and Material Handling Equipment, VPI, Blacksburg, Virginia, February 1970.

Aiken, G. E., "Bucket Wheel Excavators: How to Choose the Right One for the Job," Mining Engineering, Vol. 18, No. 1, January 1966, pp. 71-79.

AIME, "Drilling Selection Requires Value Judgement," Mining Engineering, Vol. 19, No. 10, 1967.

AIME, "Excavation and Loading - A Job for the Giants," Mining Engineering, Vol. 19, No. 10, 1967.

Allen, N., "Experimental Multiple Seam Mining and Reclamation on Steep Mountain Slopes," Proceedings of the Research and Applied Technology Symposium on Mined-Land Reclamation, Pittsburgh, Pennsylvania, NCA/BCR, March 1973.

American Iron and Steel Institute, Handbook of Steel Drainage and Highway Construction Products, 2nd Edition, New York, 1971.

Archibald, J. C., "Application of Horizontal Drilling as a Replacement of Vertical Drilling Techniques," Joy Manufacturing Company, Pittsburgh, Pennsylvania, 1976.

Ash, R. L., "The Design of Blasting Rounds," Surface Mining, Ed. E. P. Pfleider, SME/AIME, New York, 1968.

Atkinson, T., and Pryor, R., "Selection of Open-Pit Excavating and Loading Equipment discussion on," Institution of Mining and Metallurgy, Vol. 80, Bulletin No. 777, Sec. A, 1971.

Battelle Columbus Laboratories, Energy from Coal: Guidelines for the Preparation of Environmental Impact Statement, NTIS Publication PB-242-960, April 1975.

Bellum, D. P., "Open-Pit Mining," Mining Engineering, Vol. 28, No. 2, 1976.

Bengston, G. W., Mays, D. A., and Allen, J. C., "Revegetation of Coal Spoil in Northeastern Alabama: Effects of Timing of Seeding and Fertilization on Establishment of Pine-Grass Mixtures," Proceedings of the Research and Applied Technology Symposium on Mined-Land Reclamation, Pittsburgh, Pennsylvania, NCA/BCR, March 1973.

Burgos, V. T., "A Study on the Feasibility of Open-Pit Anthracite Mining," Unpublished M. Eng. Report, The Pennsylvania State University, 1975.

Caruccio, F. T., "An Evaluation of Factors Affecting Acid Mine Drainage Production and the Ground Water Interactions in Selected Areas of Western Pennsylvania," Second Symposium on Coal Mine Drainage Research, Pittsburgh, Pennsylvania, NCA/BCR, May 1968.

Day, D. A., Construction Equipment Guide, A. Wiley & Sons, New York, 1974.

Davis, G., Strip-Mine Reclamation in Appalachia (Review Draft), U.S. Department of Agriculture, Forest Service, Northeastern Forest Experiment Station, July 1971.

Dawson, V. E., "Observations Concerning On Site Brake Testing of Large Mining Trucks in British Columbia," SAE Paper 75060, Society of Automotive Engineers, Warrendale, Pennsylvania, 1975.

Dotson, G. K., Constraints to Spreading Sewage Sludge on Croplands, U.S. Environmental Protection Agency, Cincinnati, Ohio, 1973.

Duggan, C., Scanlon, D. H., and Bean, S. D., "Evaluation of Municipal Compost for Strip Mine Reclamation," Compost Science; Journal of Waste Recycling, Vol. 14, No. 3, May-June 1973.

Environmental Protection Agency, Guidelines for Erosion and Sediment Control Planning and Implementation, Environmental Protection Technology Series, EPA-R2-72-015 August 1972, [Project 15030 FMZ. Prepared by Department of Water Resources, State of Maryland, and Hittman Associates, Inc., Columbia, Maryland].

Enyedy, G., Jr., "AACE's Cost of Major Equipment (COME) Project in the Mineral Industry," AIME Council of Economic Proceedings, 1971.

Fites, D. V., "Tractor Scrapers Break New Ground," Mining Engineering, Vol. 21, No. 5, May 1969, pp. 69-70.

Ford, W. W., "Slope Reduction Methods," Unpublished Report, Division of Reclamation, Department of Natural Resources, Frankfort, Kentucky, 1968.

Ford, W. W., "Slope Reduction Method (Parallel Fill)," Unpublished Report, Division of Reclamation, Department of Natural Resources, Frankfort, Kentucky, 1970.

Gartner, E., "Trends of Development in Mining and Haulage Equipment of the Rhenish Brown Coal Open Cut Mines," Braunkohle Warme und Energie, Vol. 7, Nos. 11 and 12, June 1955, pp. 226-241.

Grant, A. F., and Lang, A. L., Reclaiming Illinois Strip Coal Land with Legumes and Grasses, Bulletin 628, University of Illinois, Agriculture Experiment Station, Urbana, Illinois, 1958.

Griffith, F. E., Magnuson, M. S., and Kimball, R. F., Demonstration and Evaluation of Five Methods of Secondary Backfilling of Strip-Mine Areas, U.S. Department of the Interior, Bureau of Mines, Report of Investigations No. 6772, 1966.

Grube, W. E., Jr., Smith, R. M., Singh, R. and Sobek, A. A., "Characterization of Coal Overburden Materials and Proceedings," Research and Applied Technology Symposium on Mined-Land Reclamation, Pittsburgh, Pennsylvania, NCA/BCR, March 1973.

Gwinn, T. A., "An Improved Environment Through Intelligent Mined-Land Reclamation," Paper presented at the fall meeting of the Institute of Mining, Metallurgical, and Petroleum Engineers, Inc., Salt Lake City, Utah, September 17 to 19, 1969, pp. 28-30.

Hartman, H. L., Principles of Drilling, Chapter 6.1, Surface Mining, Ed. E. P. Pfleider, SME/AIME, New York, 1968.

Heine, W. N., Gukert, W. E., "A New Method of Surface Coal Mining in Steep Terrain," Proceedings of the Research and Applied Technology Symposium on Mined-Land Reclamation, Pittsburgh, Pennsylvania, NCA/BCR, March 1973.

Hewes, L. I., and Oblesby, C. H., Highway Engineering, John Wiley and Sons, New York, 1960.

Hinesly, T. D., Braids, O. C., and Molina, J. E., Agricultural Benefits and Environmental Changes Resulting from the Use of Digested Sewage Sludge on Field Crops, U.S. Environmental Protection Agency (SW-30d), Cincinnati, Ohio, 1971.

Hinesly, T. D., Jones, R. L., and Sovewitz, B., "Use of Waste Treatment Plant Solids for Mined Land Reclamation," Mining Congress Journal, September 1972.

Hollingsworth, J. A., "History of Development of Strip Mining Machines," Bucyrus-Erie Company Publication, pp. 1-16.

Holzman, R. S., Executive Tax Management, Thomas V. Cromwell Co., New York, 1974.

Howarth, G. R., "How Will Major New Mines be Financed in the Future," SME Preprint No. 77-K-125, Paper presented at the AIME-Annual Meeting, Atlanta, 1977.

Hubbert, M. K., "The Energy Resources of the Earth," Scientific American, Vol. 225, No. 3, September 1971, pp. 61-70.

Jarpa, S. G., "Capital Investment and Operating Cost Estimation in Open-Pit Mining," Application of Computer Methods in the Mineral Industry, Ed. R. V. Ramani, Society of Mining Engineers of AIME, Salt Lake City, 1977.

Jelen, F. C., Editor, Cost and Optimization Engineering, McGraw Hill Inc., New York, 1970.

Kentucky Department for Natural Resources and Environmental Protection, A Manual of Kentucky Reclamation, Frankfort, Kentucky, May 1973.

Killebrew, C. E., "Tractor Shovels, Tractor Dozers, Tractor Scrapers," Surface Mining, Pfleider, Eugene P. (Ed.), The American Institute of Mining, Metallurgical, and Petroleum Engineers, Inc., New York, 1968, pp. 463-477.

Krumrey, A., "Krupp Bucket Wheel Excavators with Stacker Boom," Krupp Technical Review, Vol. 23, No. 1, 1965.

Levene, H. D., "An Unusual Coal Mine/Power Plant Complex," Coal Mining and Processing, October, 1971.

Limstrom, G. A. and Deitschman, G. H., "Reclaiming Illinois Strip Coal Lands by Forest Planting," University of Illinois Bulletin, 547, 1951, p. 50.

Linden, G., "The Bucket Wheel Excavator and the Versatility in Practice," Krupp Technical Publication. Also published in Deutsche Hebeword Fordertechnik, Nos. 3 and 4, July and October 1957.

Loftis, J. B., "The Evaluation of Horizontal Drilling In Open-Pit Mining," SME/AIME Preprint No. 75-AO-319, 1975.

Loper, F. G., "Simulation Techniques Pay Their Way as Management Tools," Mining Engineering, Vol. 19, No. 1, 1967.

Lyon, T. L., Buckman, H. O., and Brady, N. C., The Nature and Properties of Soil, 5th Ed., MacMillian Company, New York, 1952.

Mackellar, A. J., "Cost Engineering - Who Needs It?" SME Preprint 75-B-23, Paper presented at the AIME Annual Meeting, New York, 1975.

Manula, C. B., Albert, E. K., and Burgos, V., A Report on Anthracite Open-Pit Mining Part II. Engineering and Mine Cost Analysis, Special Research Report SR-104, Coal Research Section, The Pennsylvania State University, 1976.

Manula, C. B., Albert, E. K., and Ramani, R. V., "Application of a Total Systems Simulator to Coal Stripping, Volume II: OPMHS User's Manual," U.S. Bureau of Mines, 1977.

Manula, C. B., "OPMHS - Open-Pit Materials Handling Simulator," XIth APCOM Symposium, University of Arizona, 1973.

Maraboli, L. O., "Computer Simulation and Cost Analysis of Materials Handling in Chuquicamata, Chile," Unpublished M. Eng. Report, The Pennsylvania State University, 1973.

Marcus, J. J., "The Buckingham Method - An Aid in Equipment Selection," Mining Engineering, Vol. 17, No. 9, September 1965, pp. 92-94.

McClay, J., "A Dynamic Costing Model for Mining Systems," Unpublished M.S. Thesis, The Pennsylvania State University, 1974.

McFadden, D. H., "Pit Planning and Design-Coal Mines," Surface Mining, Pfleider, Eugene P. (Ed.), The Institute of Mining, Metallurgical, and Petroleum Engineers, Inc., New York, 1968, pp. 211-216.

McGraw Hill Inc., Coal Age, Vol. 75, No. 7, 1970, pp. 173-179.

Morgan, W., Fleet Production and Cost Program, Caterpillar Tractor Co., Peoria, Illinois, 1975.

Nalle, P. B., and Week, L. W., "The Digital Computer - Application in Mining and Process Control," Mining Engineering, Vol. 28, No. 9, 1976.

National Coal Association, Mined-Land Conservation Conference, Vol. 7, No. 4, Washington, D.C., November-December, 1971.

Nutter, F. B., "Mining with Front-End Loaders," Mining Congress Journal, Vol. 56, No. 6, June 1970, pp. 60-64.

Olivares, L. V., "Theory, Research and Practical Results of Inclined Drilling," An Unpublished Report in Mineral Engineering Management, The Pennsylvania State University, December 1967.

O'Neil, T. J., "Computer Simulation of Materials Handling in Open-Pit Mines," Unpublished M.S. Thesis, The Pennsylvania State University, 1966.

O'Neil, T. J., "The Minerals Depletion Allowance: Its Importance in Nonferous Metal Mining," Mining Engineering, Vol. 26, No. 10, 1974.

Peabody Coal Company, "A Model Mining Company," Coal Age, Vol. 76, No. 10, 1971.

Perez, R., "Application of a Time-Oriented Total Systems Simulator for Open-Pit Mining," Unpublished M. Eng. Report, The Pennsylvania State University, 1973.

Peterson, J. R., and Gschwind, J., "Amelioration of Coal Mine Spoils with Digested Sewage Sludge," Proceedings of the Research and Applied Technology Symposium on Mined-Land Reclamation, Pittsburgh, Pennsylvania, NCA/BCR, March 1973.

Plass, W. T., and Capp, J. P., "Physical and Chemical Characteristics of Sulfur Mine Spoil Treated with Fly Ash," Journal of Soil and Water Conservation, Vol. 29, No. 3, 1974.

Ramani, R. V., "Environmental Concerns and Decisions for the Mining Industry," Journal of the Institution of Engineers (India), MM Vol. 54, No. 2, March 1974.

Ramani, R. V., "Surface Mining," Mining Engineering, Vol. 26, No. 2, 1974.

Ramani, R. V., Iroegbu, M., and Manula, C. B., "Application of a Total Systems Simulator to Coal Stripping, Volume III: Equipment Selection Models," U.S. Bureau of Mines, 1977.

Riley, C. V., "Design Criteria of Mined Land Reclamation," Paper presented at the SME Fall Meeting and Exhibit, Birmingham, Alabama, October 1972.

Riley, C. V., "Furrow Grading--Key to Successful Reclamation," Proceedings of the Research and Applied Technology Symposium on Mined-Land Reclamation, Pittsburgh, Pennsylvania, NCA/BCR, March 1973.

Rodgers, H. C. C., "Developments in Equipment Used for Overburden Removal and Coal Winning in the Brown Coal Industry. Excavating Equipment for Brown Coal, Part I," Proceedings of the Australian Institute of Mining and Metallurgy, No. 194, June 1960, pp. 127-249.

Rose, B. A., "Sanitary District Puts Sludge to Work in Land Reclamation," Water and Sewage Works, A Scranton Gillette Publication, September, 1968.

Sathit, T., "Principles of Drilling," Mining Engineering Handbook, Cumins and Given, SME, Mudd Series, New York, 1973.

Sanford, R. L., "Stochastic Simulation of a Belt Conveyor System," Unpublished M.S. Thesis, The Pennsylvania State University, 1965.

Schellhorn, H. W., "Some Aspects of High Capacity Production with Bucket Wheel Excavators," Krupp Technical Publication. Also published in The K. A. Clark Volume, Research Council of Alberta, Edmonton, October 1963.

Schmidt, R. A., and Stoneman, W. C., A Study of Surface Coal Mining in West Virginia, Stanford Research Institute, Menlo Park, California, 1972.

Scott, R. B., Hill, R. D., and Wilmoth, R. C., Cost of Reclamation and Mine Drainage Abatement--Elkins Demonstration Project, U.S. Environmental Protection Agency, Cincinnati, Ohio, 1970.

Skelly and Loy Engineering Consultants, Surface Mine Haul Road Design Study, U.S. Bureau of Mines Report, Contract No. J0255037, Washington, D.C. 1976.

Society of Mining Engineers of AIME, Mining Engineering Handbook, New York, 1973.

Staff, Surface Mining of Coal, Coal Age Guidebook, July 1970.

Stermole, F. J., Economic Evaluation and Investment Decision Methods, Investment Eval. Corp., Golden, Colorado, 1974.

Taylor, D. W., Soil Mechanics, John Wiley and Sons, New York, 1960.

Taylor, G. A., Managerial and Engineering Economy, V. Nostrand Reinhold Co., New York, 1964.

U.S. Department of Agriculture, Kentucky Guide for Classification, Use, and Vegetative Treatment of Surface Mine Spoil, Soil Conservation Service, Lexington, Kentucky, 1971.

Weimer, W. H., "Production and Engineering in Surface Coal Mines," Surface Mining, Pfleider, E. P. (Ed.), The Institute of Mining, Metallurgical, and Petroleum Engineers, Inc., New York, 1968, pp. 229-238.

Wimpfen, F. P., "Mine Costs and Control," Section 31, SME Mining Engineer's Handbook, AIME, New York, 1973.

Woodruff, S. D., Methods of Working Coal and Metal Mines, Planning and Operations, Vol. 3, Pergamon Press, Inc., New York, 1966.

Wurtz, C. B., "Mining and Conservation," Mining Congress Journal, Vol. 56, No. 5, May 1970, pp. 40.

Zemp, J. C., "Improved Blasting Techniques Give Wider Applications for Wheel Loaders," Mining Magazine, Vol. 121, No. 3, September 1969, p. 207.

Acknowledgements

The assistance and cooperation of many mining companies and the various state and federal agencies are gratefully acknowledged.

Sincere thanks are extended to Mr. John F. Martin, Project Officer, Industrial Environmental Research Laboratory, Cincinnati, OH, and Mr. Elmore C. Grim, former Project Officer and presently Director, Division of Forestry, the Kentucky Division of Natural Resources, Frankfort, KY, for their support and guidance.

Sincere appreciation is also extended to Mr. Richard D. Ellison, Executive Vice President, D'Appolonia Consulting Engineers, Inc., Pittsburgh, PA, for permission to use information from a report "Geotechnical Investigations and Design Guidelines."

Assistance provided by Mr. Sukumar Bandopadhyay, Graduate Assistant in Mineral Engineering (Mining Section), and Dr. James M. Riddle, formerly Research Associate in Mining Engineering, The Pennsylvania State University, is gratefully acknowledged.

The materials contained were prepared by an interdepartmental and interdisciplinary group of the College of Earth and Mineral Sciences of The Pennsylvania State University. Overall management for the project was provided by the Department of Mineral Engineering. The project staff was comprised of the following personnel:

Dr. R.V. Ramani, Professor of Mining Engineering
Project Manager and Coprincipal Investigator (Surface Mining Engineering, Land Use Planning, Mine Planning)
Dr. L.W. Saperstein, Professor of Mining Engineering
Coprincipal Investigator (Federal and State Laws and Regulations, Land Use Management)
Dr. H.L. Lovell, Professor of Mining Engineering
Coprincipal Investigator (Water Quality Management)
Dr. R.R. Parizek, Professor of Geology
Faculty Associate (Geology and Hydrology)
Dr. C.G. Knight, Associate Professor of Geography
Faculty Associate (Land Use Management)
Dr. R. Stefanko, Professor of Mining Engineering
Faculty Associate (Project Consultant)
Prof. R.L. Frantz, Professor of Mining Engineering
Faculty Associate (Project Consultant)
M.L. Clar, Research Assistant in Mining Engineering
In-House Project Coordinator
L.B. Phelps, Instructor in Mining Engineering
Surface Mine Engineering
C.J. Bise, Instructor in Mining Engineering
Surface Mine Engineering
P.J. Duhaime, Graduate Assistant in Geography
Land Use Management
D. Forsberg, Graduate Assistant in Mining Engineering
Water Quality Management
K.W. Grubaugh, Graduate Assistant in Mining Engineering
Laws and Regulations
C. Murray, Graduate Assistant in Mining Engineering
Mining Engineering
D. Richardson, Graduate Assistant in Mining Engineering
Water Quality Management
J. Sgambat, Graduate Assistant in Geology
Geology
A. Weiner, Graduate Assistant in Mining Engineering
Water Quality Management

Part III

Reclamation Equipment and Techniques

The information in Part III is from *Surface Coal Mining Reclamation Equipment and Techniques. Proceedings: Bureau of Mines Technology Transfer Seminars, Evansville, Ind., June 3, 1980, and Denver, Colo., June 5, 1980* (IC 8823), compiled by the staff of the U.S. Bureau of Mines.

Reducing Costs for Recontouring Mined Land

by John M. Goris
Research Civil Engineer
Spokane Research Center
Bureau of Mines, Spokane, Washington

INTRODUCTION

The need to develop better equipment and techniques for mine spoil leveling is evident to people associated with the mining industry. Prior to the enactment of present reclamation laws, land restoration had little or no effect on production. Today, however, land restoration has become an integral part of mining, and problems associated with this activity not only can adversely affect production but can result in a shutdown of operations. The end result could be reduced coal production and extensive cost increases unless improved equipment and techniques are developed and utilized.

The mining industry's response to the recontouring requirements has been to rely on available equipment such as tractor dozers, scrapers, graders, and in some instances draglines to meet recontouring needs. This can be an expensive compromise for many operations since recontouring conditions cover a very broad range, usually exceeding the equipment's range of efficient operation. For example, spoil material relocation distances can range from as little as 20 feet to over 500 feet, whereas the effective range of a tractor dozer is approximately 100 feet, as seen in figure 1, which represents a typical production rate curve. Past this point, production rates are low and the cost per cubic yard of material moved increases. Therefore, total reliance on conventional tractor dozers for recontouring can be very expensive.

Scrapers and draglines also have limitations. Several experimental systems such as a winch-drawn blade system (fig. 2) are being researched by the Bureau of Mines; this system consists of two crescent-shaped buckets back-to-back and drawn perpendicular to the spoil bank by a winch and tailblock assembly. A prototype system has been field-tested, but it could be years before it is perfected and accepted by the industry. Even with this system, the mine operator will have to sacrifice the flexibility and mobility associated with tractor dozers. There are techniques available now, however, for improving the efficiency of tractor dozers that will result in more production at lower costs and lower fuel consumption. The same is also true for draglines employed in recontouring work. Following is a discussion of three specific areas of improvements for existing equipment that could reduce operating costs:

1. Specialized blades for tractor dozers.

2. Bulldozer work rate indicator.

3. Backfilling final highwalls with draglines.

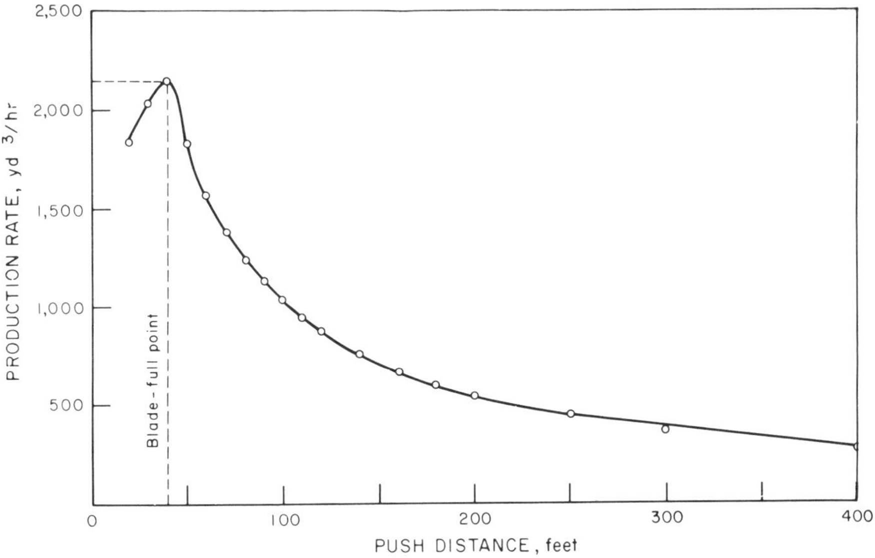

FIGURE 1. - Production curve for 410-hp tractor dozer.

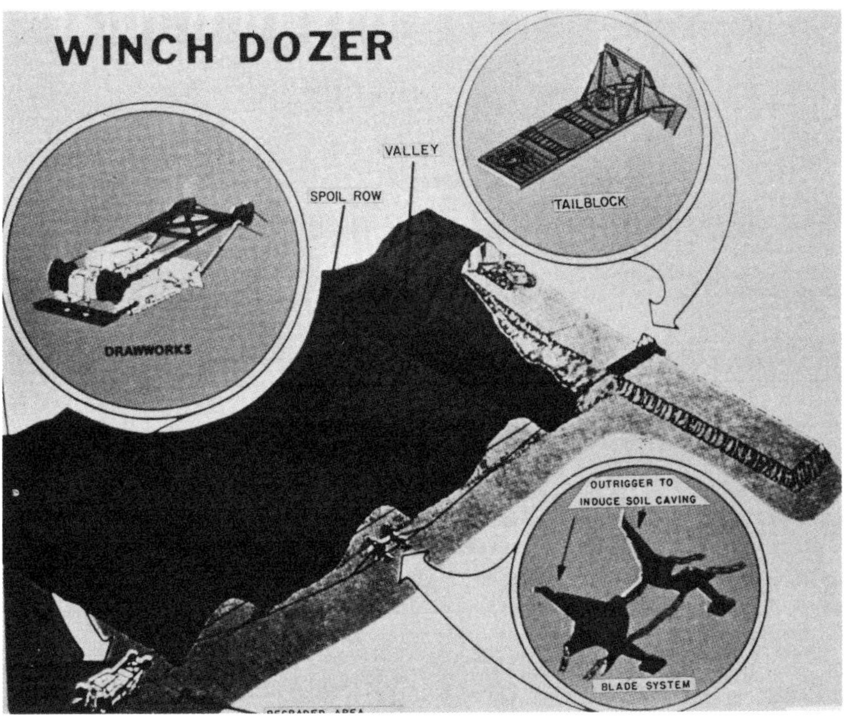

FIGURE 2. - Winch-drawn blade system.

SPECIALIZED BLADES FOR TRACTOR DOZERS

As part of the analysis of improving tractor dozer performance, it might be best to first review the requirements of leveling and recontouring area-mined land. Figure 3 shows a typical cross section of a uniform spoil bank. Tractor dozers begin leveling these banks by first pushing the peaks short distances into adjacent valleys. As the work progresses, the spoil bank height decreases and the average push distance increases. For normal spoil banks the maximum distance may never exceed 150 feet. The final work is to grade the land and prepare it for planting. The same process is generally applied for leveling box-cut spoil except that push distances can reach 500 feet or more. These are poor working conditions for conventional tractor dozers.

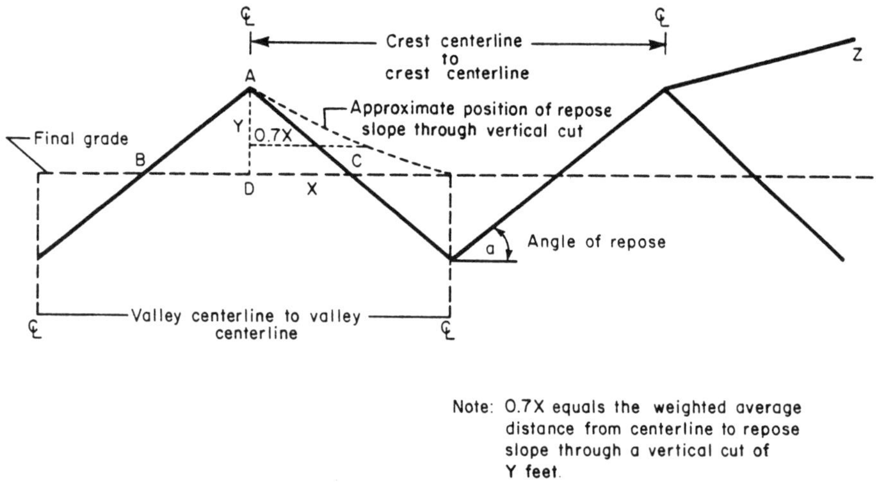

FIGURE 3. - Typical cross section of a uniform spoil bank.

The bank cubic yards (BCY) of material that must be moved depend primarily on the type of spoil bank (normal or box cut), the crest-to-crest distance of the banks, the angle of repose, and the amount of rehandle. Table 1 shows the quantities of spoil material that must be handled in normal spoil as these conditions change. In comparison, it was determined in an extensive field project on leveling spoil banks (discussed later) that an average of 27,901 BCY was handled per acre in leveling 275.9 acres of box-cut spoil. The average for normal 120-foot spoil on this project was 10,073 BCY. Therefore, the average ratio between the material handled in box-cut spoil and that moved in normal spoil was 2.77.

TABLE 1. - Theoretical estimate of spoil material handled in normal spoil

Crest-to-crest distance, ft	Bank cubic yards moved per acre					
	Angle of repose = 30°			Angle of repose = 36°		
	Rehandle 0 pct	Rehandle 25 pct	Rehandle 40 pct	Rehandle 0 pct	Rehandle 25 pct	Rehandle 40 pct
90..........	5,239	6,549	7,335	6,986	8,732	9,780
120..........	6,593	8,241	9,230	8,791	10,989	12,307

Next, consider an average dozing cycle in leveling spoil. Figure 4 shows the three primary phases, loading the blade, transferring (drifting) the material, and returning. At an average push distance of about 125 feet, the dozer will spend approximately 20 percent of its time filling the blade, 53 percent drifting the material, and 27 percent returning. Under good working conditions the production rate will be approximately 850 cubic yards per hour for a tractor dozer with about 400 horsepower and using a U-blade. The primary job conditions that affect this production rate follow:

1. Push distance.
2. Grade.
3. Type of material.
4. Operator skill.
5. Type of blade.
6. Type of dozing.
7. Job efficiency.
8. Visibility.

Each of these has some influence on production. Table 2 shows the range of correction factors applied to the production rate for most of the above job conditions as recommended by a major equipment manufacturer.

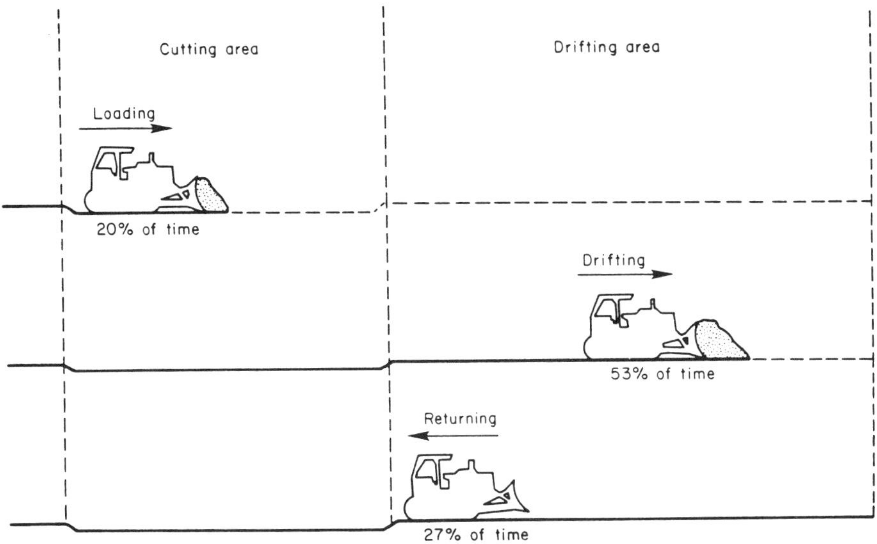

FIGURE 4. - Three primary phases of a dozing cycle.

TABLE 2. - Job conditions and correction factors for tractor dozer production rates

Job conditions	Correction factors	Job conditions	Correction factors
Push distance......	Refer to manufacturers' curves.	Type of blade:	
		Cushioned........	0.50-0.75
Grade:		Coal U-blade.....	1.20
Uphill (20 pct)..	0.65	Bowl (stockpiler)	1.30
Uphill (10 pct)..	0.95	Type of dozing:	
Level (0 pct)....	1.00	Slot............	1.20
Downhill (10 pct)	1.15	Side by side.....	1.15-1.25
Downhill (20 pct)	1.22	Job efficiency:	
Type of material:		50 min/hr........	0.84
Loose stockpile..	1.20	40 min/hr........	0.67
Hard to cut......	0.60-0.80	Visibility:	
Rock............	0.60-0.80	Darkness, rain,	
Operator skill:		snow, fog, or	
Excellent........	1.00	dust...........	0.80
Average..........	0.75		
Poor............	0.60		

These factors are multiplied by the production rates for optimum conditions, which are given in manufacturers' handbooks. It is obvious that all are significant if varied from one extreme to the next. The question is, which ones can you change and control to your advantage? Work at the Spokane Research Center has focused on four to date: push distance, type of blades, type of dozing conditions, and operator skills.

Push Distance, Type of Blades and Type of Dozing

The first three factors are actually interrelated in our research work and are therefore described together. As seen in figure 1, the shorter the push distance, the higher the production, until the blade becomes full and the operator dumps the load. This is referred to as the blade-full point. If the push distance is less than this, production will drop off because the blade is not full when the tractor dozer reaches the dump point. Consequently, the nearer the tractor dozer works to the blade-full point, the higher production rates will be. There are two techniques that an operator can employ to accomplish this:

1. Use angle blades that move material short distances laterally towards the outslope under continuous cutting and casting (blade-full) conditions.

2. Use narrow deep blades that fill in short distances and carry large volumes.

The first technique is extremely productive in normal 90- to 120-foot crest-to-crest spoil because the blade travels parallel to the spoil bank, moving the material laterally towards the outslope. The average push distances are, therefore, approximately one-half the width of the blade. Also, the blade operates in a continuous cutting and casting action as it moves along the spoil bank, which means that it is always working at the blade-full (optimum) point. High production rates result from this as well as from the fact that the nonproductive backing and positioning time of the dozer is eliminated.

The second technique is employed when push distances are short and the blade is not filled when the operator reaches the dump point; such is the case when the push distances are too great for angle dozing but short of the blade-full point.

Both of these techniques were researched and evaluated through a joint program between the Pittsburg and Midway Coal Co. (P&M) and the U.S. Bureau of Mines (USBM). The outcome of the 18-month research effort was a combination of highly productive specialized tools that can be used in combination with conventional tractor dozers and scrapers. The tools are shown in figures 5-8 and include

1. 48-foot angle blade.
2. 24-foot U-blade.
3. 13-foot narrow, deep U-blade.
4. 24-foot grading bar.

The 48-foot angle blade proved to be the workhorse of the team and is the result of many months of work designing and field-testing various types and sizes of blades. The angle blade is mounted on two 410-horsepower tractors in an offset configuration, but requires only one operator. The blade is fixed at a 45° angle. The angle blade usually begins its cycle at one end of the spoil bank, where the operator fills the blade and moves forward, casting the spoil material laterally to the outslope. This continuous cutting and casting motion continues the full length of the spoil bank at a speed of approximately 1.5 miles per hour. At the opposite end of the bank, the operator turns the system around and begins cutting and casting material into the opposite valley. The process continues until adjacent spoil banks meet or the lateral relocation distance from the centerline of the spoil bank exceeds approximately 50 feet.

Figure 5: Forty-eight foot angle blade mounted on two side-by-side tractors in an offset configuration

Figure 6: Twenty-four foot U-blade mounted on two side-by-side tractors

Figure 7: Thirteen foot narrow, deep U-blade

Figure 8: Twenty-four foot grading bar

The field evaluation of the 48-foot angle blade over a period of 520 operating hours showed an average production rate of approximately 6,000 to 6,500 BCY/hr. When used in conjunction with other specialized tools and conventional dozers as support, the system reduced overall work efforts and cost by approximately 50 percent. These reductions were achieved under very controlled field conditions, and it was estimated that the reduction would be 30 to 40 percent under actual mining conditions. This fact was later proven to be the case and will be discussed shortly.

The side-by-side offset configuration of the tractors was adopted from an existing system marketed by a major tractor dozer manufacturer which consists of two tractors side-by-side with a large single U-blade. The value of this system was recognized and adopted for the angle blade. In addition, a 24-foot U-blade was purchased for the two tractors, and the system was used as a support tool in normal spoil and as the prime mover in box-cut spoil. The 24-foot U-blade showed an approximate 25 percent increase in production and 25 percent decrease in fuel consumption over two single tractor dozers working separately. In addition, there was an immediate reduction in cost for labor because the system, like the angle blade, requires only one operator.

The value of conventional bulldozer techniques in land recontouring was not diminished in the success of the 48-foot angle blade and 24-foot U-blade. The field evaluation showed that the angle blade could handle 60 to 80 percent of material that had to be moved; the remaining 20 to 40 percent had to be handled by conventional dozers. It was also observed that these conventional dozers required 40 or more feet to get a full blade. As it turns out, this is beyond the effective relocation distance of the 48-foot angle blade as seen in figure 9. If conditions are adverse, such as in rocky material, the blade-full point for dozers could be as much as 70 feet, and yet the required push distance may only be 50 feet. The consequence, of course, is that the operator will dump the load before the blade is full, which brings us to the second technique by which an operator can increase production: using a narrow, deep blade that fills in short distances.

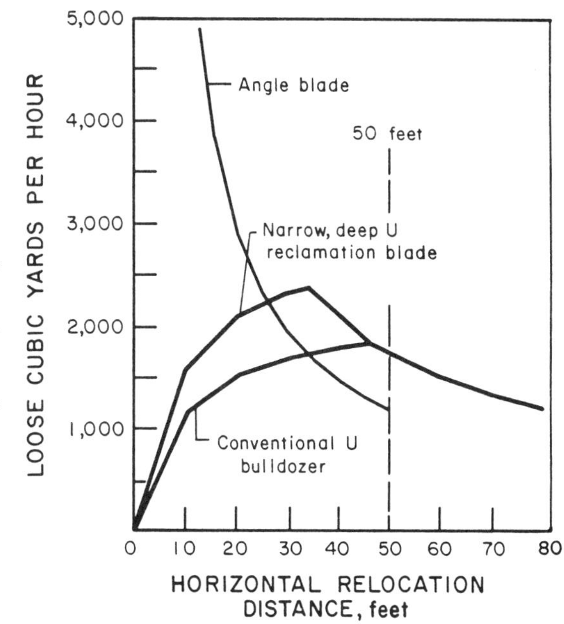

FIGURE 9. - Effective relocation ranges for specialized and conventional blades.

In an effort to study this technique a specialized 13-foot, narrow, deep U-blade was designed and tested under the P&M-USBM contract. A comparison of the blade's specifications with those of a conventional U-blade for a 410-horsepower tractor is shown in table 3.

Results of the field test of the deep U-blade show that it could fill in 30 to 40 feet by sidecutting as seen in figure 9 and was very effective in combination with the angle blade in leveling and reshaping spoil banks. The blade design does not represent a major modification in blades; it is an improvement in a tool for a specific job. It is not too specialized to be used for other routine jobs in surface mining, such as long-distance pushes in land leveling, building dragline pads, or even building roads.

TABLE 3. - Dimensional comparison of conventional U-blade and narrow, deep U-blade for a 410-horsepower tractor

	Conventional U-blade	Narrow, deep U-blade
Blade length......................ft..	15.75	13.0
Blade height......................ft..	5.92	5.75
Weight (blade only)................lb..	9,280	8,950
Width (front to back of side plate).......................in..	37	40.5
Wing angle........................deg..	25	30
Length of each wing...............ft..	4.21	3.0
Left wing...............pct of length..	26.5	23.0
Center section..........pct of length..	47.0	54.0
Right wing..............pct of length..	26.5	23.0

The grading bar, as seen in figure 8, was designed to smooth the rough-graded land in preparation for the farming phase of land reclamation. It represents another tool designed to perform a specific job at a cost savings to the mine operator.

A number of drag systems, such as I-beams, pipes, and large-diameter wire rope, have been used. However, they all have several basic drawbacks: they will occasionally hang up on large rocks, they require a large turning radius, and they are difficult to operate in limited space areas such as a point of land extending into a water impoundment.

The grading bar is based on the idea of pushing the finishing tool rather than dragging it. It is made from 24-foot channel iron, with side wings for carrying live soil, and is mounted under the heel plate of a conventional bulldozer blade using two 2-inch pins. Mounting time is about 5 minutes. Under operating conditions, the blade support arms rest on top of the channel iron, and the cutting edge of the bulldozer's blade can penetrate the ground about 1 inch. The operator has full control over lift and tilt of the grading bar, since it is attached to the blade.

The grading bar was tested under the P&M-USBM contract and smoothed 1,431 acres at rates between 2.5 and 7 acres per hour, depending on ground conditions. It was mounted on a 270-horsepower tractor dozer.

The major conclusions drawn from field-testing all of the specialized tools under the P&M-USBM contract follow:

1. The tools designed and tested to work in the short distance relocation range performed satisfactorily and resulted in a 50-percent reduction in cost and tractor time per acre of mined land leveled and reshaped. Proportional savings in diesel fuel were also experienced.

2. In leveling and shaping normal 90- to 120-foot crest-to-crest spoil banks, the 48-foot angle blade moved approximately 80 percent of the material while using only 60 percent of the total tractor time.

3. Conventional bulldozers are needed in the final stages of the land-shaping process. They will move an estimated 20 percent of the total material but will require approximately 40 percent of the total tractor time.

4. The 24-foot U-blade showed a 25-percent increase in production over two separate tractor dozers of comparable horsepower.

5. The deep U-blade demonstrated higher production rates than conventional U-blades in short-distance relocation range below the blade-full point; above this point, the deep U-blade was comparable to conventional blades.

6. The grading bar demonstrated its efficiency in smoothing rough-graded land at rates between 2.5 and 7 acres per hour depending on ground conditions.

7. No single tool was effective in all situations. It required the combination of tools to effectively and efficiently level and shape the spoil banks.

Based on the results of the P&M-USBM contract, a technology transfer project was funded to demonstrate the tools developed and to show that significant reductions in cost and work efforts could be realized by employing the proper combination of specialized and conventional tools. The mine site was in Texas, and the contractor selected to contract the demonstration was Russell and Sons Construction Co.

The team for the technology transfer demonstration was on location in mid-February 1979; field testing began around March 1 and continued until September. It included the same four specialized tools developed and/or tested under the P&M-USBM contract (figs. 5 through 8) along with conventional dozers. One exception was that a low-ground-pressure (LGP) tractor dozer was brought in as a support tool when the ground conditions were wet and flotation was a problem. The LGP had a ground bearing pressure of approximately 6.3 psi as compared to 13 to 15 psi for the other tools. It was used primarily in unstable soil conditions as well as in finish grading.

The demonstration was conducted in five test areas located throughout the mine. The five areas, totaling 312 acres, included 120-foot crest-to-crest spoil areas and box-cut spoil. The field results are shown in table 4 and equate to a 32-percent decrease in cost and fuel for normal spoil and a 27-percent decrease in cost and fuel for box-cut spoil, compared with costs using conventional equipment.

TABLE 4. - Results of technology transfer demonstration in Texas

	Location and type of spoil					
	Area 1, box-cut and irregular	Area 2, box-cut and irregular	Area 3, box-cut and irregular	Area 4, irregular	Area 5, irregular	Total or average
Approximate number of working days..	26	26	43	34	25	154
Acres reclaimed...	38	49	83	68	74	312
Acres reclaimed per working day..	1.46	1.88	1.93	2.00	2.96	2.03
Total horsepower-hours per acre...	6,904	7,912	9,049	5,363	4,717	6,778

As a result of the excellent performance of the specialized equipment, the manager of the mine adopted the equipment as his primary land-leveling and recontouring tools.

Operator Skills

This is the fourth job condition mentioned earlier which the USBM is researching for possible reduction in mined-land recontouring costs and work effort. This also leads us to the following, second major topic of this paper, as discussed in the following section.

BULLDOZER WORK RATE INDICATOR

The work rate indicator is actually a draft power sensor system designed to help the bulldozer operator to maintain high production rates without first acquiring extensive operator skill. This system continuously measures both the push of the blade against the spoil and the true ground velocity of the bulldozer. It multiplies those quantities to determine the real work rate of the machine. The system consists of a velocity sensor, a blade load sensor, processing circuits, and a work rate indicator.

The processing circuits include an analog multiplier which has an output signal representing the draft power (velocity X load = power). The draft power, or pushing power, is a measurement of the work rate at that moment.

The velocity sensor consists of an ultrasonic continuous wave transmitter and receiver operating at a fixed transmission frequency. This sensor is housed in a metal container located behind the operator's cab (fig. 10).

FIGURE 10. - Velocity sensor mounted behind operator's cab.

The signal from the velocity sensor transmitter strikes the ground behind the bulldozer and is partially reflected back to the receiver. When the bulldozer is moving forward, there is a reduction in the frequency at which the signal is received.

The load sensor is a strain gage installed on the neck of one of the push-beam trunnion's balls (fig. 11). The neck is highly stressed when the resistance of the load is transmitted through the blade and the push-beams.

An audible beeper was the indication preferred by the operators. The number of beeps per second increases as the work rate increases. Each beep is about 1/10 second long and consists of a tone burst of low sonic frequency, approximately 400 Hz. A blinking light display and meter are also available. These displays, the beeper, and the processing circuitry, are contained in a rugged metal box mounted in the cab area.

As the bulldozer moves forward with the blade up and away from the ground, the work rate will be zero. Because the product of the velocity (say 1.5 mph) multipled by the zero load against the blade is zero, the beep rate produced is essentially zero.

As the blade is lowered and digs into the ground, a load is put on the blade and beeping begins. As the blade pushes more material, the load increases and so does the beep rate until the maximum work rate is reached.

Depending on the resistance of the material being moved, the horsepower available at the tracks, and the traction conditions, the bulldozer can reach the combination of velocity and blade load that produces the maximum work rate possible for those conditions. Once the beeping reaches the fastest rate attainable for those conditions, the operator attempts to maintain that rate. This is usually done by raising or lowering the blade.

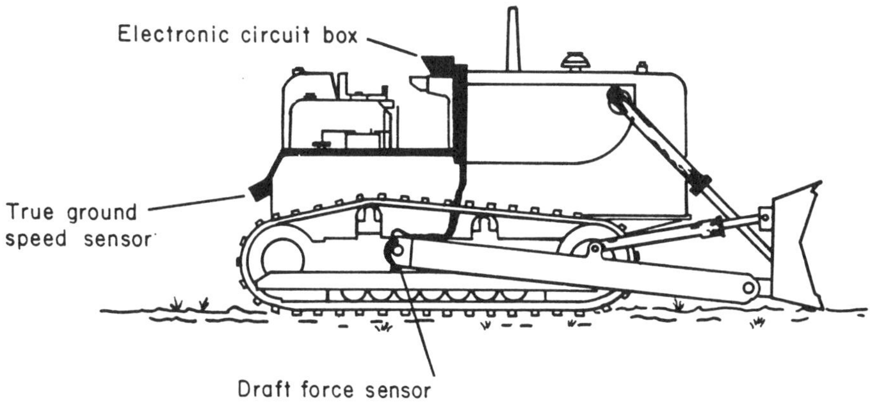

FIGURE 11. - Location of draft force sensor and electronic circuit box.

For example, if the blade is allowed to penetrate too deeply and the load against it becomes too large, the bulldozer will lose speed, either because the tracks slip or because the engine slows down. As the bulldozer work rate decreases, the beep rate will decrease proportionally. The operator must partially raise the blade to again achieve an optimum combination of velocity and blade load. Moving into an area of reduced traction would normally cause a similar indication and a lower work rate due to slippage. The correction would be similar.

On the other hand, if the beep rate dropped while the bulldozer velocity increased, this could indicate to the operator that the load against the blade had decreased. The work rate might increase if the blade were lowered. The system works regardless of the change in dozing conditions.

The prototype system, which was developed by the Southwest Research Institute under a Bureau of Mines contract, was field-tested early in 1978 at a large lignite mine in Texas.

Two 410-horsepower bulldozers were fitted with this system and used for leveling and recontouring spoil banks. A test area of 105 acres was divided into smaller areas to accommodate various tests.

Several operators ran the bulldozers both with and without the use of the work rate indicator. Logs were kept showing machine time, identity of the operator, weather conditions, fuel consumption, and other pertinent data. Tests were run for approximately 3 months.

The data showed that when the work rate indicator was used, the land was recontoured faster, providing a potential for reduced labor and capital costs. Furthermore, less fuel was used per acre. The volume of material moved per hour increased between 20 and 25 percent in the various test areas.

Although the hourly rate at which the bulldozers consumed fuel increased, this effect was more than offset because fewer hours were required to level and recontour an area. The approximate results follow: fuel consumption per hour, up 7 percent; hours of bulldozing required per acre, down 25 percent; fuel consumption per acre, down 20 to 25 percent.

The prototype design is being production-engineered as a rugged product, readily installed and used on existing reclamation bulldozers. A series of field tests of this improved system will be conducted on bulldozers of several different manufacturers and sizes. Results should be available in late 1980.

BACKFILLING FINAL HIGHWALLS WITH DRAGLINES

Public Law 95-87 mandates that all mined land be returned to approximately the original contour with all highwalls eliminated. The regulations actually specify that the backfilled slopes covering the highwall cannot exceed the approximate premining slopes.

The major problem confronting the mine operator in complying with the regulations is moving large volumes of material over long distances under conditions that will vary depending on the original ground contour (flat, rolling, etc.), overburden thickness, and the orientation of the coal seam(s) (flat or dipping).

At present, the industry is relying primarily on tractor dozers, draglines, scrapers, and trucks to move the required material; this is logical since such equipment is usually available at the mine site or at least the operator is familiar with it.

A number of studies have been, and are continuing to be, conducted by the mining industry and the Government on various methods for eliminating highwalls. Most of the methods include the utilization of the above-mentioned equipment with varying modification as to size, deployment, combinations, etc. This section of this paper will address an evaluation of several types and combinations of equipment for backfilling highwalls in modified area mining.

The mine site selected for the evaluation is in eastern Ohio and involves three different pit configurations:

1. Pit A, which was excavated by a medium-size shovel (less than 60 yd^3). See figure 12.

2. Pit B, which was excavated by a small dragline (less than 40 yd^3). See figure 13.

3. Pit C, which was excavated by a large dragline (greater than 100 yd^3). See figure 14.

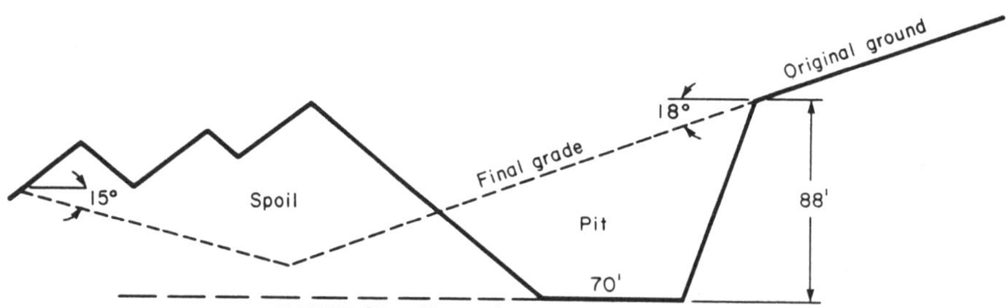

FIGURE 12. - Typical cross section of pit A.

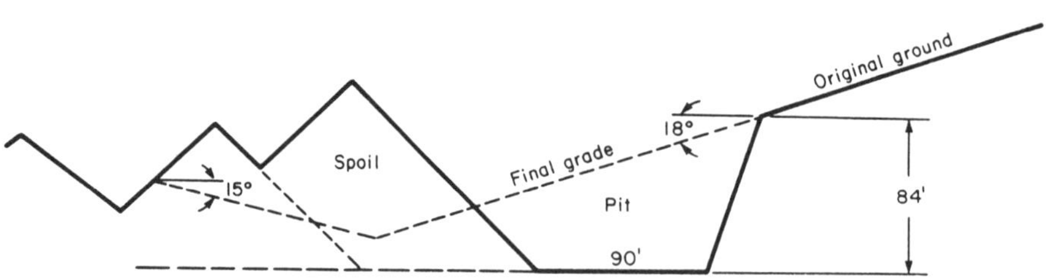

FIGURE 13. - Typical cross section of pit B.

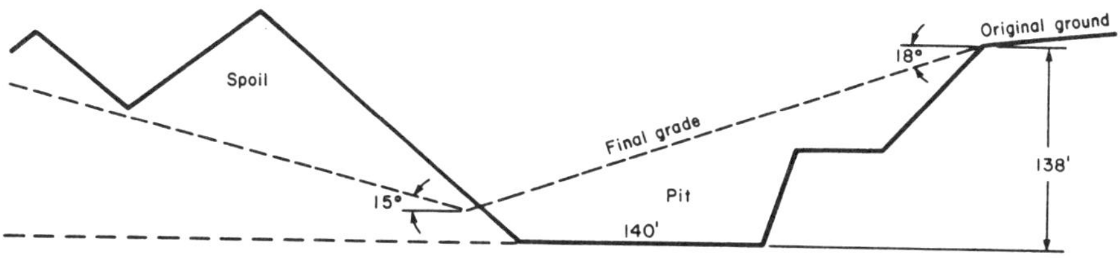

FIGURE 14. - Typical cross section of pit C.

Three basic approaches were selected for backfilling the highwalls in each of the three pits. The three approaches include

1. Conventional tractor dozers.

2. Conventional dragline working on the spoil side of the pit.

3. Conventional dragline working on the highwall side of the pit.

The specific equipment for each approach for pits A and B included

1. Conventional tractor dozers--410-horsepower units.

2. Conventional dragline--11-cubic-yard diesel with 170-foot operating radius. 410-horsepower tractor dozers were used for support work as required.

3. Conventional dragline--same as above except had a 20-cubic-yard crescent-shaped bucket and 600 feet of hoist and drag rope. 410-horsepower tractor dozers were also used for support work.

For pit C, the specific equipment included

1. Conventional tractor dozers--410-horsepower units.

2. Conventional dragline--35-cubic-yard electric dragline with 201-foot operating radius. 410-horsepower tractor dozers were used for support work.

3. Conventional dragline--same as above except had a 60-cubic-yard crescent-shaped bucket and 800 feet of hoist and drag rope. 410-horsepower tractor dozers were used for support work.

The procedure for moving the spoil in each of the three approaches was the same for all pits. In the first approach, the tractor dozers moved the material from the spoil side to the pit in a series of cuts in three primary lifts. The first lifts involved pushing approximately 25 percent of the required spoil material down a slope of 9.6 percent (5.5°) at a rehandle of 25 percent. The second lift involved pushing approximately 50 percent of the material over a zero percent slope (±2.0 pct) at a rehandle of 40 percent. The third lift involved pushing the remaining 25 percent up a slope of approximately 17.6 percent (10°) at a rehandle of 25 percent. There was sufficient material in the spoil banks to backfill the highwall and attain the final slopes, as seen in figures 12-14; consequently, material from other locations, such as the box-cut spoil, was not required.

In the second approach, the dragline is placed on a bench prepared by a dozer on the spoil side of the pit as seen in figure 15. The dragline was positioned to obtain the maximum dumping range. The dozers assisted the dragline by pushing spoil to the loading zone if some spoil material could not be reached. The tractor dozers also did the final grading in the pit spoil area and filled in any areas that the dragline could not reach.

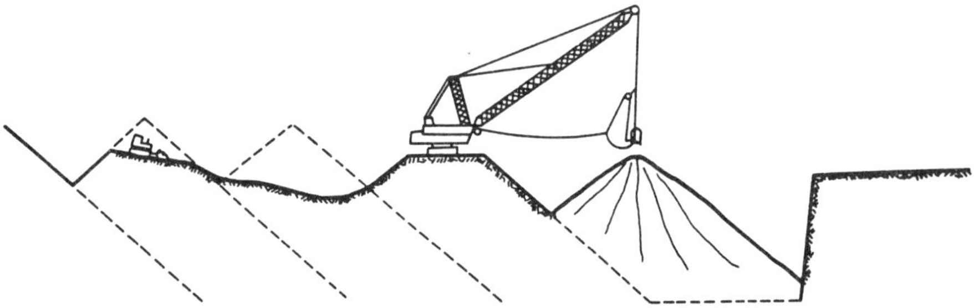

FIGURE 15. - Dragline backfilling from spoil side of pit.

In the third approach, the dragline was placed on top of the highwall (fig. 16) and equipped with additional hoist and drag rope along with a crescent-shaped bucket for dragging the material into the pit similar to a tower excavator. A 300-horsepower track-type pipelayer was used for the tail-block assembly. The dragline-tower system was able to reach all required spoil material. The final leveling and grading work was done with 410-horsepower tractor dozers.

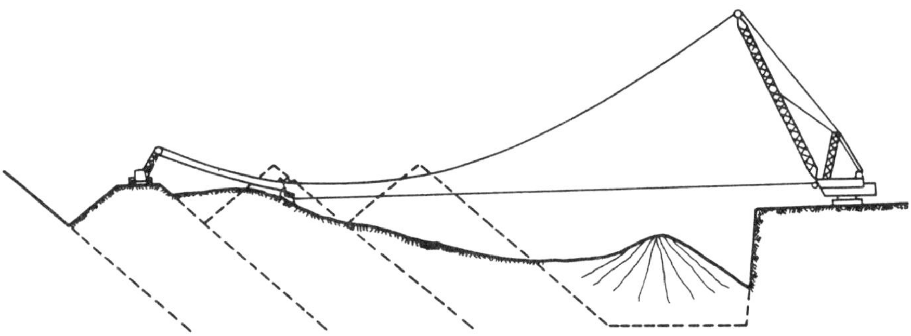

FIGURE 16. - Dragline backfilling from highwall side of pit.

Specifics about each pit follow.

Pit A

Because the pit (fig. 12) contained several inside and outside curves, its length of 2,636 feet was divided into six sections; each was evaluated separately and the results were compiled for each of the three combinations of equipment. The average highwall height was approximately 88 feet, and the average pit width was 70 feet. Each linear foot of highwall required an average of 273 BCY of spoil to backfill to the top. For this study it was assumed that no blasting would occur to facilitate backfilling. The average spoil relocation distance from the centroid of the cut area (required spoil) to the centroid of the fill area was 203 feet. Results of the analysis are shown in table 5.

Pit B

Figure 13 shows a typical cross section of pit B. Like pit A, this pit was divided into sections which were evaluated separately and the results completed. The average highwall height is approximately 84 feet, the average pit width is approximately 90 feet, and the pit length is approximately 7,690 feet.

Each linear foot of highwall averaged approximately 264 BCY of spoil to backfill to the top. As in pit A, it was assumed no blasting of the highwall would take place. The average spoil relocation distance was 236 feet. Results are shown in table 5.

Pit C

Figure 14 shows a typical cross section of pit C. In comparison to pits A and B, this pit necessitated a major deviation in some equipment because of its size and required spoil material for backfilling of the highwall; however, deployment of equipment and movement of material was essentially the same. The average overall highwall height is approximately 138 feet; the average pit width is approximately 140 feet with an average bench width of approximately 30 feet. The pit length is approximately 800 feet. Each linear foot of highwall required approximately 352 cubic yards of spoil material to backfill to the top of the bench. As in pits A and B, it was assumed no blasting of the highwall would take place. The average spoil relocation distance was 292 feet. Results are shown in table 5.

TABLE 5. - Summary of results for pits A, B, and C

Pit and primary recontouring equipment	Conventional dozers		Dragline on spoil		Dragline on highwall	
	Cost/ft highwall	Change, pct	Cost/ft highwall	Change, pct	Cost/ft highwall	Change, pct
Pit A: Small dragline with conventional 11-yd^3 bucket and 20-yd^3 crescent bucket; conventional dozers	$144	0	$108	-25	$100	-31
Pit B: Small dragline with conventional 11-yd^3 bucket and 20-yd^3 crescent bucket; conventional dozers	144	0	-120	-17	104	-28
Pit C: Dragline with conventional 35-yd^3 bucket and 60-yd^3 crescent bucket; conventional dozers	299	0	151	-49	108	-64

NOTE.--Change percent is the difference from operation with conventional dozers.

Summary

It is apparent, from at least this study, that the potential exists for gaining some significant cost savings in backfilling highwalls by using draglines either sidecasting or dragging material from atop the highwall. The next step is to conduct a comparative field evaluation study of the three systems at an active mine site. This should provide reliable data for mine operators to assess their backfilling requirements in modified area mining employing these combinations.

CONCLUSION

The primary approach Bureau research has taken to date in its efforts to reduce costs for recontouring mined land is to develop relatively inexpensive attachments for existing machines that will convert them to high-volume, low-cost tools. Emphasis is on the development of specialized tools to assist the operator until such time as (1) the industry develops systems that do not pile the spoil, or (2) enough surface coal mines increase production and their mined acreage to warrant using large expensive land-leveling equipment solely for recontouring mined land.

BIBLIOGRAPHY

1. Caterpillar Performance Handbook. Caterpillar Tractor Co., Peoria, Ill., 9th ed., October 1978, pp. 4-17 through 4-26.

2. Goris, J. M., and H. S. Benson. Maximizing Dozer Work Rate With Draft Power Sensing Systems. Pres. at Earthmoving Industry Conference, Peoria, Ill., Apr. 23-25, 1979, SAE Paper 790510, 11 pp.

3. Howland, J. W. Application of High Volume Earthmoving Methods to the Reclamation of Area Mined Spoil Banks. Final Report to U.S. Bureau of Mines, Contract No. H0252012, February 1978.

Design Considerations of Terrace-Pit Mining

by Gregory G. Miller
Mechanical Engineer
Spokane Research Center
Bureau of Mines, Spokane, Washington

INTRODUCTION

A dragline is the least expensive method to strip overburden where simple digging and casting are used. However, deep overburden, dipping seams, faulted seams, multiple seams, thin interburden, and toxic materials make dragline mining procedures very complex, and in any of these situations it is questionable whether dragline stripping is the best mining method. Other mining equipment such as shovels, bucket wheel excavators, scrapers, and backhoes may be better. Such equipment is best used where benches or terraces must be constructed to reach a deep coal seam. Figure 1 shows a typical shovel-truck equipment system terrace-pit-mining a single coal seam. Here four terraces are used to mine the overburden and one terrace to dig the coal. The benches are level and trucks haul the spoil around the pit to be back-dumped. In this way, the pit is continuously backfilled and reclaimed as the mine progresses.

RECTANGULAR PROPERTY SHAPE

If the terrain is flat and the coal seam has practically no dip, a long and narrow, rectangular property can be mined in two panels with one turnaround. The turnaround allows a better face length and places the second panel's final void close to the first panel's box-cut stockpile. The final void can then be completely backfilled with the box-cut material. In a rectangular mine, the faces should be advanced in parallel. If the mine needs to produce 20 million tons of coal per year for 40 years, a 15-square-mile property is needed with a 50-foot coal thickness. If the property is two sections wide and eight long, the above two-panel technique can be used. But if it is four sections square, four panels may be necessary to keep an optimal face length and still place the final void near the original box-cut stockpile.

CIRCULAR PROPERTY SHAPE

In the Western United States, it is rare that mine property be other than rectangular. However, surface terrain and the coal itself may make the actual mine property shape circular. Here the optimum pit shape may be circular and instead of advancing the faces in parallel, the faces are rotated around the center of the circle, with the face length equal to its radius. Sectorial panels are used again, depending on how the terrain rises and falls, and concentric panels may also be used and advanced to or retreated from the center.

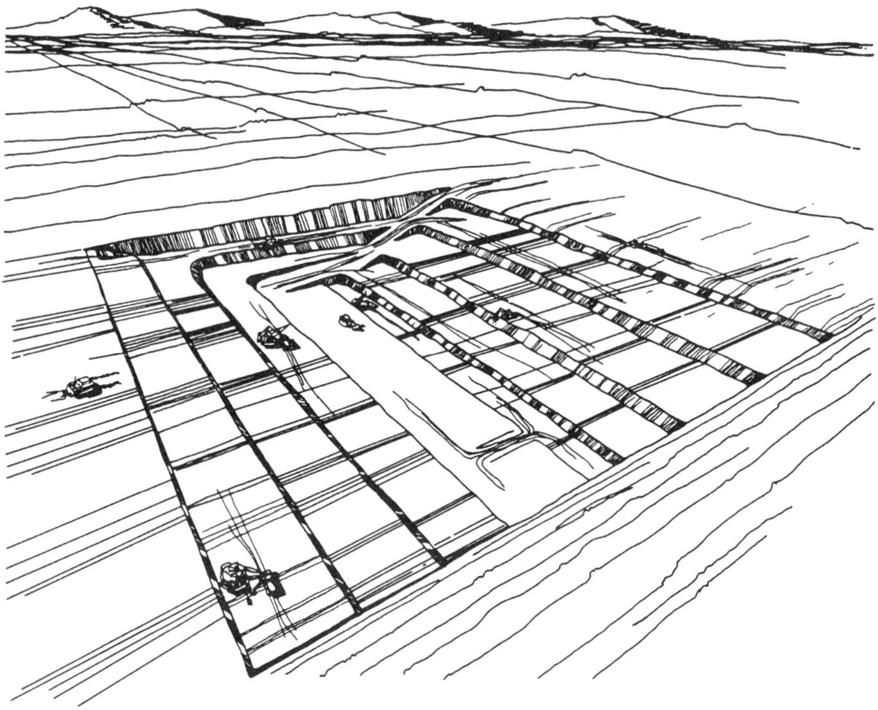

FIGURE 1. - Single-seam shovel-truck terrace-pit system.

DIRECTION OF ADVANCE

If the coal seam is lying horizontally or dips slightly (up to 3°), advance can be essentially in any direction as long as the terrain is flat. This condition is preferred in terrace-pit mining because coal production will remain uniform. However, geotechnical factors, ground water, or location of toxic zones in the overburden may dictate the direction of advance. In the case of slightly dipping seams such as those found in the Powder River Basin, if the coal does not cropout within the mine property boundary, an area should be chosen with the least amount of box-cut spoil to initiate mining. In cases of rising terrain and horizontal coal seams, it is usually preferred to advance perpendicular to the rising terrain if stability of the material permits, because of the reduced strip ratio in the first panel. If the mine property is rectangular but the seam dips about 3°, it is generally recommended that the mine advance along the strike. However, the mine can also proceed along the dip. The decision partly depends on how the property lies to the dip, and if it is long and narrow or nearly square. However, economics play a major role in deciding whether to advance downdip or along the strike. For a mine situation where the length of the mine property occurs along the strike, advancing the first panel along the strike is very economic for the first half of the mine life because of its low stripping ratio. However, there is a great danger that the coal will not be mined on the second panel because of its higher stripping ratio. Figure 2 shows the relation between the stripping ratio and a mine life of 40 years. During the first 20 years the first panel's stripping ratio average remains uniform; then as the mine turns to mine the second panel, the stripping ratio jumps up dramatically to a level to be maintained for the last 20 years. If a slot is not left between the two panels, a portion of the first panel must be rehandled in mining the second panel. This raises the second panel's average stripping ratio.

If this same property is mined along the dip, four panels will be necessary. Figure 2 shows how the stripping ratio will fluctuate over the life of the mine. At a 40-year site, the first panel will proceed downdip, the second panel updip, the third downdip, and the fourth updip. If slots are not left between adjacent panels, the average stripping ratio will be the same as for the first panel. If slots are left, the last three panels will have a reduced stripping ratio. When mining downdip, problems result because there is less room to spoil the overburden into the previous cut. When retreating updip, the trucks may encounter haulage problems, but the waste is more stable, the ground water is drained from coal and overburden, and there is more space available for spoil.

Generally, a terrace mine can operate on a 3° to 9° dip either along the strike or along the dip. However, when the coal dips more than 9°, the pit floor becomes too steep for truck haulage. Terrace-pit mining can still proceed by two methods. One method is to mine neither along the strike nor along the dip, but at an angle to them. This tends to reduce the dip to a somewhat smaller angle. In the other method, the terraces are kept horizontal, and the coal is mined selectively with hydraulic shovels as the coal crops out in the bench.

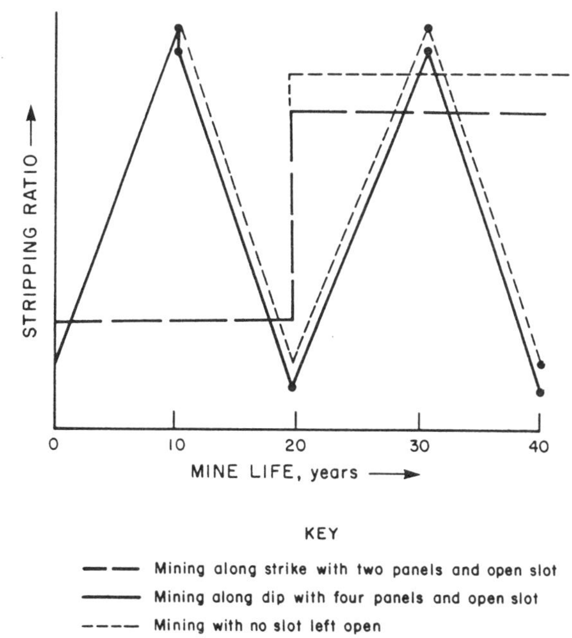

FIGURE 2. - Change in the stripping ratio over the life of a rectangular mine property with a dipping coal seam subcropping along its long side.

NUMBER OF SEAMS

The efficiency of coal mining depends on the thickness of the overburden, interburden, and coal seams. Thick coal seams and thin waste are preferred in all mining situations because of the low stripping ratio. However, multiple thin seams may be economic to mine even down to several hundred feet if the cumulative strip ratio is low and the mine is properly designed.

GROUND WATER

A small flow of ground water from interrupted aquifers can be allowed to drain into sumps cut into the pit floor and into the benches. The water can then be pumped out for uses such as dust control. If an aquifer of significant size will be interrupted, the mine must be properly designed to handle problems associated with wet overburden. Although this is not particularly a problem in the Powder River Basin or the Four Corners region, the Texas lignite region has areas of high flow rates and in some cases artesian pressures. Because artesian pressure in the underburden can cause floor heave and flood the pit, pumping wells should be placed in and around the pit to lower the water level.

GEOTECHNICAL DATA

Considerable premine research should be conducted to supply to the mine-design engineers critical geotechnical data on the properties of the site. This data will be used to determine the excavation equipment necessary and the layout of the mine that will prevent slope stability problems. Several operators have had to redesign their mines at a considerable financial loss because the excavator was unable to dig the overburden. There are cases where the excavator became spoilbound because the swell factor was not properly determined and cases where the stability of the waste and overburden was not considered important, resulting in slope failures.

TERRACE HEIGHTS

Terrace heights will vary in a multiple-seam mine because the interburden, overburden, and coal seam thicknesses change. The slope of the land or the pitch of the coal seam causes the terrace height to change as the mine progresses. Significant changes will require the addition or deletion of terraces and their side-pit haulroads. Generally, the height should be designed to be as near optimal for the loading equipment as possible.

TERRACE WIDTH

Any unnecessary bench width will increase haulage distance and costs. Therefore, all terraces, including the exposed coal seam, should be kept to a minimum width. The benches should be around 100 to 150 feet wide to permit equipment passing and turnaround, and to minimize congestion in the loading area. Generally, their width and haulroad width should be at least three times the maximum vehicle width, thereby enabling two haulage vehicles to pass maintenance equipment without interrupting production. However, the production rate or the excavator size may also be the determining factor of the final bench width.

SPECIAL TERRACES

The geological and physical characteristics of the various overburden strata encountered may require different equipment combinations or separate terraces. For example, unconsolidated overburden and easily fragmented rock, like certain shales, may be most economically excavated by scrapers, while other material, like sandstone, might be best excavated by a shovel-truck team. In addition, if a toxic layer of overburden exists in sufficient quantity, a special terrace should be included in order that the material can be excavated and deposited in the most beneficial location.

SPOIL TERRACES

Spoil terraces should be kept on the same level as the excavation terrace so that negative grades will not be encountered. If special toxic material terraces are used in excavation, a counterpart spoil terrace will be needed. This may be at a different level than its excavation terrace, thereby encountering adverse grades and reduced production. This toxic spoil terrace should be located at a level in the spoil profile to sufficiently isolate the material from contact with the environment.

FACE LENGTH

Determining the optimum face length is difficult because it depends on many variables. It is usually done in the detailed planning stage of an actual mine. Often property boundaries do not allow the use of the optimum face length, so a compromise is necessary. A whole number of panels must be planned to fit into the property, and an even number of panels is preferred because it positions the final voids near the original box-cut stockpile. In a case of an irregular property boundary that intersects the faces, a variable face length could be used to uncover all the coal. However, once mining is initiated, a significant deviation from the planned face length could require an extensive redesign of the mine.

Long face lengths are not desirable because of the numerous excavators required. In addition, Federal and State regulations may prohibit long faces because they represent an extensive source of interruption of the environment. The rate of face advance also may be less than backfilling requirements call for. However, a large production rate from a single pit may dictate the use of a long face. In such cases, coal and overburden haulage through the middle of the pit may be desirable. On the other hand, a short face would dictate a series of narrow panels. This is not desirable because the length of coal haulage roads would be continually fluctuating. Short faces require wide benches and a longer pit to maintain production levels.

METHODS OF TURNING

When a panel has advanced to the end of the property, the mine must execute a 180° turn into the next panel. This can be done three ways. First, the turn can be performed with two 90° turns. In this method, the technique of advancing the faces in parallel is maintained. When the mine has advanced to the end of the panel, the shovels are turned 90° and begin excavating the next panel. This short turning panel is advanced to the full width of the new panel while backfilling the void in the first panel; then the excavators are turned 90° again and begin retreating that panel, normally to its full length. Second, a 180° turn can be executed by rotating the pit around a point common to both panels with the radius of the swing equal to the face length. This method is similar to that used in mining properties that are nearly circular. Sectorial panels are used, and some corner coal may be lost. Third, another turn may be performed if new equipment will be purchased at the end of 20 years. Here a box-cut can be developed in the next panel before the first panel is complete. When the first panel finishes, it can backfill its void with the second panel's box-cut material. These three turning methods are quite complicated to execute and will require constant pit monitoring.

SPOIL SLOTS

When two or more panels are required for the mine property, it is beneficial not to backfill a wedge-shaped portion of the pit immediately adjacent to the next panel. This creates an open area, called a slot. By not backfilling this slotted area, rehandling of the previous panel's spoil to expose all coal is not required upon turning and retreating along the next panel. Generally, there will be spoil room available for constructing slots every panel. This results in each panel's waste being shifted over the width of the slot when backfilling takes place. The creation of a slot in the first panel may require overdeepening of its backfill area or stockpiling to dispose of the excess material caused by leaving the slot open. Savings in rehandling in subsequent panels should offset these costs. Two-panel property requires one slot whereas four-panel property requires three slots.

In two-panel property with a 40-year life, portions of the slot could be open up to 40 years. Certain regulatory permits may be required for slots because of this long-term exposure. Slope reduction and revegetation of the slot highwalls may be necessary, or slots may be banned in favor of complete backfilling to the original contour.

COAL HAULAGE

The coal haulage road can be through the middle of the pit or along the edge of the pit where the panels meet. The overburden and coal haulage costs will dictate its location. For a shovel-truck terrace-pit mine, if the face is shorter than 4,700 feet, a side panel coal haulroad is preferred because this layout allows higher average truck speeds and lower costs per ton-mile. In this situation the slot between panels can be used for coal haulage, or the coal can be taken directly in-pit to the surface and hauled to the tipple. If the slot is used, it can also become the coal haulroad in the next retreating panel.

When the face length for a shovel-truck mine exceeds 4,700 feet, it is preferable to have the coal haulroad through the middle of the pit because overburden and coal haulage distance become excessive. A coal haulroad slot can be left in the middle of the panel, but the use of two slots, the slot between adjacent panels and a coal haulage slot, may not be feasible.

Figure 3 shows coal being hauled directly to the surface. Generally coal haulage through the side slot is preferred, because coal can be hauled up a shallow incline to the surface, rather than up steep in-pit haulroads. The spoil slot or wedge should be wide enough to permit two coal haulage trucks and a road grader to pass simultaneously. Because of this required road width, the advantages of reduced rehandle and the low grade of the haulroad may not be fully realized, because the slot must be either partially backfilled or increased in overall size to obtain a wide haulroad. This increases spoil rehandling. In addition, the slot slope angles may need to be reduced for long-term stability.

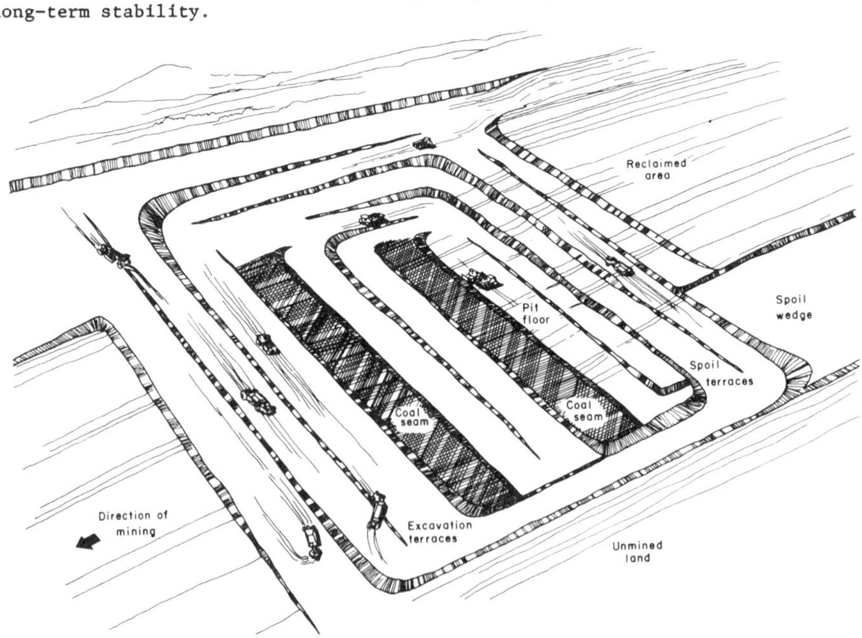

FIGURE 3. - Multiseam scraper terrace-pit system.

OVERBURDEN HAULAGE

Overburden is hauled to one side of the pit only to avoid interference with the coal haulroad; for example, when a side-pit coal haulage slot is used. Overburden haulage on the same side as coal haulage would require steep ramps down into the pit and back up to the spoil terraces. These roads are hard to maintain, lower production because of adverse grades, and are had on equipment. Overburden haulage from one side of the pit, opposite coal haulage, is preferred because level roads can be maintained. If midpanel coal haulage is used and no slot is left, coal haulage can be along both pit sides and through the middle. Terrace-pit mining requires constant overburden haulroad construction and backfilling as the mine progresses.

EXCAVATING EQUIPMENT

Bucket-wheel excavators (BWE's) appear to have potential in the Texas lignite region and possibly in North Dakota where the overburden is unconsolidated. They cannot be used effectively to dig material that needs to be blasted. They can follow the lignite on dips up to 3°, but for steeper dips, horizontal benches should be maintained. In this latter case, the lignite is selectively mined where the dipping coal crosses the highwall. In fact, a BWE can be quite effective where selective mining is necessary. A BWE can be very efficient in rebuilding the original stratigraphic sequence of the overburden. Generally, however, BWE's are relatively inflexible. They are designed for specific uses, and no significant deviation from planned mining techniques is possible. Therefore, if a BWE is to be used where flexibility may be required, the machine must be specially designed and set up.

Competent overburden should be mined with shovels. Truck haulage can permit very selective rebuilding of the spoil and produce high spoil compaction. Trucks can operate economically in a one-way-haul distance of several miles within the range needed in terrace-pit mining.

Scrapers can be used to terrace-mine unconsolidated overburden.

Topsoil Rock Removal

by

T.M. Brady
General Engineer
Spokane Research Center
Bureau of Mines, Spokane, Washington

W.W. Kaufman
Mining Engineer
Energy Division
Skelly & Loy, Harrisburg, Pennsylvania

D.N. Reynolds
Geologist
Energy Division
Skelly & Loy, Harrisburg, Pennsylvania

INTRODUCTION

An important element of surface mining reclamation is topsoil since its quality is critical for obtaining successful revegetation of reclaimed land.

In most surface mines, scrapers are used for topsoil removal, stockpiling, and redistribution. After topsoil redistribution, the soil must be prepared for revegetation or use as farmland. If the disturbed soil contains too many large rocks, farmland will not be as productive as originally. Farming in areas where no care has been taken to remove large rocks may result in damaged farm equipment and less water permeability, and in smaller, less desirable harvests.

To ensure that the soil maintains its productivity and permeability, many States have enacted laws and policies to control the size and quantity of rock in the topsoil. On the Federal level, the Office of Surface Mining (OSM) has established strict regulations governing revegetation of the reclaimed areas. Removing rocks from the topsoil is a fairly easy task; however, maintaining productivity levels is much more complex, requiring years of monitoring.

After topsoil redistribution, the soil must be prepared for revegetation. If there are too many large rocks in the topsoil, graders may be used to eliminate them prior to revegetation. Since this method does not penetrate the soil, it is often inadequate and does not comply with many regulations. When deeper penetration is desired, rock rakes attached to dozers are employed. Rakes can eliminate large rocks to depths of 18 inches and deeper, but do not remove the smaller rocks less than 6 inches in size. While the spacing of tines on rock rakes is easily adjustable, if the tines are spaced too close, the soil will be pushed in front of the dozer and rocks will not be eliminated.

Since graders and rock rakes are not totally effective for rock removal, some mining companies have elected to use rock pickers. These pickers, which are not classified as mining equipment, are generally sold as farming equipment, and to date their use by the mining industry has been limited.

Why aren't rock pickers popular within the mining industry? Often the topsoil is so plentiful and free of large rocks that there is simply no need for them. Also, since rock pickers are generally associated with the farming industry, mining companies simply may not have explored their potential for rock removal.

The following discussion outlines the extent of the rock problem throughout the United States, discusses in detail the legal requirements that affect the reclamation process, analyzes the state of the art of current rock pickers including their physical limitations, presents an analysis of rock pickers applied to mine reclamation, surveys environmental effects that may take place, and details the benefits that can result from introducing rock pickers into the mining industry.

REGULATORY ROCK REMOVAL REQUIREMENTS

For years surface mining was conducted without concern for future land uses. Consequently, the overburden strata were overturned, burying the original topsoil, and abundant rock fragments were introduced into the soils. The resultant deterioration of land utility has prompted the promulgation of regulations to insure every effort is made to restore mined areas to their original conditions.

State Requirements

As States became increasingly concerned with the adverse effects of surface mining, they began enacting and enforcing more stringent regulations governing mine operations. Among the adverse effects of surface mining that were identified was "rock pollution." The contamination of the soils with an unnaturally high percentage of rocks and the prevalence of large rocks, at or near the surface, hindered revegetation efforts, especially where farming was attempted as a postmining land use. The current tendency of regulatory legislation is to aim toward requiring mine operators to return affected land to its original condition. Thus, the States having the most rock-free soil would need the most specific regulations on rock content limitations for the topsoil layer. However, with proper handling these regulations should not impose a hardship on the reclamation efforts.

Table 1 provides a comparison of existing specific restrictions, implied regulations, and pending surface mine amendments. The States that are shown as having specified regulations are those defining limitations of rock size and/or percent in the topsoil layers of surface coal mine reclamation. Implied regulations are indefinite limitations pertaining to rock content in the topsoil layer. Such limitations are worded to the effect that "large rocks" will not hinder postmining land uses. The strictness of pending legislation was not discernible; however, almost all pending legislation was expected to reflect the intent of the new regulations promulgated by OSM.

Among the States that are monitoring topsoil rock content, the most strict regulations are found in Illinois, Indiana, Kansas, Oklahoma, and Virginia.

Federal Requirements

OSM's permanent regulatory program does not have any regulations pertaining to the size or percent by volume of rocks in a reclaimed area. However, OSM does indirectly prevent an increase in rock content by requiring that each soil horizon (distinct soil layer) be removed separately, stored separately, and replaced as it was originally. This is aimed at eliminating the possibilities of mixing sublayer rock fragments into the upper soils and of overturning the strata. When the present study was funded, the interim regulations of OSM required less than 10 percent rock coarser than 3 inches, and permeability could not be less than 0.06 inch per hour in the top 20 inches.

TABLE 1

STATE REGULATIONS ON TOPSOIL ROCK CONTENT

STATE	SPECIFIED STATE REGULATIONS	IMPLIED STATE REGULATIONS	PENDING LEGISLATION*	CURRENT POLICY
Alabama	○	●	○	Replace as was
Alaska	○	○	○	None
Arizona	○	○	●*	None
Arkansas	○	○	●*	None
California	○	○	●*	None
Colorado	○	○	●*	None
Connecticut	○	○	○	None
Delaware	○	○	○	None
Florida	○	○	○	None
Georgia	○	○	○	None
Hawaii	○	○	○	None
Idaho	○	●	○	Replace as was
Illinois	●	○	●*	No 10" rocks in top 4 feet
Indiana	●	○	○	Bury 6" rock at least 6" deep
Iowa	●	●	○	Implemented interim OSM regulations 8/78
Kansas	●	○	●	Bury 6" rocks at least 6" deep
Kentucky	○	○	●*	Replace as was
Louisiana	○	○	●*	Replace as was
Maine	○	○	○	None
Maryland	○	●	●*	Replace as was
Massachusetts	○	○	○	None
Michigan	○	○	●*	Replace as was
Minnesota	○	○	●	Requires removal of 6" rocks from surface
Mississippi	○	○	●*	None
Missouri	○	○	●*	Requires removal of excess rock
Montana	○	○	●*	Replace as was
Nebraska	○	○	○	None
Nevada	○	○	○	None
New Hampshire	○	○	○	None
New Jersey	○	○	○	None
New Mexico	○	○	●*	None

(continued)

TABLE 1 (Cont'd.)

STATE	SPECIFIED STATE REGULATIONS	IMPLIED STATE REGULATIONS	PENDING LEGISLATION*	CURRENT POLICY
New York	○	○	○	None
North Carolina	○	○	○	Requires removal from tillable lands
North Dakota	○	●	●	Requires removal from tillable lands
Ohio	○	●	○	Requires removal from tillable lands
Oklahoma	●	○	●	Adopted interim OSM regulations
Oregon	○	○	●	None
Pennsylvania	○	○	●*	Replace as was
Rhode Island	○	○	○	None
South Carolina	○	○	○	None
South Dakota	○	○	○	Replace as was
Tennessee	○	●	●*	Requires removal of 6" rocks from surface
Texas	○	○	●*	Replace as was
Utah	○	○	●*	None
Vermont	○	○	○	None
Virginia	●	○	○	Replace as was
Washington	○	○	○	Replace as was
West Virginia	○	●	●*	Must remove exposed boulders from surface
Wisconsin	○	○	○	None
Wyoming	○	●	●*	Replace as was

LEGEND: ● Yes ○ No

* Will Conform to Office of Surface Mining (OSM) Regulations.

STATE OF THE ART OF ROCK PICKERS

Characteristic Forms of Rock Pickers

Many of today's rock removal devices have evolved from the potato farming business. As a result, most of the rock pickers are designed to pick rocks above 2 inches in minimum diameter. These rocks are removed prior to potato planting in order to reduce the number of stones harvested along with potatoes.

Table 2 lists the general specifications of currently available rock pickers. In instances where companies manufacture more than one size of rock picker, only the largest models have been included in the table.

The mechanisms used in rock pickers to separate the soil and the rock can be classified into four categories: potato chain, rotating rake, rotating cage, and passive rake.

TABLE 2
ROCK PICKER SPECIFICATIONS

COMPANY	MODEL	HITCH TYPE	RAKE BAR	PENETRATION	MAXIMUM ROCK SIZE	MINIMUM ROCK SIZE	HOPPER CAPACITY Cu. Yd.
(2) Armor Metal Products	Series E	Standard P.T.O.	Rotating	2 to 4"	200 lbs.	1 1/2"	1.2
(4) Bestland	876	Hydraulic	Fixed	2 to 3"	1000 lbs	1 3/4"	1.2
(2) Degelman Industries, Ltd.	R-570-S	P.T.O.-Pull	Rotating	2 to 3"	Not Available	4"	1.0
(3) Harley Rock Picker Co.	A-High Lift	P.T.O.	Rotating	2 to 3"	20"	2"	1.1
(4) Imperial Welding & Machine, Ltd.	Mel Cam 510	Hydraulic	Fixed	0	44"	3"	0.9
(2) Leon's Manufacturing Co., Ltd.	A-7500	Hydraulic	Rotating	2"	20"	2"	2.3
(1) Lockwood Corp.	Hydra-lift L06630-00207	P.T.O. 540	Vibrating	8"	200 lbs.	3/4" ϕ	1.25
(2) McConnell Manufacturing Co. Inc.	Bin Type	P.T.O. Standard	Fixed	6"	10"	2"	1.5
(4) Melroe Division	8-D HiBoy	Hydraulic	Fixed	4"	200 lbs.	2"	0.9
(2) N. P. Nelson	40	P.T.O. Standard	Rotating	2 to 5"	9"	1 1/2"	.75
(1) Pratt Farms	Model B	P.T.O. Standard	Fixed	0 to 8"	24"	2"	5.0
(2) Rockland Manufacturing	Rotoveyer	3 Point Hitch	Rotary	12"	10"	2"	8.0
(2) Rock-O-Matic, Ltd.	8DW5	P.T.O.	Rotary	0	24"	2"	1.6
(2) Schulte	RS-H High Lift	P.T.O. or Hydraulic	Rotary	2 to 3"	28"	2"	2.4
(1) Steinman Manufacturing, Inc.	#480	Hydraulic	Fixed	2 to 4"	10"	2"	1.2
(1) Thomas Equipment, Ltd.	Hopper Type	Drawbar	Fixed	to 12"	Not Available	1 1/2"	2.0
(4) West Go	A114	Drawbar	Rotary	Not Available	24"	2 1/2"	1.1
(2) Wisconsin Rock Harvester	HD 58	P.T.O.	Rotary	2 to 4"	500 lbs.	1"	1.6

Legend: (1) Potato Chain (2) Rotating Rake (3) Rotating Cage (4) Passive Rake

(continued)

TABLE 2 (Cont'd.)

TINE ADJUSTMENT	HORSE POWER REQUIRED	SPEED OF OPERATION	WEIGHT	DUMPING HEIGHT	LENGTH	SWATH	1979 PRICE (F.O.B. Factory)
No	40	1 to 5 mph	4700#	7' 6"	13' 0"	5' 6"	$ 5,220
Yes	35	0 to 4 mph	3500#	6' 4"	10' 0"	8' 0"	$ 2,895
Not Available	70	0 to 5 mph	2500#	5' 0"	13' 6"	5' 0"	$ 5,546
Yes	50	2 to 4 mph	6810#	6' 0"	33' 0"	8' 0"	$11,260
No	Not Available	Not Available	1500#	13' 6"	12' 0"	5' 0"	$ 2,200
No	60	2 to 3 mph	6000#	2' 8"	16' 0"	4' 10"	$ 8,135
No	60-80	4 to 5 mph	4250#	6' 3"	21' 4"	8' 0"	$ 6,837
No	45	0 to 4 mph	3830#	6' 0"	16' 0"	8' 0"	$ 5,978
No	45	0 to 4 mph	2800#	7' 6"	17' 0"	6' 0"	$ 2,565
Yes	7	0 to 6 mph	1500#	4' 0"	10' 0"	6' 0"	$ 3,340
No	250	0 to 6 mph	3750#	12' 0"	30' 0"	10' 7"	$25,000
No	125	2 to 4 mph	13,000#	10' 0"	34' 0"	8' 0"	$31,400
No	Not Available	3 to 4 mph	6900#	8' 0"	20' 0"	20' 0"	$ 9,897
No	50	2 to 3 mph	5000#	8' 6"	14' 4"	5' 0"	$ 7,600
No	30	> 5 mph	1350#	3' 4"	12' 0"	3' 6"	$ 1,625
No	45	0 to 8 mph	2800#	7' 0"	20' 6"	7' 7"	$ 6,082
No	>40	2 to 4 mph	2510#	6' 8"	14' 3"	8' 0"	$ 3,459
No	60	1 to 4 mph	4850#	8'-0"	15' 0"	5' 3"	$ 6,200

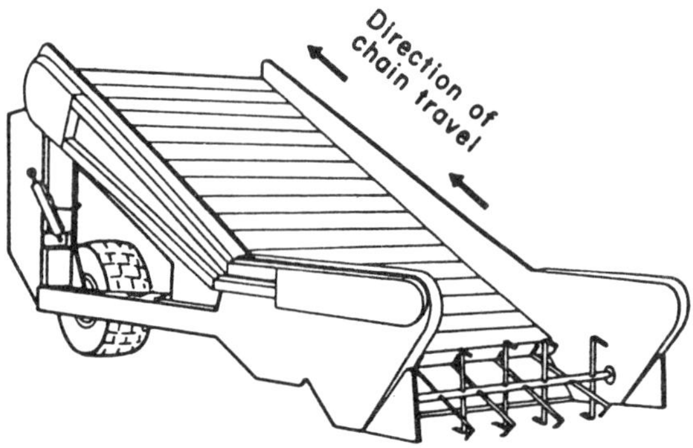

FIGURE 1. - Potato chain separation mode. *(Adapted from Rockland Product literature.)*

Potato-Chain Rock Picker

The potato chain (fig 1.) is the most common separation mechanism used to separate rocks from soils. These rock pickers are towed behind a tractor mounted on the draft arms of a three-point hitch. Power is obtained from the tractor's power takeoff (PTO) shaft. The rock picker consists of three basic components: the digging head, the conveying potato chain, and a storage hopper.

The digging head can take two forms. It is normally either a fixed blade or a rotating rake. The fixed-blade pickup method uses a steel blade to lift a layer of soil and rocks from the surface. Forward motion of the rock picker forces this material into the separation mechanism. The cutting depth is hydraulically variable on some models to permit adjustment while in operation. Soil conditions and operating speed affect the depth of penetration. In general, the deeper the cut, the slower the rock picker must travel to avoid overloading the separating mechanism. The fixed-blade pickup method is used by many models in conjunction with potato-chain, rotating-cage, or passive-rake separation modes.

Rotating-Rake Rock Picker

The rotating rake is generally best suited to surface and shallow rock picking. The rocks must be loose and free, or else the spring-loaded rake teeth will ride over them. As the rake turns, it lifts rocks up onto a screen which allows the soil to drop out while the rock is directed into a collection box. These machines require dry, vegetation-free soil since vegetation and damp soil bind the separation mechanism and make it inoperative.

Rotating-rake rock pickers operate on the principle of a rotating, spring-loaded picking reel passing over a heavy grill-bar-type apron (fig. 2). The depth at which the leading edge bar of the apron runs through the soil is controlled hydraulically, on most models, from the tractor. The action of the reel teeth loosens the rock, helps to break up clods, and rapidly moves the rock back into the hopper. Manufacturers claim that continuous picking operation even during tight turns is possible. They also claim a speed range from 1 to 5 mph, depending on soil conditions. Self-cleaning action and very little jamming should be attained using rake rock pickers. The design of the aprons is generally such that, when used for surface work on pasture or sod, the machine will take all the surface rock with little tearing of the sod.

Rotating-Cage Rock Picker

The rotating-cage rock picker (fig. 3) does not penetrate the soil; rocks must be windrowed before this machine is employed. A rotating shaft pushes the rocks onto a conveyor belt that transfers them into the rotating cage. As the cage turns, tumbling rocks break up dirt clods, liberating soil particles that can fall through the cage openings and return to the ground. The inclined cage directs the rocks through the length of the cage to the rear conveyor, which deposits them into a storage hopper or following trucks.

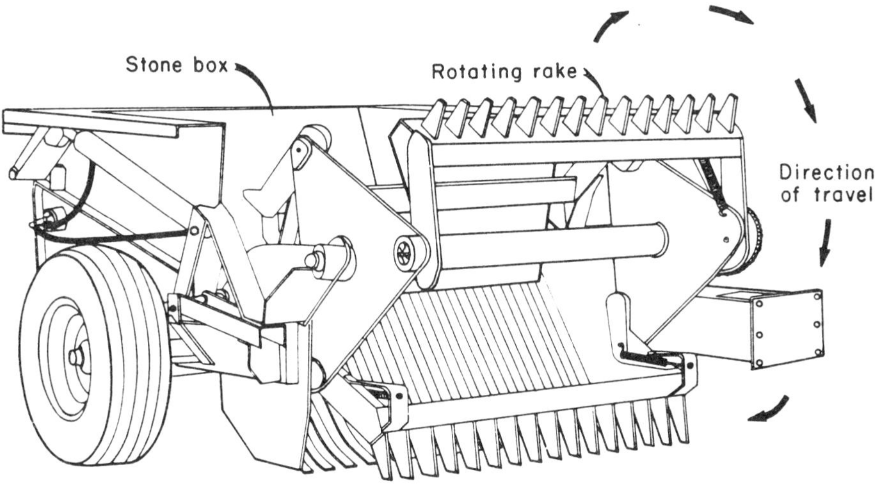

FIGURE 2. - Rotating-rake separation mode. *(Adapted from Degelman Product literature.)*

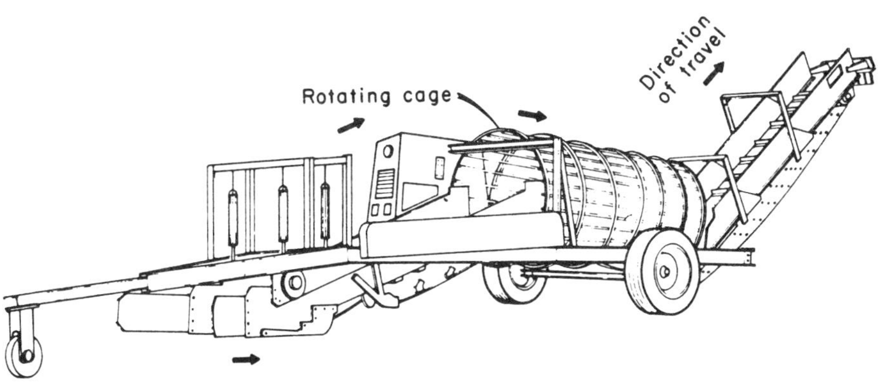

FIGURE 3. - Rotating-cage separation mode. *(Adapted from Harley Product literature.)*

For best performance by this type of rock picker, the rocks should not be mixed with excessive amounts of soil and vegetation. The separation area of this machine is the smallest of the four types. If plant material is encountered, it tends to clog the cage openings and reduce the available separation area even more.

Passive-Rake Rock Picker

Passive-rake rock pickers (fig. 4) are the least complicated of all the available rock pickers. Most passive-rake rock pickers have a fixed rake with adjustable tines. They are not designed to penetrate the soil more than a few inches. As the passive rake is towed behind a tractor, the rocks are maintained on the L-shaped tines, while the fine particles fall through. When the tines are full of rocks, the collecting head is hydraulically lifted and dumped into the rear storage hopper. For effective operation, the soil must be loose and dry.

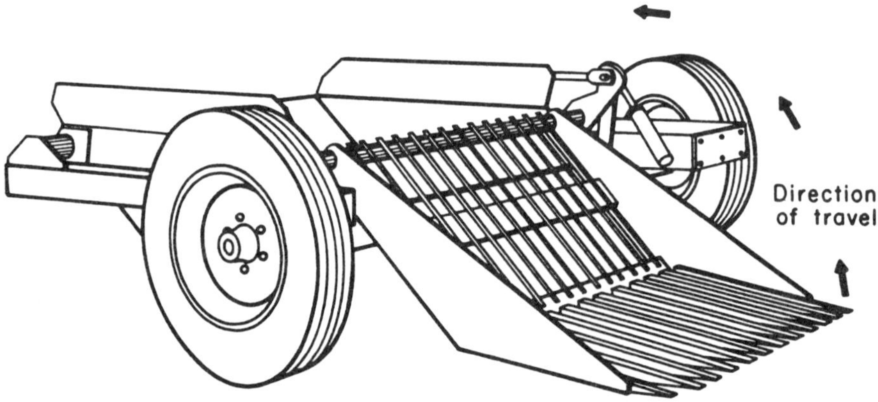

FIGURE 4. - Passive-rake separation mode. *(Adapted from Leon Product literature.)*

Rock Rakes

Another form of rock removal device is the rock rake. Table 3 lists the general specifications of some popular models. Rock rakes can be either dragged behind a tractor or pushed by a dozer. The dragged varieties are used for light-duty application and will not penetrate the surface. Dragged rakes only windrow surface rock and debris. Other methods must be used to pick up the windrowed material.

TABLE 3

ROCK RAKE SPECIFICATION COMPARISON TABLE

COMPANY	MODEL	HORSE POWER REQUIRED	OPERATING SPEED (MPH)	TINE SPAC (IN)	PENETRATION (INCHES)	WEIGHT (LBS)	HEIGHT	WIDTH	PRICE (1979 FOB FACTORY)
DRAG TYPE									
AUSTIN PRODUCTS INC.	LITTLE RHINO	55 HP	0–6	2	DRAGGED ON TOP OF EARTH	370	3'–4"	8'–0"	$825
BRILLION IRON WORKS	LR–8	30 HP	0–5	2		455	4'–6"	8'–0"	$854
DEGELMAN INDUSTRIES LTD.	ROCK RAKE W/BRUSH KIT	50 HP	3–5	2		3200	3'–8"	14'–0"	$4,860
YORK MODERN CORPORATION	R.E.	25 HP	2–6	2		425	3'–2"	8'–0"	$1,550
PUSH TYPE									
FLECO CORPORATION	9–S	USE WITH D–9 DOZER	VARIES WITH EQUIPMENT	12	21	10,520	5'–4"	13'–8"	$18,100
ROCKLAND	FF–3 FOR FIAT ALLIS	USE WITH FIAT– ALLIS DOZER		12	20	5,800	5'–8"	11'–6"	$8,710
ROCKLAND	RF–3 FOR CAT	FOR D–9H DOZER		11	20	7,500	5'–11"	11'–11"	$8,710

The push-type rock rakes are attached to the front of a dozer. They can be effectively used to remove large rocks and vegetation from the soil. Again, since the rocks are not picked up, they must be pushed into a pile or burial pit. The tines on the rake are generally about 12 inches long.

TOPSOIL ROCK CONTENT INVESTIGATIONS

To explore the feasibility of transferring topsoil rock removal technology in its present state from its agricultural origin into surface mine reclamation, nine representative mine sites were visited. Figure 5 shows the location of the sites visited. By sampling rock contents of soils before and after they were affected by mining, an indication of the need for incorporating rock removal into the reclamation plans could be ascertained. In addition, the capability of current rock-picker technology to function in mining environments was assessed. Each site chosen provided a unique combination of topsoil quality, topography, climate, and method of mining. Further considerations were made of the States that had rock limitation regulations. In addition, mines using rock removal devices to improve reclamation were sought to assess successes and problems.

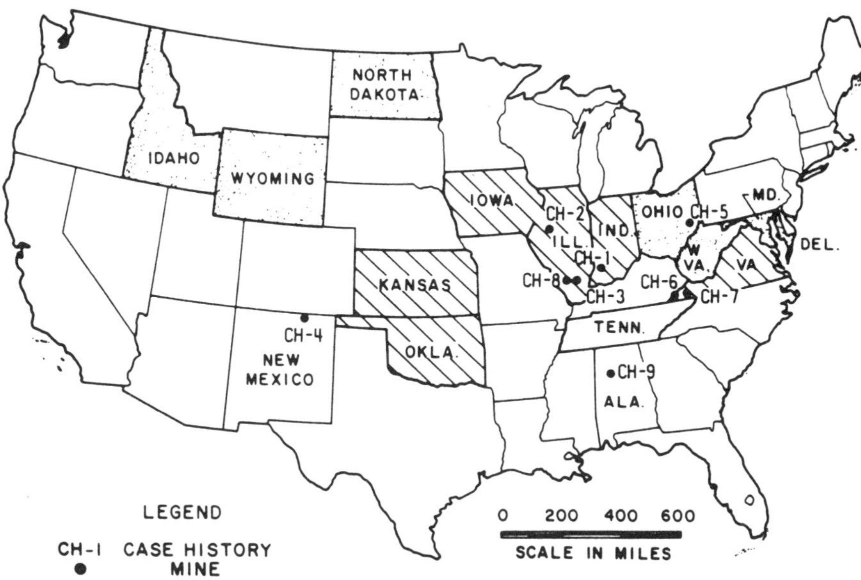

FIGURE 5. - Location map—mine visitation sites.

Each mine visit consisted of a discussion with mine operators of mining methods, availability of topsoil, overburden characteristics, reclamation equipment, amount of land reclaimed annually, average amount of time spent on reclamation each year, and any particular reclamation problems experienced at the mine. The operator would then provide a tour of the operation for a first-hand view of mining practices and rock content evaluation of the soils.

Rock content was evaluated by digging small test holes and then separating and weighing the rock and soil components. Assuming a maximum mechanical picking depth (for optimum rock removal operation) of 12 inches, the samples were examined in two stages. The percent by volume of rock to soil was calculated at 0 to 6 inches and for 6 to 12 inches. Segregation of rocks from soil was made by considering all material that passes through a 1-inch sieve to be soil and all that is retained to be rock. Those rocks in excess of 3 inches in their smallest dimension were weighed separately; however, in most cases this size fraction was a small portion of the total rock content. Test results for each mine are summarized in table 4.

TABLE 4

TEST HOLE RESULTS

MINE SITE	ORIGINAL ROCK CONTENT (%)			RECLAIMED ROCK CONTENT (%)		
	HOLE DEPTH		HOLE AVG.	HOLE DEPTH		HOLE AVG.
	0-6"	6-12" (1)		0-6"	6-12" (1)	
CH-1*	0	0	0	0	0	0
CH-2*	0	0	0	0	0	0
CH-3*	0	0	0	0	0	0
CH-4	5	10	7.5	5	6	5.5
CH-5	0	0	0	10 (2)	10 (2)	10 (2)
CH-6	7	13	10	9	14	11.5
CH-7	7	12.5	9.75	9	14	11.5
CH-8*	0	0	0	0	0	0
CH-9	7	10	8.5	5	7	6

* DENOTES MINE SITES IN PRIME FARMLANDS

(1) HOLE DEPTHS VARIED BETWEEN 8-12" DUE TO DIGGING DIFFICULTIES AND TOPSOIL THICKNESS

(2) PRIOR TO ROCK PICKING STEP: AFTER PICKING IT IS RETURNED TO 0 FOR ALL CATEGORIES

Sensitivity Analysis

To determine the effects of incorporating a rock picker into the reclamation plans, a sensitivity analysis was performed on two of the mines visited. While keeping certain criteria constant and varying those items that would most likely fluctuate, sensitivity graphs were constructed. Criteria held constant throughout this analysis were

1. 6-inch effective penetration depth of rock picker.

2. 8-foot effective picking width.

3. 5-mile-per-hour ground speed during dumping cycle.

4. 60-second dump cycle.

5. $30,000 cost for rock picker and tractor.

6. 7-1/4-hour shifts.

7. UMW operator for tractor (Grade 1).

8. 90-percent availability of rock picker and tractor.

Items considered as variable were

1. Operating speed--1 to 5 miles per hour.
2. Hopper capacity--4 to 16 cubic yards.
3. Haul distance to dump--0 to 5,000 feet.
4. Hopper versus direct loading.

Each of the four variable items was graphed as a function of the percent of rock contained in the topsoil. These analyses of the relationship between the rock content of the topsoil and the horsepower required to remove that rock (expressed in horsepower-hours per acre) provide a visual display of the sensitivity of the reclamation effort to these variables.

Horsepower-hour per acre units were chosen to portray the incremental increases in the rock removal effort as the rock content increased in order to provide a relationship of universal and timeless application. These units reveal the physical energy required to accomplish given tasks by summing the products of the total available potential flywheel horsepower outputs of all contributing pieces of equipment multiplied by the total hours over which they were applied. This figure is, in turn, divided by the number of acres reclaimed. These units can be converted to cost per acre in a single step, at any future point in time, by simply multiplying times a conversion factor.

Projected Rock Picker Efforts

Table 5 presents the topsoil removal and reclamation efforts with projected rock picker effort for all nine mine sites visited. Effort data obtained from the sensitivity analysis were used to compute the projected rock picker efforts at the other seven sites visited. Also included in table 5 are topsoil removal and reclamation efforts per ton of coal with and without the addition of a rock picker.

TABLE 5

TOPSOIL REMOVAL AND RECLAMATION EFFORTS WITH PROJECTED ROCK PICKER EFFORTS

MINE SITE	EFFORT WITHOUT ROCK PICKER		PROJECTED EFFORT WITH ROCK PICKER	
	H.P.-HRS/ACRE	H.P.-HRS/TON	H.P.-HRS/ACRE	H.P.-HRS/TON
CH-1	44.900	5.0652	46.462	5.0668
CH-2	72.700	15.9942	74.262	15.9958
CH-3	35.100	9.3535	36.662	9.3558
CH-4	123.100	11.3315	124.662	11.3363
CH-5	7.500	3.6608	9.062	3.6761
CH-6	217.100	15.5063	218.662	15.5175
CH-7	25.700	11.4131	27.262	11.4247
CH-8	113.900	16.0300	115.462	16.0312
CH-9	148.300	30.2588	149.862	30.2695
AVERAGE	87.600	13.1793	89.151	13.1860

Table 6 translates effort into costs based on 35 cents per horsepower-hour.

TABLE 6
TOPSOIL REMOVAL AND RECLAMATION EFFORTS
WITH PROJECTED ROCK PICKER EFFORTS

MINE SITE	EFFORT WITHOUT ROCK PICKER	EFFORT WITH ROCK PICKER	PERCENT INCREASE
CH - 1	1.77/TON	1.77/TON	0
CH - 2	5.60/TON	5.60/TON	0
CH - 3	3.27/TON	3.27/TON	0
CH - 4	3.97/TON	3.97/TON	0
CH - 5	1.28/TON	1.29/TON	.78
CH - 6	5.43/TON	5.43/TON	0
CH - 7	3.99/TON	4.00/TON	.25
CH - 8	5.61/TON	5.61/TON	0
CH - 9	10.59/TON	10 59/TON	0
AVERAGE	4.61/TON	4.62/TON	.22

CONCLUSIONS AND ENVIRONMENTAL IMPACT

Rocks are a natural component of soil, yet at times they can be detrimental to revegetation efforts. Large rocks hinder attempts to seed an area. A prevalence of rock on or near the surface retards the establishment of vegetative cover by restricting root growth and limiting moisture distribution.

As a side effect of this impedance to vegetation, there will be an increase in the amount of sedimentation. Erosion, transportation, and deposition of soil particles impair environmental quality. The source area is robbed of nutrient-providing soil, the transporting water becomes turbid and affects the aquatic life of the receiving stream, and the deposition can result in a clogged stream channel, lake, or pond, or in suffocation of productive land.

There can also be harmful effects on the wildlife from erosion and sedimention in the form of destruction of their refuge areas. These are often damaged or destroyed without being noticed, since they are remote from the "affected area."

Past mining practices led to mixing of consolidated and unconsolidated strata within the same spoil pile. In the process, the strata were inverted, making it impossible to replace the topsoil in its original condition. To alleviate the destruction of the natural soil conditions, States began requiring separate handling of the topsoil.

The segregated removal, storage, and replacement of soil horizons during coal mining virtually eliminate the introduction of bedrock fragments into the reclaimed topsoil layer; thus the rock content should be unchanged and reclaimed areas should be of equal quality to their premining conditions. Recently, implementation of the Federal surface mine law forced the segregation of soil horizons to be implemented nationwide. This practice has had a noticeable effect in States without previous rock content restrictions, where an extreme decrease in the contamination of topsoils by rocks is apparent.

Although national implementation of rock pickers by the mining industry is neither necessary nor probable, for those instances when land use improvement is desired, a rock removal system may prove to be surprisingly cost effective. Of the currently available designs, the direct-acting rock pickers (figs. 1-3)--those that continuously loosen and sieve a layer of soil--perform better and more efficiently than cyclic-acting machines (fig. 4). Passive rakes often bounce off large rocks or force them deeper into the soil, and are thus less effective for rock removal. In addition, they are slower owing to the required stops for emptying the rake into the storage hopper and for dumping the storage hopper. These two nonproductive cycles of a three-cycle operation greatly limit productivity rates as compared to the continuous direct-acting types of rock pickers.

For the operator who plans to improve the land use from its premining status, the employment of a rock removal system is effective and relatively inexpensive. The economic impact on both the small and large operators is slight, as shown by the previous sensitivity graphs and unit cost estimates. In terms of the cost versus benefit, any area whose land use has been improved will be more valuable than it was. The cost of utilizing a rock removal system that is properly sized for the volume of rock removal necessary can be more than offset by the increased land value. Surface mine operators with a vested interest in the properties they mine could be shown that, for a nominal investment of capital and effort, a sizable valuation increase could be derived.

In areas where the former optimum land use was woodlands owing to an inherently excessive rock content, the use could conceivably be upgraded to grasslands. Likewise, in areas where cropland is scarce, a suitable substitute might be produced from previous grasslands or even woodlands. The most impressive cost-benefit results, however, would be in upgrading rocky wasteland into developable property. There are many minable areas with rocky or no topsoil and abundant bedrock outcrops. When these areas are mined, an increased amount of soil materials is produced. The utilization of a rock picker to decrease rock content would manufacture land of a quality that surpasses that of its surroundings for developmental purposes.

The financial advantage of a rock removal system begins even before the reclamation is completed on a severely rocky site. The removal of boulders and/or the reduction of the rock content can decrease the planting cost by permitting the mechanical preparation of a proper seedbed, mechanical tree planting, and improved revegetation success. Thus, the initial cost of planting, as well as the probability of partial or total replanting (to meet the bond release requirements), will be reduced.

In areas like the semiarid Southwest, where topsoil is of limited quantity and poor quality and is underlain by decomposed sandstones, the probability of successful establishment of vegetative cover may be improved by special rock reduction methods. The productivity of a reclaimed area might be improved by employing a machine that crushes the rock it engulfs and returns the reduced rock particles to the ground. If this type of rock reduction was applied to the upper subsoil layers, it would increase the amount of soil material, improve the availability of soluble minerals for plant nutrients, increase the moisture-retention capabilities of the root medium, and accelerate the development of soil horizons.

A possible disadvantage of using rock pickers in the reclamation of surface mines would occur when the removal of rocks accelerates the erosion of restored slopes. Removal of the resistance to erosion that rocks provide may allow excessive gullying to occur. This disadvantage of rock removal could be eliminated by frequent terraces, or contour plowing, raking, or disking. These practices retard surface flow velocities, reducing the water's erosive power.

Areas that are reclaimed for farming uses will experience the reoccurrence of rocks periodically, due to the combined effects of repetitive plowing patterns, siltation, and frost action. Plowing overturns the topsoil, and if the same plowing pattern is repeated year after year, the result is a lateral shift of soil, which is most pronounced on sloping ground. Siltation can carry away thin layers of topsoils, which decreases the rock cover. Finally, the frost action physically pushes rocks toward the surface. Separately these are minor factors, yet collectively they will allow rocks below the cleared level to migrate up into the tillage zone. This return of rocks will take varying amounts of time, depending on the depth of rock removal and the depth of frost penetration. Overall, the infiltration of rocks into the tillage area is extremely slow and is not a formidable problem; current restoration practices will provide a reoccurrence rate similar to that of adjacent, unmined lands of similar strata.

Transplanting Native Vegetation

by

Earl M. Frizzell
General Engineer
Spokane Research Center
Bureau of Mines, Spokane, Washington

James L. Smith
Professor of Agricultural Engineering
Colorado State University
Fort Collins, Colorado

Kent A. Crofts
Manager of Reclamation and Environment
Energy Fuels Corp.
Steamboat Springs, Colorado

INTRODUCTION

Transplanting native vegetation (trees, shrubs, and other species) is a viable alternative for reclamation of surface-coal-mined lands. Transplanted native vegetation both provides mature growing plants immediately to the reclaimed area and creates a nucleus from which seeds and other organisms can reinvade the area. Transplanting also provides immediate cover for wildlife, can be used for erosion control, and improves landscape aesthetics of recontoured areas. The transplant system described in this paper, consisting of a front-end loader and a transporter, is intended to provide equipment with which transplanting can be accomplished effectively and efficiently with minimal costs since current mechanized transplanting systems are expensive as compared with reseeding or planting of nursery stock.

Most of the current transplanting equipment has evolved from nursery and landscape applications and is not sufficiently durable for the operating environment and steep slopes encountered in surface mines. Further, with this equipment, it may be possible to move only one moderately sized plant at a time, and the removal operation generally destroys virtually all the vegetation around the plant. Other specialized pieces of equipment, although suitable for reclamation, have high acquisition costs, and since the reclamation period may be of short duration, restricted usage may make their purchase difficult to justify. However, if the reclamation equipment can serve as a backup to regular mine production equipment when not in reclamation service, economic analysis may show the investment justifiable. Examples of such equipment are tree spades and mechanical transplanters.

FRONT-END LOADER STUDY

The front-end loader (FEL) is a piece of mine production equipment that has been used, on a limited basis, for transplanting native vegetation in surface mine reclamation. However, the FEL is expensive, in terms of both capital and operating costs. Therefore, if an FEL is used in reclamation, its functions and capabilities should be optimized to achieve maximum productivity during the time it is available.

Conventional FEL buckets are designed to maximize the volume of material that can be loaded from piles or embankments. However, for transplanting, the FEL bucket should be designed to maximize the area of material (soil plus vegetation) that can be removed. Further, because of operating costs and production efficiency, the FEL should be used only for removing clumps of vegetation from undisturbed soil and transporting them short distances to the transplanting area.

Many mine reclamation operations require moving native vegetation several miles. Also, to maximize use of available vegetation, it may be necessary to remove clumps of vegetation and store them in a holding area for several weeks before they are moved to the transplant site. In these situations, the transplant system requires a transport capability in addition to the removal capability of the FEL bucket. The transporter should be capable of picking up several clumps of vegetation, moving them several miles rapidly, and placing them on the ground at the transplant site.

Transplant Bucket Operating Procedure

Conventional FEL buckets are designed to load from piles or embankments. Therefore, they are deep and have relatively high sides, and the horizontal distance that the bottom extends beyond the bucket top is relatively small. When this type of bucket is filled with clumps of vegetation, this configuration disturbs taller plants such as trees, and after transplanting they may be crooked.

The general design of the transplant bucket requires understanding the operating procedure used for transplanting. This procedure, which is markedly different from typical FEL operation, follows:

1. A vertical bank or step, approximately 2 feet high, is cut around the source of trees and shrubs. This must be done with the front-end loader and transplant bucket.

2. With the bucket bottom parallel to the ground surface and between 12 and 24 inches below the top of the vertical bank, the bucket is pushed into the bank.

3. When the bucket is full or driven to refusal, it is lifted vertically, thereby removing a clump of vegetation and soil.

4. The bucket is tilted back (towards the loader) about 10° to 15°, and the clump is transported by the loader to the transplant or storage area.

5. At the transplant area, the bucket is tilted forward (30° to 45° or more) and the loader is reversed, leaving the clump on the ground.

6. The loader is returned to the source of vegetation, and the cycle is repeated. Any access roads that may be required for transplanting are constructed prior to picking up the next clump, and any necessary surface leveling is done. All roadbuilding and leveling operations must be done with the reclamation bucket.

Bucket Design Criteria

The reclamation bucket design criteria follow logically from the above operating procedure. The criteria developed for the reclamation bucket are

1. The force required to push the reclamation bucket into the soil bank was determined. To accomplish this, a series of "plate penetration tests" was conducted. The plate penetration apparatus consisted of a flat plate and a cutting edge, representative of a segment of the bottom of the reclamation

bucket. The force required to push the plate into the soil bank was determined as a function of the distance the plate was pushed into the bank.

2. The tractive capability of a Terex* model 72-71A FEL was obtained from the loader manufacturer and standard construction equipment resource information. A similar procedure could be used for any size or make of FEL.

3. The size of the reclamation bucket was determined by matching the tractive capability of the FEL to the force required to push the bucket into the soil bank.

4. The force required to lift the clump of vegetation vertically from the soil bank was determined and compared with the lift capability of the FEL as given in the manufacturer's specifications.

5. The safe transport load for the reclamation bucket was determined from the manufacturer's specifications and standard construction equipment resource information.

6. Soil must be held on the plant roots with minimal disturbance during loading and unloading. In this regard, a suitable liner or coating should be provided on the surface of the reclamation bucket to reduce friction and decrease the tendency for soil to stick. The latter problem becomes particularly serious in cold weather because of the tendency of soil to freeze to the surface of the bucket. Once a soil clod sticks to the surface of the reclamation bucket, it disturbs the clump of vegetation until it is physically removed.

7. The force required to push a bucket into a soil bank can be reduced by mounting teeth on the cutting edge of the bucket. This was confirmed during the plate penetration tests. However, it is also necessary to confine soil in the clump of vegetation, particularly if the clump includes trees. Confining soil in the clump is a problem if the clump is transported in the bucket. Teeth allow soil to fall out of the clump and were, therefore, not included on the initial bucket.

Reclamation Bucket Specifications

1. Size will be 15 feet wide, 5 feet deep, and 3 feet high. This size is within the operating capability of the Terex model 72-71A loader.

2. Lucite will be used to line the inside bottom of the reclamation bucket. This material proved superior in field tests and will significantly reduce the sticking problem and problems of soil freezing to the bucket.

3. The front cutting edge of the reclamation bucket will be straight and balanced (the cutting edge will have an equal angle and length bevel on top and bottom). This will aid in controlling the depth of cutting and help confine soil in the clump.

TREE CLUMP TRANSPORTER

Analysis of the FEL vegetation clump transplanting operation indicated two factors as limiting the operation. The first was that the FEL was very efficient in removing clumps and planting them, but was relatively inefficient as a means of transportation, particularly if the distance was more than 300 yards. The second limitation was that some mines may not have an FEL available when needed for the transplanting operation. The FEL is a very common piece of production equipment in surface mines, but the total time an FEL was available for transplanting might be relatively small.

*Reference to specific equipment or trade names does not imply endorsement by the Bureau of Mines.

These problems can be solved by including a separate transporter in the transplant system. The vegetation clump transporter must pick up clumps of vegetation removed from undisturbed premined land by an FEL, transport the clumps, and place them in a transplant area. Proposed design criteria for the transporter are

1. Capable of picking up clumps of vegetation, each having an area of 75 square feet, and including small trees.

2. Capable of transporting approximately 200 to 300 square feet of vegetation clumps per load at speeds up to 20 miles per hour.

3. Capable of placing the vegetation clumps in the transplant area either as individual clumps or as groups of clumps up to the total area on the transport.

A used Hesston stack mover is presently being modified and tested to evaluate the functions required for the transporter. The machine was originally manufactured to load, transport, and unload haystacks. It consists of the follwoing mechanical elements:

1. A tilt-bed trailer having a surface area of approximately 170 square feet.

2. A chain conveying system capable of sliding the stack across the top of the trailer.

3. A pickup roller for separating the stack from the ground.

4. A set of tracks located at the rear of the trailer to pull the trailer under the stacks.

The rear tracks are synchronized with the conveyer so the stack is pulled up into the trailer at the same rate the tracks pull the trailer under the stack. When unloading, the stack is moved off the trailer at the same rate the rear tracks push the trailer forward. The maximum tilt angle of the trailer is approximately 15° (about 20 percent of the angle used between the FEL transplant bucket and the ground surface when unloading). When the clumps of vegetation are handled by the transporter, they are subjected to minimal bending, and the plant survival rate should be satisfactory.

The stack mover was converted to a vegetation transporter by the following modifications:

1. The subframe and tilt bed were strengthened for increased load-carrying capability.

2. The pickup roller was removed, and a short incline was located on the tail of the trailer to slide the vegetation clumps up on the conveyor chain.

3. The tilt-bed trailer was covered with Lucite to reduce friction and facilitate moving vegetation clumps across the top of the trailer.

4. A self-contained hydraulic system was mounted on the trailer, so only a prime mover (truck or farm tractor) will be required for operation of the transporter.

It should be noted that although the converted stack mover is narrower than the transplant bucket (9 feet compared to 15 feet), it will be possible to move clumps having the full 5-foot lateral dimension of the transplant bucket. This is the critical dimension, since it represents the length over which the clumps are subjected to bending as they are moved on and off the transporter.

In the final design concept, the vegetation clump transport system would be operated as follows:

1. An FEL would be used for a short time to remove a large number of clumps of vegetation from the undisturbed premined land and store them conveniently nearby. The FEL would then be free to return to mine production activities.

2. The transporter would pick up the stored clumps of vegetation, move them to the transplant area, and place them on the ground.

FIELD DEVELOPMENT TEST

During the growing season of 1980, a developmental test of the vegetation clump transplant system will be conducted to study its effectiveness. Field test studies will be conducted in the following areas:

1. Survival rate by species.

2. Growth rate by species.

3. Survival of species as affected by the time of year transplanted.

4. Equipment operating efficiency and costs.

5. Provision for wildlife cover in transplant areas.

6. Buildup of moisture in transplant areas due to retention and buildup of snowpack.

7. Provision of ground cover and/or erosion control in the transplant areas.

8. Cold weather transplanting procedures.

9. Method analysis to reduce number and/or complexity of operations.

The prime contractor for the field test is Colorado State University, Fort Collins, Colo. The cost-sharing cooperator is Energy Fuels Corp., which will provide senior staff, machinery, equipment, and facilities for evaluation of the tree transplant system. Field test development will be conducted at the Energy Fuels Corp. mine, near Steamboat Springs, Colo.

SUMMARY

Advantages of the complete transplanting system, including the transplant bucket and transporter, are

1. Native vegetation can be moved efficiently and economically over longer distances, perhaps several miles, than is possible with the FEL alone.

2. An FEL, a piece of conventional mine production equipment, would be readily available but would be used only on a limited basis so as not to impede production activities.

3. The transporter could be used continuously without interfering with mine production. Larger areas could be thus transplanted in the same time period.

4. Native vegetation could be removed from an unmined area, placed in a holding area by the FEL, and moved by the transporter to the transplant area

when convenient. This will allow maximum productivity with the FEL when it is available and provide increased flexibility in scheduling the FEL.

5. The transporter would be lower in initial cost than an FEL and less expensive to operate. Transplanting costs would therefore be reduced.

6. A truck, large farm tractor, scraper tractor, or other piece of reclamation equipment, not directly required for mine production, could be used to pull the transporter. All power for the transporter operation other than mobility is self-contained, so the prime mover could be changed easily.

Selective Overburden Placement

by Gregory G. Miller
Mechanical Engineer
Spokane Research Center
Bureau of Mines, Spokane, Washington

INTRODUCTION

The Bureau of Mines sponsored a study of selective overburden handling at a coal mine in Colstrip, southeastern Montana. The Montana Agricultural Experiment Station in Bozeman, Mont., performed the work. They determined a recommended drilling intensity needed to accurately identify salt and trace metal materials that would be inhibitory to plant growth and performed a selective handling demonstration.

DRILLING INTENSITY

Current practice of identifying toxic ions of salts and trace metals involves drilling on a 1-mile grid in unfaulted areas and on a 1/2-mile grid in faulted areas. This guideline is based on a commonly accepted assumption that if inhibitory materials are present, they will be located in a physiochemically uniform geologic strata. Consequently, if an inhibitory zone is found, selective overburden handling of this single strata is necessary. To evaluate this guideline, 300 boreholes were drilled on an approximate 150-foot grid over a 100-acre premine site. Samples were taken on 5-foot intervals, and approximately 100,000 separate physiochemical laboratory tests were performed at a mobile laboratory at the mine site.

Data on the physiochemical properties at the site were graphed showing a three-dimensional distribution of each parameter. Figure 1 is a three-dimensional view of the distribution of clay at the site where the first four dragline passes would occur. This diagram and other diagrams of soluble salts, lead, and nickel show that inhibitory physiochemical parameters are distributed in overburden zones independent of the geologic strata and generally independent of each other. It was also determined that the horizontal extent of the toxic zones was considerably less than 1 mile, indicating that the current guideline is not effective. To detect each parameter at a 90-percent accuracy level required an overburden drilling grid of 100 to 200 feet, considerably less than the guidelines suggest.

Therefore, to properly identify inhibitory material, a two-step drilling program is proposed. First, the overburden should be drilled on a 1/2-mile grid. This would identify which physiochemical parameters exist in inhibitory concentrations. Then the site should be drilled on a 100- to 200-foot grid to identify the vertical and horizontal extent of those parameters determined to be present in inhibitory concentrations. This two-step procedure will reduce the costs of laboratory analysis by eliminating repeated analysis of parameters found in step 1 to be insignificant. Another conclusion is that, generally speaking, mines that selectively bury a certain strata based on a drilling grid identification in the thousands of feet are likely to entirely miss considerable zones of inhibitory material.

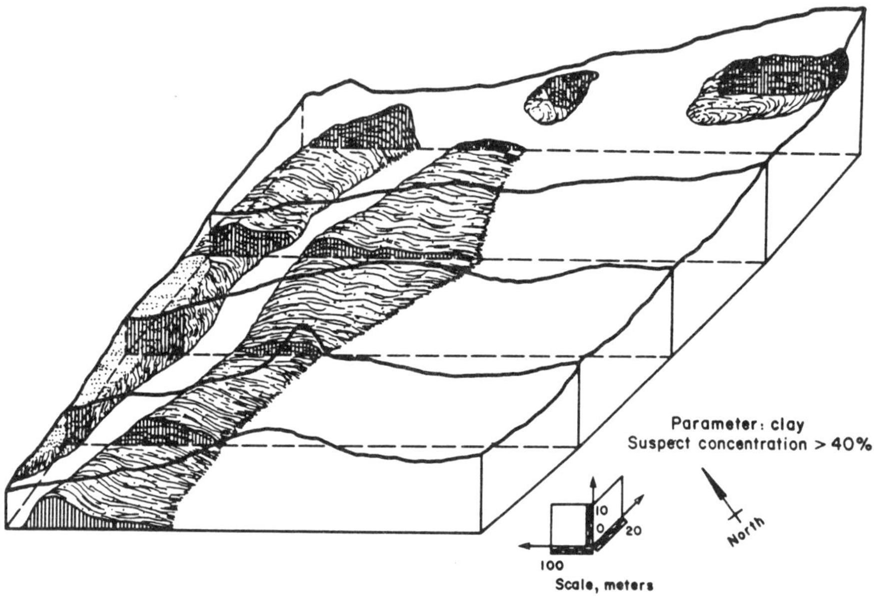

FIGURE 1. - Three-dimensional view of clay distribution.

INHIBITORY MATERIALS

In the Western States, inhibitory materials contain concentrations of salts and trace elements. Salts of calcium, magnesium, and sodium associated with sulfates (SO_4) and carbonates (HCO_3) are present. Trace metal ions of nickel, cadmium, zinc, molybdenum, and lead were a problem at the study site, and levels of copper, manganese, mercury, selenium, and boron should be checked. Suspect levels of pH, ammonium, sand, and clay were found at the site.

High concentrations of salts adversely affect plant-water relationships by upsetting the plants' osmotic pressure. High concentrations of trace metals cause metabolic interference in plant cell physiology. Spoils containing more than 40 percent clay prevent plant root penetration and water infiltration. Clays are usually associated with high levels of positive trace metals ions which cling to negatively charged clay. Spoil containing more than 70 percent sand retains little water or nutrients essential to plant life. Drinking water contaminated with salts and trace metals causes metabolic problems in animal life.

THE LEACHING PROCESS

Leaching takes place if the spoil water begins to flow by gravity, carrying with it soluble constituents of the spoil. The water content must be above the field capacity for leaching to be significant. The field capacity is the amount of water held in the spoil after excess water has drained away and the rate of downward movement has substantially decreased. Leaching is most significant as the water content approaches saturation.

It is desirable to suppress the leaching process in the presence of spoil containing inhibitory concentrations of trace metals or salts. This may be done in four ways: (1) The inhibitory material can be sealed with an impermeable layer; (2) the inhibitory material can be diluted through mixing; (3) a topographic dome can be made above the inhibitory material to divert surface water away; and (4) subsurface water can be consumed by plantlife. Clay seals and dilution were investigated in this project.

CLAY CAPPING

Figure 2 shows selectively handled inhibitory material that was buried and capped with clay. It is desirable to use spoil located at the mine as a source of clay to reduce haulage costs. A clay cap intended to divert water is constructed with spoil dominated by a smectite clay. Smectite clay has a 2:1 lattice structure of a tetrahedral-octahedral-tetrahedral arrangement. Two-to-one lattice clays do not have a hydrogen bonding effect, so the units can readily expand and contract with wetting and drying, creating a formation that tends to be impermeable when wet. If the clay source is dry, it must be irrigated to achieve an optimal moisture content of 14.6 percent by weight. The clay should be laid over the inhibitory zone with scrapers. The scrapers should also attempt to compact the clay to a standard Proctor density of 112 pounds per cubic foot. The cap should be formed with a 5:1 grade in all directions to create an umbrella effect to shed percolating water. The clay cap should be thick enough so that it will not rupture when the dragline spoils overburden on it.

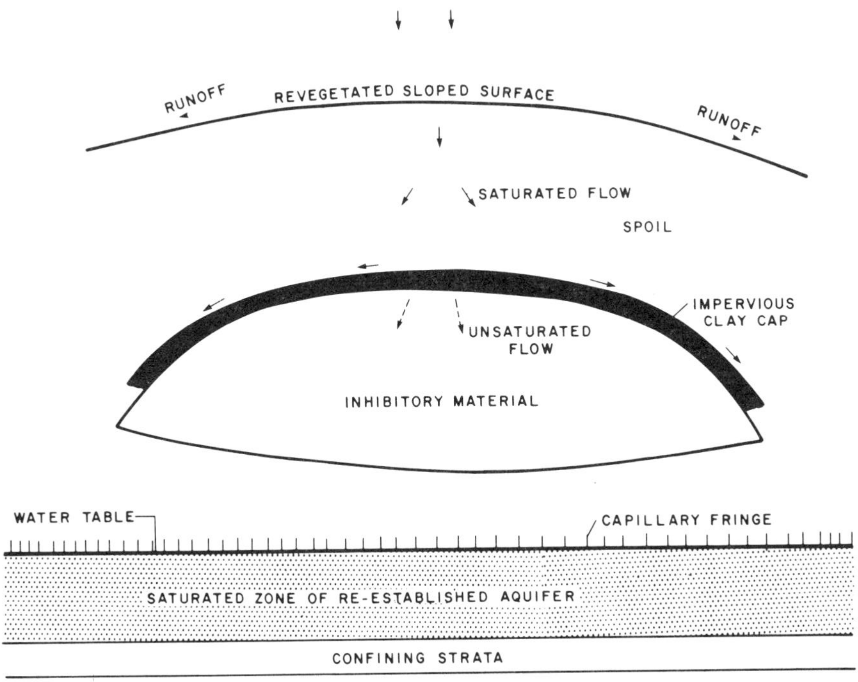

FIGURE 2. - Infiltrating water is diverted by clay cap away from the inhibitory material.

A 60-cubic-yard dragline was able to build the foundation material and selectively place saline inhibitory material for capping. This was performed on approximately two-thirds of an acre and 14,000 cubic yards of a saline material were buried. The saline material was originally located in the top

15 feet of the overburden. The stripping ratio was 3:1. The cost of the operation was 50 percent more than normal dragline operations.

The cost and complexity of selective overburden handling rest on whether the inhibitory materials are at the top, middle, or bottom of the overburden profile, and on the availability of clay. More efficient selective overburden handling procedures could significantly reduce costs.

The inhibitory overburden material should be buried deep enough to prevent future contact with the surface environment caused by man or nature. If the mine has interrupted an aquifer, the material should be buried above the predicted reestablished aquifer. The new aquifer capillary fringe must not intercept the buried inhibitory material. (The capillary fringe is a zone above the new aquifer into which unsaturated water enters due to capillary rise.) If the capillary fringe intercepts the inhibitory material, fluctuation of the water table level causes a corresponding fluctuation in the capillary fringe. When the aquifer level drops, the fringe drops. When it later rises, the chemical constituents are now in the aquifer zone. The fringe can then carry dissolved chemical constituents with it.

Because of the semiarid climate at the site after 2 years, unsaturated flow of water through the spoil has not been enough to evaluate the effectiveness of the clay cap. However, a small (50-foot-diameter) experimental cap was also built which could be flooded with water to determine its effectiveness in shedding the downward flow. Over a period of 2-1/2 months, about 9-1/2 feet of water was applied in a diked area above the cap. This water application rate was extremely intense in comparison to the amount of water normally expected from precipitation. The change in the water content above and below the clay cap was monitored. The wetted front mushroomed out laterally over the cap. Although the cap itself became quite wet, the inhibitory material beneath it never reached saturation. The cap was determined to be an effective barrier to saturated flow.

DILUTION OF INHIBITORY MATERIALS

Mixing of inhibitory spoil material with other "clean" spoil tends to dilute the concentration of inhibitory material to acceptable levels. Mixing can be performed three ways with a dragline through digging and spoiling. First, when an inhibitory zone intersects the digging face, the dragline can drag the bucket at an incline, mixing clean and inhibitory material. Second, the dragline can dump the spoil as it swings, spreading and thus mixing the material. Third, the dragline can dump on spoil peaks, causing mixing as spoil rolls down.

Three types of dragline spoiling techniques were investigated: normal, dump, and scatter. Normal spoiling with a 60-cubic-yard dragline with a 325-foot boom generally results in a bucket load of material being cast over an area 65 feet long by 35 feet wide. This was found to be true for several dragline operators, although their styles differed. Dump spoiling is used for initial spoiling on the pit floor. This involves stopping the dragline swing before dumping the load. It is used to avoid entrapments of pit water in the spoil because entrapped water causes subsidence and slope failure. Dump spoiling covers an area 35 feet square. Scatter spoiling involves dumping the load on the swing. The material is cast over a large area about 130 to 200 feet long. Scatter and dump spoiling increase cycle time. Three volumes of spoil were created with each method. Dump spoiling generally results in less mixing of inhibitory zones, and the technique should be used only when necessary. Scatter spoiling seemed to produce the best degree of mixing; however, normal spoiling could not be judged to be significantly different.

Analysis of the resultant spoil yielded the following conclusions. When the inhibitory material constitutes less than 5 percent of the total overburden volume, the inhibitory zone was diluted to acceptable levels. When the inhibitory zone constitutes 5 to 15 percent of the overburden volume, only partial dilution occurs. When the inhibitory material exceeds 15 percent, only a small portion is diluted. In this case, selective overburden placement is necessary. No mixing occurs if the inhibitory material is spread throughout the entire overburden. Additionally, surficial overburden material may not be mixed as well as subsurface material. This may be due to a tendency to spoil surface materials onto the pit base in a dumping fashion, whereas deeper overburden materials are often cast normally or scattered over a larger area and slide down an existing spoil pile.

It is always preferable to mix clays and sands. Although mixing of inhibitory concentrations of trace metals and salts will increase the surface area available for deep leaching, mixing reduces the amount of inhibitory material in the plant root zone and the amount of inhibitory material available for leaching in the reestablished aquifer zone.

GROUND WATER

Because of the limited depth of surface strip mining, only the shallow ground-water system was studied. Three aquifers were present in this system. The first is located in the permeable sandstone portion of the overburden, and the other two in the two coal seams. The second coal seam was not mined. Its aquifer was confined, and research found that surface strip mining has little effect on it. The top two aquifers were disturbed by mining, significantly affecting their level and direction of flow. A fourth aquifer now exists in the shallow ground-water system located in the spoils. How and where this new aquifer forms will impact water supplies down gradient. Water availability in this semiarid region is critical to local agricultural activities.

The spoils aquifer should be constructed of pervious, uncontaminated material so that leaching of soluble salts and trace metals will not take place. Wells placed in the spoil confirm that the new spoils aquifer is forming and is exibiting confinement characteristics similar to those that exist in the unmined coal seams.

A network of monitoring wells were placed in the overburden and spoil. Water levels and water quality measurements were taken. Data suggest that mining activities have had little effect on the overburden water levels at distances greater than 2,000 feet from the mine pit. This may be due, in part, to variations in lateral hydraulic conductivity, lack of sufficient time for adjustment to new conditions created by continued mining, and greater amounts of precipitation during the monitoring period. The top coal seam water level was reduced up to 1 mile from the mine pit. The water level tended to drop more in the coal seam aquifer than in the overburden aquifer. This seems to be due to the relatively continuous and homogeneous nature of the coal seam and the highly discontinuous and heterogeneous nature of the overburden. Consequently, dewatering of the coal seam is evident at a greater distance from the mine pit than is dewatering of the overburden aquifer.

Surface water from streams, ponds, impounded water, and the mine settling ponds can be expected to be a source of recharge of the aquifers. The mine hydrologist must gather enough data to estimate the aquifer level in the spoil. This estimate is used in specifying the depth an inhibitory material can be safely placed.

A comparison of 1923 water quality data with 1973 analyses indicates that no major changes in water quality have occurred owing to commercial mining at the site. Our analyses of water quality coming from the new spoil squifer indicate the water is generally more mineralized and of higher salinity. Mining altered the ground-water flow pattern in the overburden and coal towards the pit. However, this change in flow direction will cease once mining in the area has been completed and the spoil recharge is complete.

Part IV

Environmental Aspects

The information in Part IV is from *Proceedings of the Second U.S.-Polish Symposium—Coal Surface Mining and Power Production in the Face of Environmental Protection Requirements* (EPA 600/7-79-159), prepared by J. Libicki, symposium director, of the Central Research and Design Institute for Openpit Mining, Wroclaw, Poland, for the U.S. Environmental Protection Agency, October 1979.

Coal Mining and Ground Water

by John Hardaway
Physical Science Administrator
Office of Energy
U.S. EPA, Denver, CO 80203

INTRODUCTION

Coal mining has the potential to measurably disrupt local ground water systems. In addition, coal mining on a regional scale may disrupt regional ground water systems. Both surface mining, which requires removal and thus extensive disturbance of overburden, and underground mining have this disruption potential. In-situ extraction also has a high potential for disruption of ground water systems. The extent and tenure of such disruptions will vary, often in proportion to the size of the mine and the degree of hydrologic complexity of the ground water system. Documented impacts to the ground water system have included lowered water tables and deterioration of water quality.

The objectives of this paper are to discuss the potential effects of coal mining on ground water systems and to report on the measurement of such effects by others. It should be noted that the documentation of such efforts as they relate to coal mining in the United States is limited. Perhaps one reason for this limited documentation is the fact that until the last four years, affected ground water systems have not been thoroughly evaluated. To date the majority of reported data has dealt with acid and iron production from the oxidation of sulfide minerals associated with coal mining activities in the Eastern and Midwestern United States. It is hoped that the recently passed Surface Mining Control and Reclamation Act of 1977 (Public Law 95-87) will serve to improve future efforts to evaluate the ground water impacts of coal mining in the United States.

Potential disruptions of the ground water system brought about by coal mining may be divided into "physical impact" and "chemical impacts." Hamilton and Wilson (1977) prepared a generic analysis of the effects of the surface strip mining of coal which emphasized the theoretical impacts of this mining on the ground water flow system. McWhorter et al (1977) also reported on the possible effects on the flow system. This paper makes use of Hamilton and Wilson's analysis in addition to textbook presentations of ground water flow systems such as those contained in standard texts.

PHYSICAL IMPACT

The geohydrologic setting in the vicinity of coal seams is varied. (See Figure 1.) Coal seams may lie anywhere within the regional recharge-discharge system. The coal may be overlain or underlain by significant ground water aquifers. The coal may be part of an aquifer. In most cases, the older the coal, the less will be its water transmission capabilities since it is denser and less likely to contain open fractures. The majority of the ground water found in coal seams flows through the fractures in the coal, rather than through the coal matrix itself since the internal permeability is low.

Underground mining and the first stages of surface mining, namely the excavation of overburden and coal, both create a void in the geologic system which then affects the ground water system by forming a "sink." If saturated strata are encountered in a mine, the active mine will lower the piezometric surface at the mine itself, usually to the bottom of the active mine area. Seldom are conditions in the United States such that dewatering wells are required and thus the active mine pit itself serves as a large well and causes a cone of depressions in the piezometric surface.

Figure 1: Generalized Geohydrologic Settings

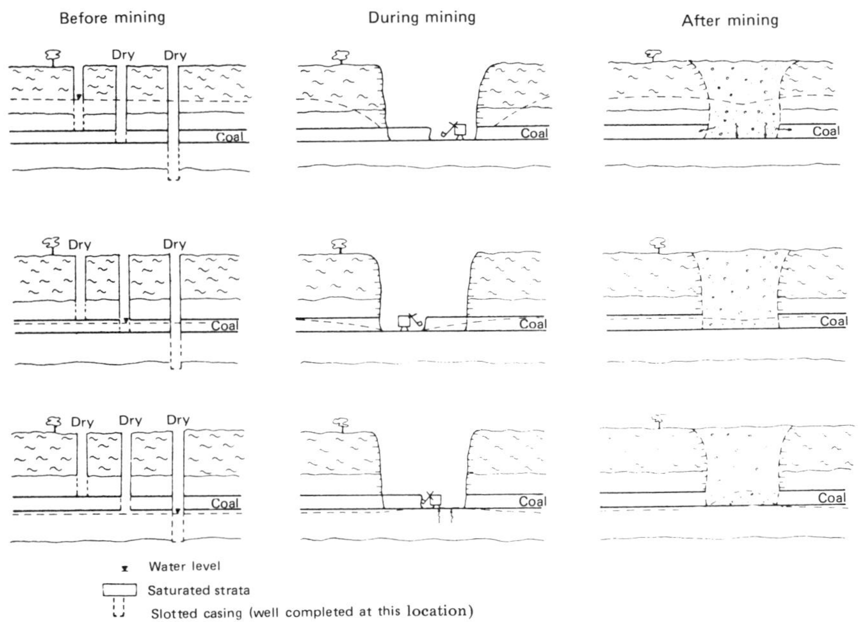

Unconfined aquifer above coal mining causes drop in ground water levels if leakage to coal and below occurs.

Unconfined aquifer in coal mining causes no significant change in ground water levels if no downward leakage. If coal was confined water level would rise.

Confined aquifer below coal mining disturbs strata so as to remove confinement and allow upward leakage.

The horizontal extent of this depression is predictable and has ranged from a few hundred to a few thousand meters from the active mine. This distance is a function of the permeability and the storage coefficient of the surrounding geologic strata as well as the premining ground water gradient through the mine area.

Three hydrogeologic situations are identified for purposes of examining the physical impacts of mining saturated strata. These are (1) saturated strata above the coal, (2) saturated strata including the coal, and (3) saturated strata below the coal. In the first two cases, the drawdown caused by active mining may be approximated by an appropriate potentiometric equation. The accuracy of the prediction depends on the degree to which premining hydrologic measurements have described the hydraulic complexity of the ground water flow system.

In the case of saturated strata above the coal, mining will generally cause these strata to be drained. Reclamation may allow reestablishment of the piezometric surface if the disturbed area is small in relationship to the size of the ground water flow system. If, however, the disturbance is relatively large, and the transmissivity of the reclaimed area is high, the piezometric surface of the overlying strata may be permanently lowered. In the case where saturated strata occur at the level of the coal seam, the piezometric surface will generally be lowered during mining. However, following mining a trend toward reestablishment of the piezometric surface should take place. If saturated strata exist below the coal seam, they will be affected if they contain water under sufficient pressure to cause the water to flow upward into the mined area.

It would appear from the literature available that only in limited cases in the interior western United States has there been adequate ground water monitoring during coal mining to document the drawdown effect of coal mining. In addition, all of monitoring reported in the literature has been performed at surface mine sites. Hamilton and Wilson (1977, page 22) report that the large surface mining operation conducted along Caballo Creek in eastern Wyoming (The AMAX Belle Ayr Mine) has not produced drawdown past 300 meters from the mine edge. Van Voast and Hedges (1975, Plate 9) report drawdown out to at least 3.2 kilometers from the edge of the Decker mine (in the up-gradient direction).

As surface mining progresses, disturbed and broken overburden is returned to the mine and the ground water system begins to approach some physical equilibrium with the disturbed material. In the case of most underground mining, the hydrologic equilibrium reached after the mining is likely the same as that produced during mining since no material is back-stowed into the mined-out seam to fill the void and since adits will in many cases permit continued drainage to the surface. While active surface mining causes voids with infinite permeability, subsequent backfilling again gives a finite value to permeability in the mined area. It is postulated that the postmining permeability of backfilled areas in the majority of cases will be greater than that existing prior to mining. This does not mean to infer that the surface infiltration rate in most cases will be greater since impermeable surfaces may be created if the spoiled overburden surface is inadequately treated. However, unless spoiled overburden is selectively placed, overburden moisture is controlled, and heavy equipment is used for compaction, it is believed that permeability cannot be reduced to that created over geologic time by pressure and chemical action. An exception occurs in those areas of the Midwest and West of the United States where the undisturbed overburden is not indurated. Van Voast et al (1975) have graphically portrayed ranges in hydraulic conductivity for coal and spoil aquifers at surface coal mines located near Colstrip and Decker, Montana. These are repeated here as Figure 2. Rahn (1976 pp 26-31, 54) has indicated that hydraulic conductivities of spoil measured in the vicinity of Sheridan, Wyoming probably will never approach the values of natural formations (i.e., hydraulic conductivities are always higher).

McWhorter et al (1977, pp 91) presented a streamflow equation that was used to project postmining flow patterns. The equation may be used to project streamflow affected by surface coal mining as follows:

Stream functions outside the disturbed area are described by

$$\Psi = q_{\infty} y \quad 1 - \left(\frac{k_o - k_i}{k_o + k_i}\right) \left(\frac{R^2}{x^2 + y^2}\right)$$

Stream functions inside the disturbed area are described by

$$\Psi_i = \left(\frac{2q \; k_o}{k_o + k_i}\right) Y$$

The notations are:

q_{∞} = solution flux

k_o = hydraulic conductivity of undisturbed strata located outside the mine

k_i = hydraulic conductivity of spoil

R, x, y = the coordinates

McWhorter et al used a circular mine area and assumed a simple hydrologic flow system of two-dimensional, steady flow. The stream function outside the disturbed area is controlled by the following term:

$$\left| \frac{k_o - k_i}{k_o + k_i} \right| \quad \frac{R^2}{x^2 + y^2}$$

which is set equal to "C" where C is treated as a decimal measure of the significance of the differences between premining and postmining streamflow functions.

The perimeter of the area within which the flow paths are significantly modified forms a circle of radius "r" given by:

$$r = x^2 + y^2 = \frac{R}{\sqrt{C}} \left(\frac{k_o - k_i}{k_o + k_i} \right)^{1/2}$$

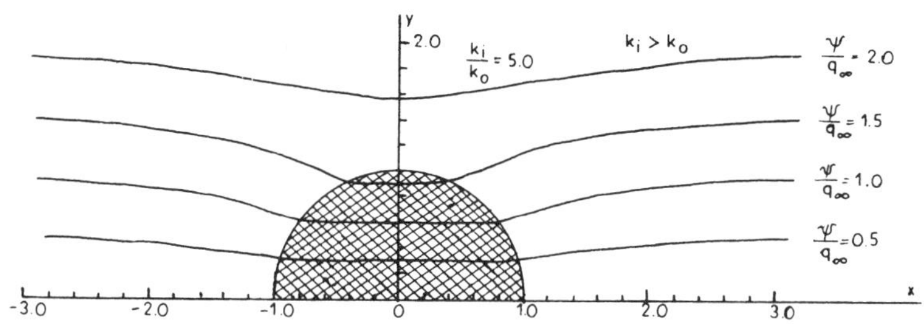

Flow pattern in a disrupted aquifer, $k_i/k_o = 5$

k_i = inside mine area
k_o = outside mine area

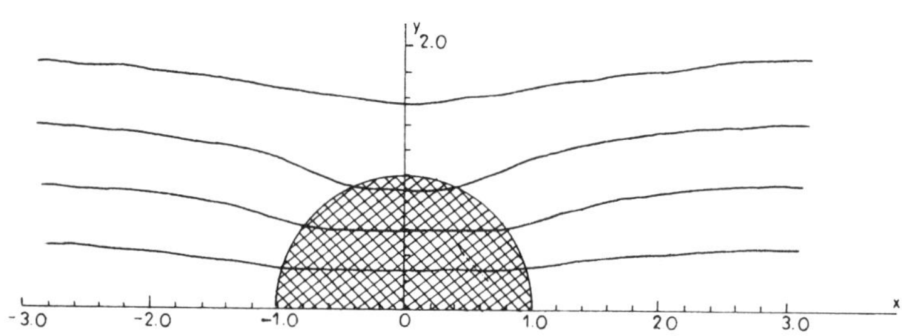

if ki = 2 ko and we assume

that C=5% or C=0.05,

then r=2.6R where R = radius of mine

if $k \rightarrow 0$ or ∞, $r \rightarrow \frac{R}{\sqrt{C}}$

Figure 2: Hydraulic-Conductivity Ranges and Means for Coal and Mine Spoils in Southeastern Montana. After Van Voast

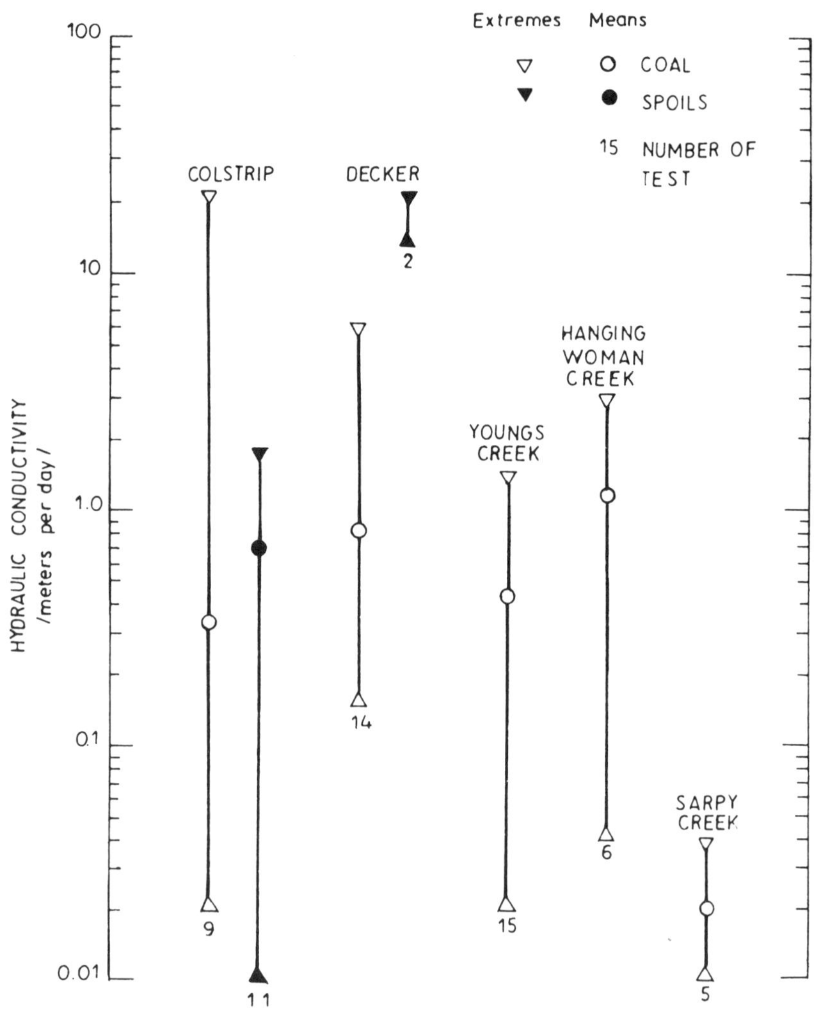

Thus the effects of mining are projected as minimal part of radius of 2.6 times the radius of the mine. Rogowski et al (1977) observe that many air pockets (i.e., large voids) are evident in mine spoils. They point out that these voids may contribute to non-Darcian flow behavior of the spoil. This heterogeniety of the spoil may be further complicated by the existence of respread topsoil on top of coarser spoil. Thus the wetted front that occurs as a result of infiltration may be unstable (Hill and Parlange, 1972, cited in Rogowski et al, 1977).

Data provided by Gilly et al (1977) show how water movement into spoils can be significantly and positively affected by tilling (i.e., cultivating) the spoil.

Thus water movement through spoils may not be as simplistic as assumed, and anomalous flow patterns may be found.

CHEMICAL IMPACTS

Major consideration has been given by the research community to the effect of coal mining and reclamation on ground water quality (at least as that quality subsequently affects the surface water system). McWhorter (1977), Dollhopf et al (1977), Van Voast and Hedges (1975), Van Voast (1974), and Rahn (1976) have addressed, in varying degrees, potential and measured effects. Ground water quality impacts during mining are not considered to be particularly significant due to the fact that the mine acts as a sink for ground water flow and the chemical changes in the quality of the interrupted water are relatively small since the active mine is kept as dry as possible during operations and discharged water is treated to meet water quality effluent limitations. However, when mining is completed, water infiltration and saturation of the mined area will likely reoccur and ground water quality will be changed. The reaction of sulfide minerals with oxygen and water is well documented (Ohio State University Research Foundation, 1971). In the case of pyrite, an idealized chemical reaction is: $2FeS_2 + 7O_2 + 2H_2O \rightarrow 2FeSO_4 + 2H_2SO_4$.

This is a characteristic reaction in coal-bearing formations with a pyrite content of more than a few percent, where neutralizing carbonates are in short supply, and where water is readily available. Thus, chemical reaction is persuasive with respect to coal mining in much of the eastern and midwestern United States. Interestingly, acid formation had been either ignored or overemphasized in the coal mining regions of the western United States until the past year. At this time certain researchers including Croft, et al (1977) have identified oxidation of sulfide minerals and formation of sulfuric acid as a key step in the commonly observed increase in alkalinity caused by coal mining in the interior western United States. The chemical reactions, as identified by Croft, and others in ground water associated with a lignite mine in North Dakota, are caused by rain water (containing $Ca(HCO_3)_2 + O_2$) as this water passes through the lignite are:

$$2FeS_2 + 7O_2 + 2H_2O \rightarrow 2FeSO_4 + 2H_2SO_4$$

and

$$Ca(HCO_3)_2 + H_2SO_4 \rightarrow CaSO_4 + 2CO_2 + 2H_2O$$

and

$$2CaCO_3 + H_2SO_4 \rightarrow Ca(HCO_3)_2 + CaSO_4$$

An additional source of SO_4 is through dissolution of gypsum. The water then becomes saturated with $CaSO_4$. However, if it flows through a sodium clay, the calcium is exchanged for sodium (cation exchange) resulting in water with a high Na_2SO_4 content (and high in HCO_3).

Rogowski et al (1977) note that a considerable amount of clay must be present and that it must possess a high exchange capacity to significantly alter the acid character of acid drainage from pyrite-rich strata. The hydrogen ions react with the crystal lattice to produce charged aluminum species. The aluminum will then be dissolved to some degree.

Pagenkopf et al (1977) have tabulated water quality data for two mine areas in Montana. Table 1 shows increased concentrations for a number of parameters. As indicated in Table 1, iron and pH tend to change much less than do the other common elements shown. Dollhopf, et al (1977) have found somewhat high concentrations of manganese in some ground water in spoils at one mine site. However, the natural levels of manganese also appear high suggesting that the coal mining may not be the source of high manganese water.

McWhorter et al (1977, pp 41) also noted sporadically high concentrations of manganese in Colorado streams and suggested, in the case of high manganese in ground water passing through spoils, that the source was older spoils (about a year or more) and time delays in manganese dissolution.

Table 1: Comparison of Natural Ground Water and Spoil Waters in Southeastern Montana

	Decker		Colstrip	
	Wells	Ground Water in Spoil	Wells	Ground Water in Spoil
Ca	3.2	31	35	429
Mg	0.9	13	70	—
Na	445	1,355	27.4	346
HCO_3	1,143	1,996	366	655
CO_3	0	19	0	0
SO_4	4.7	1,320	118	3,306
SiO_2	8.6	8.6	15	21
Cl	15.3	16	4.3	63
TDS	1,626	4,801	641	5,380
SPC	1,590	5,010	776	5,300
Fe	0.02	0.02	0	0.85
pH	8.1	8.3	8.1	7.6

Pagenkopf, Whitworth, and Van Voast

Thorstenson, Fisher, and Croft are reported by Feder and Saindon, 1976 to have found that as ground water moves naturally downdip from recharge areas, a combination of chemical reactions with clays, sulfate-reducing bacteria, and other minerals gradually causes a change in the water of one of the Fort Union aquifers of Montana (the Fox Hills) from a Ca^{++} Mg^{++} HCO_3SO_4 type with a pH of about 8.0 to a $NaHCO_3$ type with a pH above 8.5. Feder et al (1977) utilized data from nineteen randomly selected wells in supporting this generalization showing statistically that Na, pH, HCO_3 and TDS tend to increase together in natural ground waters of the Fort Union, while SO_4, Ca and Mg tend to decrease in relative relationship.

The data developed by Feder et al are repeated here as Tables 2 and 3 to characterize the geochemistry of ground water from the Fort Union Formation in eastern Montana and western North Dakota.

Table 2: Chemical Analyses of Samples Representing a Typical Recharge Water (Sample A) and a Typical Discharge Water (Sample B) in the Powder River Coal Region After Feder et al 1977

Parameter	Sample A	Sample B
Ca, mg/ℓ	180	1.9
Mg, mg/ℓ	82	0.6
Total hardness (as $CaCO_3$), mg/ℓ	790	7
HCO_3, mg/ℓ	362	1,150
Na, mg/ℓ	250	450
SO_4, mg/ℓ	1,100	5.5
SAR*	3.9	73
pH	6.8	8.3
TDS, mg/ℓ	1,830	1,110
SiO_2, mg/ℓ	11	7.8
Cl, mg/ℓ	6.1	47
K, mg/ℓ	9.3	1.5
F, mg/ℓ	0.5	14
I, mg/ℓ	<0.01	0.07
Br, mg/ℓ	0.1	0.7
Al, μg/ℓ	20	20
As, μg/ℓ	<1.0	<1.0
Ba, μg/ℓ	<6	100
B, μg/ℓ	80	240
Cd, μg/ℓ	<1.0	<1.0
Cu, μg/ℓ	6	<5
Fe, μg/ℓ	2,500	30
Hg, μg/ℓ	<0.1	<0.1
Pb, μg/ℓ	180	15
Li, μg/ℓ	40	20
Mn, μg/ℓ	2,200	10
Mo, μg/ℓ	<12	12
Se, μg/ℓ	<1.0	<1.0
Sr, μg/ℓ	2,000	160
Zn, μg/ℓ	40	200
U, μg/ℓ	0.30	0.09
Ra, pCi/ℓ	0.1	0.2

*Sodium adsorption ratio.

Table 3: Geochemical Summary of Ground Water from the Powder River Coal Region, Montana and Wyoming. After Feder et al (1977)*

Parameter	Detection Ratio	Geometric Mean	Geometric Deviation	Maximum	Minimum
Al, μg/ℓ	19/19	17	1.9	60	6.0
Ba, μg/ℓ	9/15	24	2.7	128	6.0
B, μg/ℓ	19/19	148	2.2	422	32.0
Cu, μg/ℓ	4/19	—	—	14	1.0
Li, μg/ℓ	20/20	36	2.4	180	10.0
Sr, μg/ℓ	15/16	444	4.1	2,754	19.0
As, μg/ℓ	6/20	—	—	6	1.0
Se, μg/ℓ	5/20	—	—	12	1.0
Cd, μg/ℓ	3/20	—	—	1	1.0
Hg, μg/ℓ	4/20	—	—	0.2	0.10
Zn, μg/ℓ	20/20	50	5.5	1,800	0.70
Fe, μg/ℓ	20/20	170	10.5	28,000	7.0
Mn, μg/ℓ	20/20	21	9.5	4,800	0.70
Br, mg/ℓ	13/20	0.15	2.0	0.7	0.10
F, mg/ℓ	20/20	0.68	2.8	14	0.10
I, mg/ℓ	4/20	0.01	1.4	0.02	0.01
Cl, mg/ℓ	20/20	8.7	2.1	47	1.9
SO_4, mg/ℓ	20/20	292	5.1	1,800	5.5
HCO_3, mg/ℓ	20/20	504	1.6	1,400	195.0
Ca, mg/ℓ	20/20	24	6.5	530	1.9
Mg, mg/ℓ	20/20	13	7.0	150	0.60
K, mg/ℓ	20/20	3.9	2.0	12	1.5
Na, mg/ℓ	20/20	173	3.0	1,000	24.0
SiO_2, mg/ℓ	20/20	11	1.6	26	5.8
THD, mg/ℓ	20/20	112	6.9	1,900	7.0
pH**	20/20	7.7	0.6	8.5	6.5
Specific conductance, mhos/cm	20/20	1,494	1.8	4,000	582.0
TDS, mg/ℓ	20/20	1,076	1.9	3,190	345.0
NO_3 + NO_2, mg/ℓ	18/20	0.08	8.3	3.2	0.01
SAR***	20/20	6.9	6.0	73	0.40
Ra, pCi/ℓ	17/18	0.23	2.0	0.80	0.10
U, μg/ℓ	16/18	0.25	6.0	7.3	0.01
Beta, pCi/ℓ as Cs-137	12/18	6.6	1.8	14	4.0

*Detection ratio: number of samples in which constituent was determined to total number of samples analyzed.

**Arithmetic mean and standard deviation.

***Sodium adsorption ratio.

These data give background conditions for the area studied, and show the natural changes in water quality that must be separated from changes caused by coal mining and reclamation. Pietz et al (1974), through a ground water study conducted at a surface mining site in Illinois, identified changes in ground water quality similar to those found in the western United States. Ground water collected from the mine spoils contained significantly higher concentrations of sulfate, sodium, calcium, magnesium, manganese, and iron than did ground water collected from undisturbed sites (see Table 4).

Regardless of the climate or geographical region in which coal mining is conducted, mining processes, both surface mining and underground mining, expose fresh rock surfaces to an oxidizing environment. After some as yet undetermined period of time, this oxidation potential should be satisfied. From Pagenkopf et al, it's estimated that certain freshly broken spoils collected in the western United States will require four volumes of pore water before the oxidation potential is satisfied to the degree that the additional waters passing through the spoils show little changes in dissolved solids content. The period of time for the oxidation potential to be satisfied then depends on the rate that water is available. McWhorter et al (1977, pp 5) suggest that the effects of leaching of spoils will not become apparent until about one pore volume of water passes through the spoil, thus in the more arid climates, where recharge to saturated aquifers does not exceed a few centimeters a year, a long time period would occur before the effects of leaching are noticeable in terms of a measureable increase in total dissolved solids.

Laboratory experiments conducted by McWhorter et al (1977, pp 59) on spoil specimens collected in Colorado gave the results shown in Figure 3. Samples were collected for total dissolved solids analysis after passing 100 ml samples. Fresh water was continually added to maintain a head of 5 cm during the test. The spoil was aerated three times during the test, and the results suggest that 6.8 m^3 of deionized water must pass through 1 m^3 of spoil material to achieve a 95% reduction in conductivity. McWhorter et al use an example of spoils 20 M thick requiring 136 m^3 of water/m^2 for leaching to achieve a 95% reduction in the original (high) conductivity. At an infiltration rate of 10 cm, it would then take 680 yr to reduce conductivities by 95% and longer if the effects of weathering are considered (McWhorter et al, 1977, pp 60). These leaching studies produced at least 2.4 kg of SO solubilized elements/m^3 of spoil.

Table 4: Means and Ranges of Ground Water Chemical Characteristics for Monthly Samples Collected During 1971-1973 from Selected Placed Land and Mine-Spoil Monitoring Wells at the Fulton County, Illinois Land Reclamation Site.* (Pietz et al, 1974)

Chemical Characteristic	MDL**	. . . Placed Land Wells. . . . Range	Mean	. . . . Mine-Spoil Wells . . . . Range	Mean	T-Test for Means***
pH	—	6.40-8.90	7.48±0.03†	6.20-8.90	7.19±0.03	xx
Total P, mg/ℓ	0.01	0.00-2.10	0.14±0.02	0.00-0.41	0.08±0.01	x
Cl^-, mg/ℓ	1.0	0.00-87.0	11.9±1.0	2.0-44.0	17.5±0.6	xx
$SO_4^=$, mg/ℓ	1.0	0.0-1,253	127±15	21.0-1,812	609±32	xx
Kjeldahl N, mg/ℓ	1.0	0.00-6.50	1.22±0.10	0.00-7.30	1.65±0.21	xx
NH_3-N, mg/ℓ	0.1	0.00-4.00	1.04±0.07	0.00-6.90	1.25±0.06	ns
NO_3 + NO_2-N, mg/ℓ	0.01	0.00-30.5	0.82±0.30	0.00-0.43	0.05±0.01	xx
Alkalinity, mg/ℓ	1.0	110-700	363±10	100-1,600	661±19	xx
E.C., μmhos/cm	—	200-1,500	786±90	1,000-4,000	2,406±63	xx
K, mg/ℓ	1.0	0.0-20	1.5±0.2	2.0-18.7	8.3±0.3	xx
Na, mg/ℓ	1.0	7.0-131	30.4±1.9	19.0-657	246.8±14.4	xx
Ca, mg/ℓ	1.0	38.5-226	103.2±3.3	35.0-707	260.3±13.6	xx
Mg, mg/ℓ	1.0	23-102	56.8±1.6	86.0-625.0	207.4±8.8	xx
Zn, mg/ℓ	0.1	0.0-140	8.2±1.2	0.40-100.0	14.5±1.5	xx
Cd, mg/ℓ	0.01	0.00-0.03	0.002±0.001	0.00-0.20	0.010±0.002	xx
Cu, mg/ℓ	0.01	0.00-0.82	0.04±0.01	0.00-0.52	0.06±0.01	ns
Cr, mg/ℓ	0.02	0.00-0.04	0.003±0.001	0.00-0.05	0.008±0.001	xx
Ni, mg/ℓ	0.1	0.00-0.30	0.02±0.005	0.00-1.10	0.08±0.01	xx
Mn, mg/ℓ	0.01	0.21-2.79	0.86±0.04	0.39-9.00	2.53±0.22	xx
Pb, mg/ℓ	0.03	0.00-0.44	0.08±0.01	0.00-0.66	0.15±0.01	xx
Fe, mg/ℓ	0.1	0.0-78.8	17.3±1.0	2.70-193.0	52.1±3.4	xx
Al, mg/ℓ	1.0	0.0-5.8	0.8±0.1	0.0-8.20	0.8±0.2	ns
Hg, μg/ℓ	0.1	0.00-1.40	0.17±0.02	0.00-2.40	0.22±0.02	ns

*Groundwater data from six placed land and six mine-spoil monitoring wells were used.
**Minimum laboratory detection limit.
***x denotes significance at the 0.05 level, xx denotes significance at the 0.01 level, and ns is not significant.
†Standard error of mean, $S_{\overline{x}} = S/\sqrt{n}$, where S is the standard deviation and n is the sample population.

Figure 3: Results of Leaching Test with Aeration of Sample to Produce Weathering After McWhorter et al

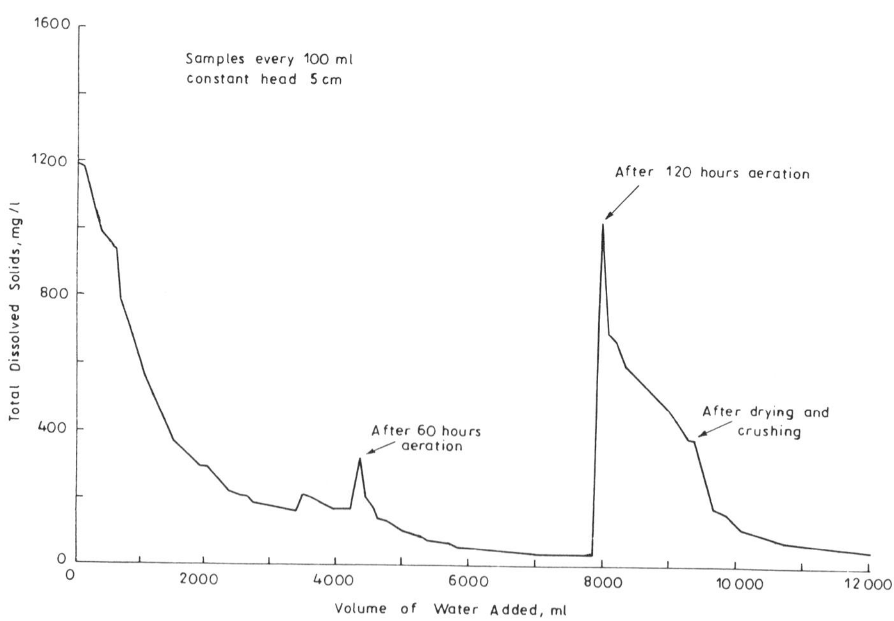

Pagenkopf et al (1977) also performed laboratory experiments for the purpose of analyzing the leachates from spoil. They concluded, from the data presented in Figures 4 and 5, that the majority of the leaching took place in the first hour in weathered spoil, but was delayed in unweathered spoil. The weathered spoil also produced less leachable salts to the solution. These experiments were steady state mixtures of 12.5 grams of spoil in contact with 172.5 ml of distilled water (i.e., 10% spoil). The spoil sample designated D-3 is concluded to have produced 8.5 kg of dissolved solids per cubic meter during the test. Other samples produced the amounts of dissolved solids shown in Table 5.

Table 5: Soluble Salts in Spoil (Pagenkopf et al, p 118)

	Kilograms per Cubic Meter
C-1	0.84
C-2	1.10
C-3	1.89
D-1	2.73
D-2	3.14
D-3	8.50
D-4	4.29

SUMMARY

In conclusion, both surface and underground coal mining activities have a high potential to impact the quantity and quality of both local and regional ground water systems. Such impacts may be short-term and/or long-term in duration. As identified in this paper, the local hydrogeologic system plays the principal role in the extent of quantity-related imports to the ground water system. Ground water quality effects as they relate to coal mining activities are primarily related to the chemical characteristics of the local geologic system.

Figure 4: Leaching of Colstrip Spoil with Time, 10% Solution, C-1, C-2, C-3 Designate the Particular Spoil (Pagenkopf et al)

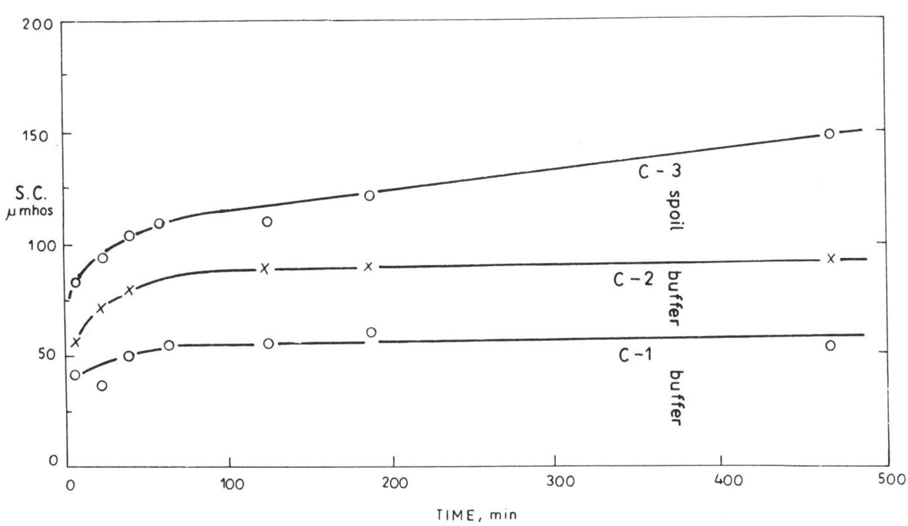

Figure 5: Leaching of Decker Spoil with Time, 10% Solution, D-1, D-2, D-3, D-4 Designate the Particular Spoil (Pagenkopf et al)

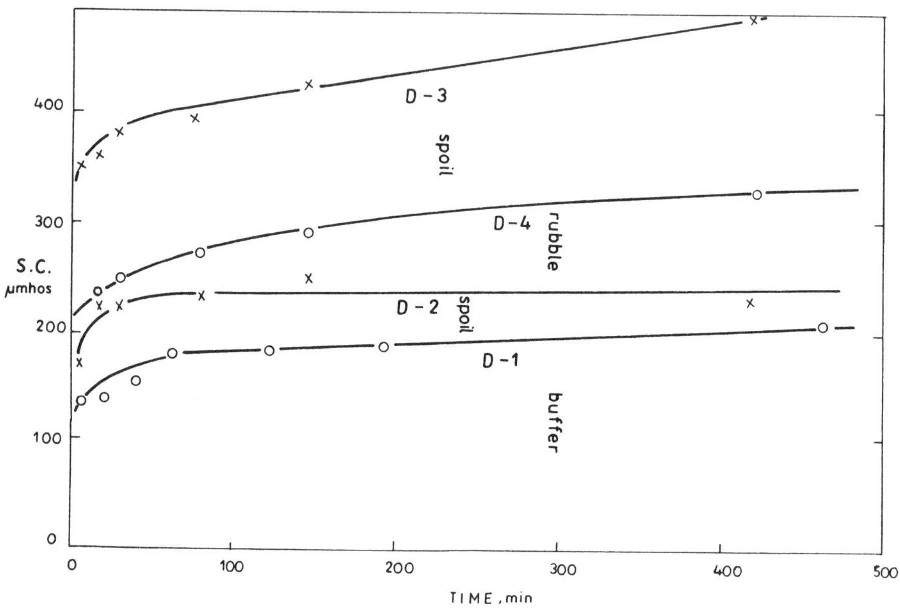

REFERENCES

(1) Van Voast, Wayne A. and Robert B. Hedges, 1975. *Hydrogeologic Aspects of Existing and Proposed Strip Coal Mines Near Decker, Southeastern Montana.* Montana Bureau of Mines and Geology Bulletin 97.

(2) Van Voast, Wayne A., 1974. *Hydrologic Effects of Strip Coal Mining in Southeastern Montana—Emphasis: One Year of Mining Near Decker.* Montana Bureau of Mines and Ecology, Bulletin 93.

(3) Croft, M.G., Donald W. Fisher, and Don Thorstenson, 1978. *Preliminary Report Hydrologic Effects of Strip Mining in the Gascoyne Area, Bowman County, North Dakota.* U.S. Geological Survey, Bismarck, North Dakota.

(4) Van Voast, Wayne A., Robert B. Hedges and John J. McDermott, 1976. "Hydrologic Aspects of Strip Mining in the Subbituminous Coal Fields of Montana." *Proceedings Fourth Symposium on Surface Mining and Reclamation—National Coal Association and Bituminous Coal Research, Inc.* (pp 160-172).

(5) Pennington, Dennis, 1975. "Relationship of Ground Water Movement and Strip Mine Reclamation." Natural Coal Association and Bituminous Coal Research, Inc. (pp 170-178).

(6) Pietz, R.I., J.R. Peterson and C. Luatling, 1974. "Ground Water Quality at a Strip-Mine Reclamation Area in west central Illinois." *Proceedings. Second Symposium on Surface Mining Reclamation—Natural Coal Association and Bituminous Coal Research Inc.* (pp 124-149).

(7) Rogowski, A.S., H.B. Pionke, and J.G. Broyan, 1977. "Modeling the Impact of Strip Mining and Reclamation Processes on Quality and Quantity of Water in Mined Areas: A Review." *Journal of Environmental Quality*, Volume 6, July 1977. No. 3.

(8) Feder, Gerald L., Roger W. Lee, John F. Busby, and Linda G. Saindon, 1977. *Geochemistry of Ground Waters in the Powder River Coal Region in Geochemical Survey of the Western Energy Regions, Fourth Annual Progress Report, July 1977.* U.S. Department of the Interior, Geological Survey, Denver, Colorado, Open-File Report 77-872.

(9) McWhorter, D.B., J.W. Rowe, M.W. Van Liew, R.L. Chandler, R.K. Skogerboe, D.K. Sunda, and G.V. Skogerboe, 1977. *Surface and Subsurface Water Quality Hydrology in Surface Mined Watersheds.* U.S. Environmental Protection Agency Research Report.

(10) Pagenkopf, Gordon K., Clarence Whitworth, and Wayne A. Van Voast, 1977. "Influence of Spoil Material on Ground Water Quality." *Energy Communications*, 3 (2), 107-126.

(11) Dollhopf, D.J., I.B. Jenson and R.L. Hodder, 1977. *Effects of Surface Configuration in Water Pollution Control on Semi-Acid Mined Lands.* U.S. Environmental Protection Agency Research Report.

(12) Hamilton, David A. and John L. Wilson, 1977. *A Generic Study of Strip Mining Impacts on Ground Water Resources.* Ralph M. Parsons Laboratory for Water Resources and Hydrodynamics Department of Civil Engineering, Massachusetts Institute of Technology, Report No. 229 (R-77-28).

(13) Rahn, Pery H. 1976. *Potential of Coal Strip-Mine Spoils as Aquifers in the Powder River Basin.* Project Completion Report prepared for the Old West Regional Commission by the Engineering and Experiment Station, South Dakota School of Mines Rapid City, South Dakota.

(14) Ohio State University Research Foundation. *The 1971 Acid Mine Drainage Formation and Abatement Report.* DAST-42, 14010FPR 04/71, Environmental Protection Agency, Water Quality Office, Washington, D.C.

(15) Feder, Gerald Land, Lynda G. Saindon, 1976. "Geochemistry of Ground Waters in the Fort Union Coal Region." *Geochemical Survey of the Western Energy Regions, Third Annual Progress Report, 1976.* U.S. Department of the Interior Geological Survey, Denver, Colorado, Open File Report 76.

The Impacts of Coal Mining on Surface Water and Control Measures Therefor

by Ronald D. Hill
Director, Resource Extraction and Handling Division
Industrial Environmental Research Laboratory—Cincinnati
U.S. Environmental Protection Agency Cincinnati, OH 45268

INTRODUCTION

During 1971, coal provided 19% of the 68,975 trillion Btu's of energy consumed. 17% of the projected 1985 demand of 133,396 trillion Btu's is to be supplied by coal. In 1972 an estimated 582 million tons of coal were consumed. The forecasted coal production in 1985 is 1,088 million tons of which 245 million tons will be used to make synthetic gas (1). These figures indicate the important role that coal plays in the energy picture.

Two important trends are occurring, i.e., increased surface mining and Western mining. In 1972 for the first time, surface mining produced over half the coal. Surface mining will probably hold this position in the near future. With the above projected increase in demand, the amount of land disturbed each year will nearly double by 1985.

Another trend is the shift to western coals. Approximately 60% of the coal mined in 1970, mostly eastern coals, will not meet the EPA air quality standards (1). The low sulfur content of the western coal and improved coal prices has offset the high transportation cost. Mine mouth power plants and coal gasification are projected added incentives to utilizing western coal. Peabody Coal Company estimated that by 1980 the states of Montana and Wyoming will rank with the top five states in surface mining. Arizona, Colorado, Utah and New Mexico will also be up in the rankings (2).

WATER PROBLEMS ASSOCIATED WITH COAL MINING

Acid Mine Drainage

One of the most troublesome mine drainage problems is caused by acidity. Although the exact mechanism of acid mine drainage formation is not fully understood, it is generally believed that pyrite (FeS_2) is oxidized by oxygen (equation 1) or ferric iron (equation 2) to produce ferrous sulfate and sulfuric acid (4).

$$2FeS_2 + 2H_2O + 7O_2 \rightarrow 2FeSO_4 + 2H_2SO_4 \quad (1)$$
$$FeS_2 + 14Fe^{3+} + 8H_2O \rightarrow 15Fe^{2+} + 2SO_4^{2-} + 16H^+ \quad (2)$$

The reactions may proceed to form ferric hydroxide and more acid:

$$4FeSO_4 + O_2 + 2H_2SO_4 \rightarrow 2Fe_2(SO_4)_3 + 2H_2O \quad (3)$$
$$Fe_2(SO_4)_3 + 6H_2O \rightarrow 2Fe(OH)_3 + 3H_2SO_4 \quad (4)$$

A low pH water is produced (pH 2-4.5). At these pH levels, the heavy metals such as iron, calcium, magnesium, manganese, copper and zinc, are more soluble and enter into the solution to further pollute the water. Water of this type supports only limited water flora, such as acid-tolerant molds and algae; it will not support fish life, destroys and corrodes metal piers, culverts, barges, etc., increases the cost of water treatment for power plants and municipal water supplies, and leaves the water unacceptable for recreational uses.

As estimated, 10,000 miles of streams have been degraded in Appalachia alone (3). Additional acid problems in Western Kentucky, Indiana, Illinois, and Montana have been documented (4)-(7).

Alkaline Mine Drainage

Alkaline mine drainage may result where no acid-producing material is associated with the mineral seam or where in situ neutralization of that acid which is produced has taken place. Alkaline mine drainage may be, but is not usually, as bad as acid mine drainage. Drainage from freshly exposed strata usually has a higher mineral content than that from undisturbed land because the strata has high levels of readily leachable materials.

Some alkaline waters have high concentrations of ferrous iron and upon oxidation and hydrolysis, form acid which lowers the pH and changes the drainage to the acid type. These types of discharges are more common to underground mines than surface mines.

The Western mines present new problems. In many areas the overburden material is highly alkaline and sometimes saline (8). Improper handling of this material can lead to pollution of surface water, stock-watering ponds and maybe even ground water.

Documentation of the alkaline and saline problem is almost nonexistent.

Sedimentation and Erosion

The removal of the vegetation and the loosening and breaking up of the overburden by blasting, shovels, draglines, and dozers creates materials and conditions conducive to erosion. Curtis (9) and Collier, et al (10) have documented the increased sediment and suspended solid loads from surface mined areas.

Sediment fills creek beds destroying fish habitat and creating flooding conditions. Suspended solids increase treatment cost for industrial and municipal supplies.

Erosion is influenced by the soil type (sand, clay, salt, infiltration rate and percolation rate), the climate (temperature and amount and type of rainfall), the topography (steepness, length of slope, land configuration and exposure), and the vegetation cover. Each of these factors should be considered in the planning, development and reclamation of a surface mine to prevent erosion.

SOURCES OF MINE DRAINAGE

Surface Mines

Surface mines fall into two broad types: area and contour. Area surface mining is practiced on relatively flat terrain and usually encompasses a large tract of land, up to several hundred or more acres. It can be found mainly in Ohio, Western Kentucky, Illinois, Indiana, and west of the Mississippi River. Contour mining is practiced on rolling to very steep terrain. Usually only a few cuts are made into the hillside and the coal seam is followed along the edge of the hill or mountain, leaving a trace which appears as a "contour line."

In general, the pollution from area mines is not as severe as that from contour mines. Overburden can more easily be segregated during mining and the acid producing or saline material buried. Silt from erosion can often be confined to the mining area. The overburden can be graded to a slope that is less erosive. Final cuts can be filled with water to prevent oxidation of the pyrite material. Areas that have been area mined have been shown to have lower peak flows and increased base flows (4)(11).

The common practice in the past was the placement of the overburden from a contour mine onto the downslope of the adjacent area. This material was subject to severe erosion and landslides. Plass (12) made a survey of eastern Kentucky and found that 12% of the outslope area had failed (landslides). Because of the landslide problem, West Virginia has limited the bench width on steep slopes greater than 65% (17). Even with these precautions, landslides still occur. Many water problems arise from this type of mining. The highwalls and spoil piles, because of their long, bare, steep, uninterrupted slopes are subject to severe erosion. Sediment coming off these slopes has been known to clog stream channels, cover highways, settle on cultivated land and kill crops, cover fish spawning beds, fill water courses, and thus, subject adjacent land to flooding, etc.

A study of water sheds draining to Beaver Creek, Kentucky (9)(13) showed that a partially stripped watershed (6.4% of area) had an average erosion rate of 5.9 tons per acre per year as compared to 0.7 ton per acre per year for an unmined area.

Curtis (14) has found that contour mines in Kentucky in nonacid areas still cause changes in water quality in the adjacent streams. Among the elements showing greatest increase are sulfate, calcium and magnesium. Aluminum, manganese, iron and zinc also increase some. In general the increase did not reach a serious magnitude.

Even when spoil piles are graded, problems occur. If the grading is toward the highwall, the water may accumulate in an area of poor quality spoil, i.e., a spoil high in pyritic material, and may therefore be degraded in quality. Where underground mines are located behind the highwalls, grading toward it often causes water to flow into the underground mine. This flow often flushes toxic material from the mine.

When spoil piles are graded away from the highwall, erosion may be as bad or worse, unless good water management practices are followed, such as diversion ditches across the top of the highwall, ditches or terraces across the slope to break the slope length and control structures to remove the water from the mining area.

Underground Mines

During the period when an underground mine is active, water must be removed either by gravity or by pumpage. The characteristics of this water will vary from seam to seam and even within the same mine. Following active mining, below drainage mines usually fill with water and are often no longer a source of pollution. However, those above drainage continue to discharge polluted water for decades.

The Appalachian Study (3) reported that active underground mines produced 18.8% of the acid mine drainage and inactive underground mines 52.5%. An additional 9.2% came from a combination surface-underground. Thus, underground mines were involved with 80.5% of the acid mine drainage.

Siltation is not usually associated with underground mines; however, the waste "gob" removed and piled outside the mine is a major source of silt, acid mine drainage and often air pollution from ignition.

Whereas control technology for surface mines is highly developed, the technology for underground mines is severely limited.

Refuse Piles and Slurry Ponds

Associated with many coal mines is a cleaning or processing plant where the coal is processed to remove dirt and impurities present in the coal. Two types of wastes are discharged from a cleaning plant—refuse and slurry. The refuse is the coarse portion consisting largely of coal intermixed with pyrites, sandstones, clays and shale. Refuse has historically been piled adjacent to the cleaning plant and has created major silt and acid water pollution problems and air pollution. Slurry is the fine reject material which contains mostly coal, shale and clay. The slurry is usually placed in lagoons adjacent to the plant. Slurry can create major sediment problems in streams when it is discharged directly to the stream or escapes from lagoons because of poor design, construction or maintenance. Upon abandonment, slurry lagoons continue to be a menace unless they are stabilized. Often slurry ponds are not acid.

CONTROL OF MINE DRAINAGE

Acid Problems

The ultimate solution to the acid mine drainage problem is preventing its formation. As noted in equations (1) through (4), the reaction is dependent on air (oxygen) and water coming in contact with the sulfides. In addition to water being a reactant, it serves as the transport medium for the metals. All prevention techniques are based on excluding water and/or air from the mining environment.

Air: Air is necessary during the active operation of an underground mine. After the coal has been extracted, excluding air should be an integral part of the mining plan. As areas are worked out, blockages should be made to prevent air from entering that section of the mine. Mines should be developed downdip so that they flood when abandoned. In this manner, water serves as an oxygen barrier. Injecting solid material into the mine, such as stope filling, should be encouraged and expanded. Not only can the mine drainage problem be decreased by blocking the flow of air, but a solid waste disposal problem can be resolved simultaneously.

Sealing adits to prevent air from entering an underground mine has been in practice in the coal fields since the 1920's. This procedure at its best, has only been marginal. Seals can be built to prevent air from entering the adit, but air reaches the mine readily through the fractured outcropping and overburden. With each change in barometric pressure, air is pumped in and out of the mine. Air sealing should not be considered as an acceptable method.

Bulkhead or hydraulic mine seals have been shown to be successful (15). For this method, a seal is built in the adit and the outcrop is grouted to prevent water from leaving the mine. In time, the mine workings are flooded and oxygen is excluded from the sulfides.

Excluding oxygen from the active surface mine is not feasible. Usually there is a delay between the time the sulfides are exposed to air and acid mine drainage is formed. High-sulfide-bearing material should be exposed for the shortest time possible. As part of the mining operation, this material should be covered with material containing little or no sulfides. The saving and replacement of top soil should be practiced. The cover material acts as an oxygen barrier. Wind and diffusion are the only driving forces for moving air through the cover material to the sulfides. The cover material should be vegetated with grasses, legumes, shrubs, forbs, etc. Not only does the vegetation serve to stabilize the cover material and prevent it from eroding, but upon drying, it decays and acts as an oxygen absorber.

Refuse dumps at mines are major sources of mine drainage and should be constructed to prevent air movement into them. Sound techniques are compacting waste material as it is placed and sandwiching layers of compacted clay between layers of waste. The slopes of the dumps should be kept to 33% or less to prevent wind from driving air into the pile. Upon abandonment, the surface of the pile should be sealed with an air barrier. Although many different materials have been evaluated, soil still remains the best. As in the case of surface mines, the surface should be vegetated.

Water: Enough water is usually available in the moist air within an underground mine or within the material itself in a surface mine or waste dump to meet the requirement for the formation of mine drainage. The other function of water is as a transport medium. It flushes the oxidation products from the sulfide and carries them into the environment. By controlling the water, these products can be contained.

Provisions should be made to prevent water from entering the mining environment, except where it is to be used for flooding and an oxygen barrier. Diversion ditches, dewatering of the working and sealing of fractures are just a few of the methods available. Water that does enter the workings should be removed as rapidly as possible.

Silt Problems

The control of silt from surface mines starts in the mine planning stage and follows through mining and final reclamation. During the planning state, water handling throughout the operation should be carefully planned. This stage includes the location of diversion ditches to keep water out of the mine working, protection of natural drainageways, location of silt traps and basins, and the selection of a mining method to minimize erosion and landslides. The States of Kentucky (16) and West Virginia (17) have developed drainage handbooks as guidelines. In recent years several new mining methods such as slope reduction, box cut and block cut have been developed to reduce silt and slides (18)(19). The removal, storage and replacement of topsoil are other valuable tools in erosion control. Backfilling and planting should follow mining closely to minimize the period that bare materials is exposed. Grading should be planned to keep slope length and steepness to a minimum. Terraces and benches should be included to reduce runoff rate. Final grading should be across the slope.

Grasses should be planted on all areas for rapid erosion control. Trees may follow if needed or desired. Before planting, soil analysis should be made and the proper fertilizer and lime added. On steeper slopes mulches may be needed.

In some cases the silt basins may not be capable of reducing the sediment load to an acceptable level because of the colloidal nature of the material. The addition of lime, alum and other flocculating chemicals might be required in conjunction with the silt basin.

Treatment

Treatment has three places in the acid mine drainage scheme: (1) during active operations to produce a water acceptable for discharge to a stream, (2) in those situations where the production of acid mine drainage cannot be prevented, and (3) in those cases where water is needed for industrial or domestic use.

Neutralization

The most commonly used method for treating acid mine drainage and removing heavy metals is neutralization. A typical system would include adding an alkaline reagent, mixing, aerating, and removing the precipitate. Alkaline reagents that may be used are ammonia, sodium carbonate, sodium hydroxide, limestone, and lime. In most cases, lime is used because of its lower cost and higher reactivity.

Most heavy metals tend to precipitate as the pH is raised. Metals such as copper, zinc, iron, aluminum, manganese, nickel and cobalt can be reduced to low levels, less than half mg/ℓ, with pH adjustment, precipitation and solids separation.

In removing iron, aeration is used to convert the ferrous form to the ferric form to take advantage of the precipitation of ferric iron at a lower pH.

In most cases, lime is used for neutralizing. Except for limestone, it is the cheapest reagent, and also reacts rapidly. The disadvantage is the voluminous sludge produced. Although limestone is cheaper and produces a denser sludge, it is not effective above a pH of 6.5 because of the slow attack of limestone by acidity at higher pH's. Two-stage treatment with limestone followed by lime appears to offer the advantages of both reagents. EPA has studied its use on coal acid mine drainage (20).

Although neutralization can be used to remove heavy metals and a portion of the sulfate, the resulting water is high in dissolved solids and sulfate. The disposal of the sludge is also a problem.

Ion Exchange

Various ion exchange schemes had been applied to the treatment of coal acid mine drainage (10). These schemes can upgrade the water quality to potable use. Modification can be incorporated to recover heavy metals. Ion exchange has not received wide acceptance because of difficulties encountered with resin fouling, interfering ions, limited loading capacity, cost of operating and disposal of regenerating solutions. Further development is needed.

Reverse Osmosis

Reverse osmosis (RO) has effectively removed multivalent ions from mine drainage. The following table presents data from the treatment of coal mine drainage. All heavy metals should be removed at about this same level (99% or better). Reverse osmosis is a concentrating process in which the pollutants are contained on one side of a membrane while the water passes through. Water recoveries (percent of water fed to unit that is available as treated water) have been as high as 90%. Water recovery is limited by the precipitation of material on the membrane when they have been concentrated beyond the materials' saturation point. Calcium sulfate is usually the first material to be precipitated in mine drainage. Equations have been developed to predict the highest water recovery allowable under any influent conditions (22)(23).

Since RO is a concentrating system, the disposal of the waste stream is a major problem. EPA has developed a system whereby the waste stream is neutralized, the sludge removed, and the neutralized water returned to the influent of the RO unit. This system has been named "Neutrolosis" and has obtained water recoveries in excess of 99% (24).

Treatment of Mine Drainage by Reverse Osmosis

	Removal Percent
Ca	98-99.8
Mg	98.5-99.8
Fe, total	98.5-99.9
Al	91.7-99.2
Mn	97.8-99.1
Cu	98.7-99.5
SO_4	99.3-99.9
Acidity	81.0-91.7
Specific conductance	95.0-99.9

SUMMARY

Coal will continue to provide a significant amount of the energy of this country. The very concept of disturbing the earth to remove any material, whether by surface or underground methods will cause changes in the environment. We must strive to make these changes create a minimum of temporary or long-term damages.

With our present technology, most acid mine drainage problems can be controlled during active mining and upon abandonment of surface mines and refuse piles. This statement cannot be made of underground mines. In the majority of cases, acid mine drainage cannot be controlled from abandoned or orphaned underground mines.

We are still short on technology for controlling silt and landslides from surface mines. Great strides are being made in these areas and we can be optimistic. The opening of the Western fields has raised many issues for which we do not have solutions or which we do not fully understand.

REFERENCES

(1) Beall, John V. "Muddling Through the Energy Crisis," *Mining Engineering* 24, 10, October 1972.

(2) Anonymous, "Peabody Looks at the Future of Surface Coal Mining," *Mining Engineering* 24, 10, Oct. 1972.

(3) Appalachian Regional Commission. *Acid Mine Drainage in Appalachia.* Appalachian Regional Commission. Washington, DC June 1969.

(4) Grubb, H.F. and Ryder, P.D. "Effects of Coal Mining on the Water Resources of the Tradewater River Basin, Kentucky." Geological Survey Water—Supply Paper 1940, U.S. Geological Survey, Washington, DC 1972.

(5) Corbett, D.M. "Acid Mine Drainage Problem of the Patoka River Watershed, Southwestern Indiana," Water Resources Center Report of Investigation No. 4, Indiana University, Bloomington, 1969.

(6) "Lime/Limestone Neutralization of Acid Mine Drainage," EPA Grant 14010 DAX, Environmental Protection Agency, National Environmental Research Center, Cincinnati, Ohio, 1968.

(7) Ferris, Orrin. Sand Coulee Creek Mine—Acid Pollution Abatement, Montana Department of Natural Resources and Conservation, Kelena (personal letter).

(8) Hodder, R.L., "Surface Mined Land Reclamation Research in Eastern Montana," Montana Agricultural Experiment Station, Bozeman, Montana, Research and Applied Technology Symposium on Mined-Land Reclamation, National Coal Association, Pittsburgh, Pennsylvania, 1973.

(9) Curtis, Willie R., "The Effects of Strip-Mining on the Hydrology of a Small Mountain Watershed in Appalachia," U.S. Forest Service Paper, Berea, Kentucky, 1969.

(10) Collier, C.R., et al, "Influence of Strip Mining on the Hydrological Environment of Beaver Creek Basin, Kentucky," U.S. Department of the Interior, Geological Survey, Prof. Paper 427-B 1964 and Geological Survey, Prof. Paper 427-C, 1966.

(11) Corbett, D.M. *Water Supplied by Coal Surface Mines, Pike County, Indiana,* Water Resources Research Center Report of Investigation No. 1, Indiana University, Bloomington, December 1965.

(12) Plass, William T. "Land Disturbances from Strip Mining in Eastern Kentucky," U.S. Forest Service Research Notes, Berea, Kentucky, 1967.

(13) Musser, John J. "Description of Physical Environment and of Strip-Mining Operations in Parts of Beaver Creek Basin, Kentucky," U.S. Department of the Interior, Geological Survey, Prof. Paper 427-A, 1963.

(14) Curtis, W.R. "Chemical Changes in Streamflow Following Surface Mining in Eastern Kentucky, "U.S. Forest Service, Berea, Kentucky, Fourth Symposium on Coal Mine Drainage Research, April 1972.

(15) Foreman, J.W. "Evaluation of Mine Sealing in Butler County, Pennsylvania," Fourth Symposium on Coal Mine Drainage Research, Bituminous Coal Research, Incorporated, Monroeville, Pennsylvania, April 1972.

(16) *A Manual on Kentucky Reclamation,* Kentucky Department for Natural Resources and Environmental Protection, Frankfort, 1973.

(17) *Drainage Handbook for Surface Mines,* West Virginia Department of Natural Resources, Division of Reclamation, Charleston, 1972.

(18) *Green Lands, Quarterly,* West Virginia Surface Mining and Reclamation Association, Charleston, Vol. 3, No. 1, pp 17-28, spring 1973.

(19) Grim, E.C. and Hill, R.D. "Surface Mining Methods and Techniques," National Environmental Research Center, Environmental Protection Agency, Cincinnati, Ohio, 45268.

(20) Wilmoth, R.C., Scott, R.B., and Hill, R.D. "Combination Limestone-Lime Treatment of Acid Mine Drainage," Fourth Symposium on Coal Mine Drainage Research, Bituminous Coal Research, Incorporated, Monroeville, Pennsylvania, April 1972.

(21) Holmes, J. and Schmidt, K. "Ion Exchange Treatment of Acid Mine Drainage," Fourth Symposium on Coal Mine Drainage, Bituminous Coal Research, Incorporated, Monroeville, Pennsylvania, April 1972.

(22) Gulf Environmental Systems Company, *Acid Mine Waste Treatment Using Reverse Osmosis,* EPA Water Pollution Control Research Series #14010 DYG, August 1971, Washington, DC 1971.

(23) Wilmoth, R.G., Mason, D.G., and Gupton, M. "Treatment of Ferrous Iron Acid Mine Drainage by Reverse Osmosis," Fourth Symposium on Coal Mine Drainage, Bituminous Coal Research, Incorporated, Monroeville, Pennsylvania, April 1972.

(24) Hill, R.D., Wilmoth, R.C., and Scott, R.B. "Neutrolosis Treatment of Acid Mine Drainage," 26th Annual Purdue Industrial Waste Conference, Lafayette, May 1971.

Coal Refuse Disposal Practices and Challenges in the United States

by John F. Martin
Extraction Technology Branch
Industrial Environmental Research Laboratory
U.S. Environmental Protection Agency, Cincinnati, OH 45268

INTRODUCTION

On August 3, 1977, the Surface Mining Control and Reclamation Act of 1977 was signed into law. The Surface Mining Reclamation and Enforcement Provisions supporting the law define coal waste as "earth materials, which are combustible, physically unstable, or acid-forming or toxic-forming, wasted or otherwise separated from product coal and are slurried or otherwise transported from coal processing facilities or preparation plants after physical or chemical processing, cleaning, or concentrating of coal" (1). The regulations also state that "toxic-forming materials means earth materials or wastes which, if acted upon by air, water, weathering, or microbiological processes, are likely to produce chemical or physical conditions in soils or water that are detrimental to biota or uses of water." The handling of waste, refuse, slurry, and gob is a practical problem for the coal industry and one set apart for special consideration by law. Proper disposal of these by-products or utilization of them for construction or production of other products remains as a challenge to the United States and to all coal producing countries.

Statistics for the coal industry in the United States show that 1975 production for bituminous coal and lignite was 588 million metric tons (2). This figure can be further divided into underground production of 266 million metric tons and surface production of 323 million metric tons. In 1971 approximately 50% of the domestic lignite and bituminous coal produced was the result of underground mining methods. The production emphasis has since moved to surface mining. Of the coal produced in 1975, 41.2% was mechanically cleaned yielding 97 million metric tons of waste material.

The mechanical coal cleaning plants in the United States are clustered in the eastern and interior coal provinces east of the Mississippi River. The 1975 statistics show that the largest number of cleaning plants are in West Virginia, and the eastern states as a whole account for more than 360 of the total 388 plants recorded. Wet jigging and dense medium processing continue to be the most popular mechanical coal cleaning techniques.

A summary of production trends indicates that the waste handling problem is increasing. Bituminous and lignite coal cleaning wastes rose from 81 million metric tons in 1971 (3) to 97 million metric tons in 1975. As a result of increased production of low sulfur coal and more careful and selective mining procedures, the percentage of coal that was cleaned in 1975 (41.2%) was less than that in 1971 (49.1%). Selective mining or washing of all coals, however, is becoming more important as environmental concern increases (4).

ENVIRONMENTAL CONCERNS

The coal waste disposal problem bears equally on air pollution, water pollution and land use. In addition, the disposal site is often considered to be aesthetically offensive.

A loosely packed refuse pile that contains a large amount of combustible material is a prime candidate for spontaneous ignition and a refuse pile fire is extremely difficult and dangerous to quench. The resulting fumes have a high sulfur oxide content and present a serious health hazard to the nearby population. Of 135 active refuse piles recently visited in Pennsylvania, 19 had signs of fire. In addition to fumes from refuse fires, fine refuse particles carried by the wind are a source of considerable dust.

A source assessment performed by Monsanto Research (5) attempted to summarize the severity of the air pollution problem resulting from burning coal refuse piles. Estimates of the yearly national mass emissions from burning coal refuse were: particulates, 190 metric tons; sulfur oxides, 39,000 metric tons; nitrogen oxides, 34,000 metric tons; total hydrocarbons, 34,000 metric tons; carbon monoxide, 4,500,000 metric tons. Other atmospheric pollutants detected from burning waste areas included polycyclic organic materials, hydrogen sulfide, ammonia, and mercury.

Water is polluted by coal waste in basically two forms: acid drainage and siltation. The acid drainage formed by mine waste varies according to location, seams, compaction, cover material, etc. Although coal seams vary locally and by region, there appears to be less of an acid drainage problem in southern West Virginia, eastern Kentucky, and the western coal provinces of the United States than in other coal producing areas.

Siltation is directly influenced by the steepness, compaction, drainage control structures, and cover material of a waste or refuse pile. Work at the Truax Tracer refuse pile (EPA Grant 14010 DDH) in Illinois has shown that even the moderately sloping sides of a pile are susceptible to severe erosion (6). A stable grass cover, often difficult to establish on refuse, is essential if erosion is to be controlled.

Water moving through coal waste picks up the sulfuric acid that has been generated by oxidation of pyritic materials. The acid, in turn, enables the water to dissolve extremely large quantities of iron, aluminum, and other heavy metals. Previous studies have indicated that effluents from coal waste areas contain significant quantities of zinc, copper, nickel, iron, lead, boron, magnesium, and manganese. An Environmental Protection Agency survey (7) found that the effluent from an Indiana waste site had a pH of 2.4, acidity of 6,500 mg/ℓ, and iron concentration of 2,600 mg/ℓ. In addition, the zinc and manganese concentrations were 7.2 mg/ℓ and 120 mg/ℓ respectively. The suspended solids loads may range into the thousands of milligrams per liter, depending on the amount of flow, stream gradients, and characteristics of the waste pile.

Studies at the University of South Carolina (8)(9) have shown that only the framboidal form of pyrite (smaller than 0.5 micron) in coal and coal wastes produces abundant acidity. In disposal or reclamation efforts it would be of benefit to be able to predict the acid potential of the waste material, so research was undertaken to present a point counting technique for quantitatively estimating acidity production.

Very few studies have been carried out in the United States to document ground water pollution problems associated with mining, especially regarding the effects of coal waste disposal.

One short-term monitoring effort of an abandoned coal waste pile was undertaken by Argonne National Laboratory at Consolidated Mine No. 14 in Illinois (10). The mine waste was highly eroded and generated an acidic surface discharge. Twenty-seven wells were placed in and around the disposal site in order to monitor the shallow ground water system. Thirteen residential wells in the area were also sampled. The water table was about 3 m below the ground surface near the waste pile and sloped away from the pile in all directions. At wells within 130 m of the waste pile the pH values were low, while acidity, sulfates, and several metal ions were extremely high. At one particular well near the waste pile, the pH was 2.4 and acidity, sulfate, and aluminum concentrations were 8,370 mg/ℓ, 2,722 mg/ℓ, and 980 mg/ℓ, respectively. Values of iron, manganese, and zinc were recorded at 394 mg/ℓ, 7.0 mg/ℓ and 123 mg/ℓ, respectively. Water quality greatly improved at distances greater than 130 m from the pile. Combinations of dispersion, adsorption, cation exchange, and precipitation were cited as possible processes involved in improving the water quality.

A joint study by Poland's Central Research and Design Institute for Opencast Mining Poltegor and the United States Environmental Protection Agency, carried out in Poland, showed that coal waste disposal pile had a negative effect on ground water quality (11). The study, involving disposal of coal waste and fly ash, showed there was significant contribution of sulfates, chlorides, sodium, potassium, boron and other ions from the waste material to the ground water.

Streams affected by drainage from poorly constructed coal waste disposal sites are rendered useless for many miles. The most obvious effect of such pollution is the destruction of aquatic life. Even after a waste pile is reclaimed, the effects of silt accumulation on the bottom impact the spawning areas of fish populations for many years. Such streams, polluted by acid drainage and suspended materials, afford no recreational use. They are basically unfit for cooling water, of no good for irrigation, and require extensive treatment before they can serve as a domestic or industrial water supply.

The impact of coal waste disposal on land use is basically one of reducing future alternatives for development. In some cases reclamation efforts have turned waste piles and slurry ponds into pasture land, building sites, and recreational areas, but these examples are in the minority. It is expensive to correct a problem area after the damage has been done. Proper planning and construction, aimed toward pollution control and post-reclamation land use, is a proven, effective tool for solving the land area problems associated with disposal of waste or toxic-forming materials. Often the flat surface of a coal waste disposal area can serve as an economic asset in mountainous or steeply sloping terrain where flat building sites are scarce.

CURRENT PRACTICE

The culm bank, refuse pile, or coal waste heap was, in the early coal mining days, the easiest spot for random dumping. It may have been adjacent to the preparation plant, over the nearest hillside, or in a stream bed. Our failure to properly plan and engineer these waste sites has caused many of them to be environmental hazards. Today, the careful construction of refuse piles is essential.

Various methods have been employed to transport material to the waste dump (12)(13). Each system was developed to take advantage of the terrain and to apply to the type and quantity of waste material produced. Small size wastes in slurry form are readily pumped to the disposal site, while coarser fractions must be transported in some other manner. The most popular haulage technique today for coarse wastes is by truck. It is essential that the disposal site be well shaped and compacted. Ground machinery should be used in conjunction with trucks to spread and compact material. A sheep's-foot roller greatly increases the effectiveness of the dozer in compacting thin layers. On many active disposal sites the truck drivers are being encouraged to drive to the dumping location on a slightly different course each trip to aid in compaction. Pans or scrapers also can be used effectively to spread the material in thin layers, to form the waste pile, and to compact it.

The aerial tram is a popular haulage tool for solid wastes in mountainous regions, but it does a rather poor job of forming a stable pile. As the material drops, it tends to segregate into coarse and fine fractions. The side slopes rest at a steep natural angle of repose, and there is relatively no compaction. A dozer, when used in conjunction with a tram is helpful in building the pile, but can accomplish less compaction than when trucks are used for haulage. Belt haulage of waste also requires some type of ground machinery to spread and compact material at the site.

Many of the huge anthracite culm banks in Pennsylvania were formed by dumping material over the end of the pile, and raising the road or mine car tracks as the pile grew. Often these piles burned because of their porous structure and low density; they were riddled by erosion and polluted the surface drainage with silt. Excessive surface acidity or heat from the sun kept most of them devoid of vegetation (14).

The angular character of coal waste particles or refuse often allows it to rest at a steep slope. One anthracite refuse pile in Eastern Pennsylvania held 35 degree side slopes; bituminous coal refuse on a hillside dump in Eastern Kentucky held 30 degree side slopes. These slopes are often subjected to high pressure from winds, and allow air to be forced through the pile contributing to pyrite oxidation and spontaneous combustion.

Since combustion and acid mine drainage require oxygen in their processes, oxygen most logically is the factor to control at the waste disposal site. With regard to runoff and leachate, oxygen is necessary to convert pyritic material into ferrous iron and sulfuric acid. The acid in turn increases the ability of the water to dissolve heavy metals and to carry them to the receiving stream. The density of the pile, size of material, water content, and side slopes contribute to the rate of oxygen entry.

For stabilization and control of the refuse pile, it is necessary to compact the material as it is placed. This reduces the air voids, increases the bearing capacity, and reduces the possibility of slides and slope failures due to excessive water content. To aid in halting the movement of oxygen, a sealant (such as plastic, cementing agents, clay) can be placed over the pile (15). Vegetation on a refuse bank further aids in trapping oxygen, because the humus that develops forms a moisture barrier and demands oxygen itself. In certain instances it may prove feasible to seal only the sides of a compacted pile, thereby cutting off most of its oxygen supply. Various research projects are currently investigating a host of surface sealants, including gypsum, wood chips, fly ash, waste cement, and various types of vegetation and earth materials (16)(17)(6).

In response to the Buffalo Creek Disaster of 1972, the Mining Enforcement and Safety Administration published design and inspection manuals (18)(19) to establish uniform construction and regulatory procedures for coal refuse disposal facilities. The design manual covers such topics as the origin of coal refuse and waste, classification of disposal techniques, general planning and design techniques, geotechnical investigations, hydrologic considerations, environmental factors, construction procedures, and monitoring systems. The Administration classifies disposal facilities as impounding (cross-valley, side-hill, diked pond, incised pond, other) or nonimpounding (valley-fill, cross-valley, side-hill, ridge, heaped, other) structures to facilitate monitoring and inspection. The current regulations of the Surface Mining Control and Reclamation Act of 1977 do not cover disposal of coal refuse except in surface mine pits or other backfilled and regraded areas. The final regulations are expected to cover environmental and safety considerations in off-site disposal of waste in ponds, piles, or various types of valley-fills.

Recently, many mining companies have begun to return waste materials, refuse, and toxic-forming materials to the surface mine pit for disposal when feasible. They have made efforts to integrate disposal and reclamation plans to safely and effectively solve their coal waste problem. A pronounced advantage of returning waste to the surface pit is the fact that no pile or fill is constructed and no other land disturbance is required beyond the actual mine pit. Operations of this type are extremely well suited to area mining operations with onsite coal preparation or washing facilities.

Returning coal wastes to the worked-out voids in underground mines is much less common. Backfilling has long been common practice in metal mines, but it is rare for coal mines (20). Some hydraulic backfilling is accomplished, however, in abandoned mines for subsidence control. In this case, utilization and disposal of coal waste materials is combined into one operation.

The initial regulations (21) published by the Department of the Interior in support of the Surface Mining Control and Reclamation Act of 1977 set forth partial guidance for future waste disposal operations. The regulations basically provide for covering all toxic-forming, acid-forming, combustible, and waste materials with a minimum of 1.2 m of nontoxic material as a part of the backfilling and grading operation of a mine site. In addition, it must be demonstrated to the regulatory authority that the use of waste materials will not adversely affect water quality, flow, vegetation, or stability in the backfilled area, and that the use of waste will not present a public health hazard. All activities associated with the mining operation are to be conducted in such a manner as to minimize impacts to the prevailing hydrologic balance. The section of the regulations dealing with disposal of spoil and waste materials in areas other than the mine workings or excavations does not currently deal with waste or toxic-forming materials except to specifically prohibit disposal of combined spoil and waste in valley or head-of-hollow fills.

Water quality standards set forth in the regulations apply to all surface drainage from the disturbed area. This provision would necessarily include waste materials being returned to the mine pit for backfilling and disposal. After being routed through a sedimentation pond or series of ponds, the surface drainage must meet effluent limitations as indicated in the following table.

Point Source Coal Mine Effluent Limitations

Effluent Characteristics	Maximum Allowable	Average of Daily Values for 30 Consecutive Discharge Days
Iron, total, mg/ℓ	7.0	3.5
Manganese, total, mg/ℓ	4.0	2.0
Total suspended solids, mg/ℓ	70.0	35.0
pH	6-9	

There are exceptions to the limitations on a case-by-case basis for some western states and in cases where high manganese levels are encountered. In addition, discharges from the disturbed area within the permit area resulting from precipitation in excess of the 10 year, 24 hour event will not be subject to the above effluent limitations.

The use of coal waste material in dam construction, and any dam to impound waste materials must be directly approved and supervised by the regulatory authority to ensure stability, safety, and operating characteristics.

DISPOSAL RESEARCH

In order to develop more advanced and economical methods of coal refuse disposal and abandonment of disposal areas, various research and demonstration projects have been undertaken by state and federal agencies and private industry. One such project was accomplished as a joint venture between EPA and the Consolidation Coal Company at the New Kathleen Mine in Illinois (23). An abandoned refuse pile was regraded and covered with soil material to determine the effectiveness of different thicknesses of earth cover in reducing acid runoff from the refuse. In general, it was found that acid production was controlled by covering and revegetating the pile and that no significant difference was observed among 0.3, 0.6, or 0.9 m soil covers. Before reclamation, runoff from the pile had a high iron content with a pH value of 2.0. After completion of the project, the effluent from the pile was of good quality with but a few tiny acidic seeps along the toe.

Two problems related to the surface soil covering were encountered in the reclamation of the pile. First, although no significant water quality or vegetation difference was detected among the various soil cover thicknesses there was a problem in controlling the 0.3 m soil application. With such a thin soil blanket, the construction equipment gouged some spots of the pile and overcovered others. Secondly, several of the side slopes on the pile were so steep that gullies eroded into the waste material. Later, remedial action was necessary to repair the damage.

From a feasibility study conducted by EPA and the State of Indiana at the Greene-Sullivan State forest, it was pointed out that leveling and covering large refuse areas is extremely expensive (23). The project area consisted of a large refuse pile and adjacent acid lake. The restoration program was to include grading, covering, revegetating and permanently inundating portions of the refuse. These procedures were designed to break the cycle of erosion, exposure of pyritic material, oxidation, acid formation, and erosion. The lake had a pH value of 2.8, with sulfate, iron, and magnesium concentrations of 1,700 mg/ℓ, 100 mg/ℓ, and 100 mg/ℓ respectively. The demonstration phase of the project was not accomplished because of the prohibitive cost.

In 1971, the Peabody Coal Company and Illinois Department of Mines and Minerals began a cooperative study for direct seeding of coal refuse after lime was added to soil (24). Seed and fertilizer were added at two study areas but growth at the end of the first growing season was almost negligible. Additional plots were vegetated with increased lime application rates and these showed more favorable results. One area also received a 15 cm soil layer in addition to a lime treatment of 48 tonnes/ha. This area showed what seemed to be superior vegetation compared to the others that received lime only. The study showed that vegetation could survive on refuse material with proper lime neutralization. Initial results showed that about 11 tonnes/ha of lime, incorporated to a 15 cm depth, was necessary for each 1% of sulfur in the refuse material.

Disposal of fine coal refuse is also receiving some research and demonstration efforts. Fine refuse in slurry form may be dewatered mechanically and placed in a waste landfill, dewatered in a holding pond for future landfill disposal, or permanently impounded in slurry form. With many preparation plants now recycling the slurry water, excavation of settling or storage ponds is becoming more common. The fine coal refuse is, however, difficult to excavate, difficult to haul by truck, and nearly impossible to compact in a fill.

An approach to the problem, which seems to be gaining popularity, is to add a stabilizing agent to the slurry material in order to help it gain shear strength and reduce its permeability. The Island Creek Coal Company and Dravo Lime Company have demonstrated that treated refuse can be excavated and used in a landfill (25). The slurry material, treated with Calcilox, was handled, spread and compacted with conventional equipment immediately after being excavated from the holding lagoon.

COAL WASTE UTILIZATION

Utilization of coal refuse and waste materials offers several advantages over mere disposal. As a by-product material, coal waste can be used as a fill or construction material, processed for use as an aggregate material, or supplied as a raw feed for extraction of coal and other elements carried in it.

Use of coal mine waste material for backfilling of mine workings in the United States dates back to at least 1864 when a church in Pennsylvania was saved from coal mine subsidence by stowing preparation waste in the void beneath the building. The practices of controlled placement and blind flushing of refuse into abandoned mines has continued to be a control technique for subsidence. Because of the porous and compressible nature of the backfilled material, the use of a binding agent such as portland cement, would be necessary to consider a coal waste backfill as a total subsidence prevention technique.

In 1972 and 1973, the Bureau of Mines, U.S. Department of the Interior, undertook a full-scale demonstration of coal waste backfilling (26). Nearly 344,900 cubic meters of refuse were pumped in water-slurry form into mine voids below Scranton, Pennsylvania. Five injection boreholes were used with a network of observation holes to monitor progress of the operation. Pressure at the slurry pump delivering material to the holes was maintained at a gauge reading of 4.2 to 6.7 g/cm^2. The waste material ranged in size up to 12.7 mm.

The work demonstrated that coal mine waste could not only provide some measure of protection against subsidence, but that utilization of waste material eliminates the need for surface disposal. This secondary environmental benefit also extends to a reduction in air pollution from burning of abandoned waste piles and reduction of surface water pollution emanating from acid drainage. Hopefully, the backfilling process will effectively seal any acid generating materials from oxygen contact and thereby eliminate any future water leachate problems.

At the Orient No. 4 mine in Illinois, the Freeman United Coal Mining Company conducted a similar demonstration of stowing mine waste underground (20). Their work differed from that previously cited in that the mine was active and not abandoned. Since 80% of the refuse material at Orient No. 4 was less than 1 mm in size, it was pumped directly from a preparation plant to a worked-out section of the mine. The refuse slurry was fed to the underground void by gravity at 5.3% solids. It appears that the total extra cost of injecting the slurry underground as opposed to ponding it on the surface was about $0.60 per metric ton. Even with the extra crushing required, the costs of returning both coarse and fine refuse to the mine void may be economically feasible where surface disposal systems are either impractical or expensive.

Various other coal refuse utilization projects have been undertaken in the United States. In Pennsylvania, refuse has been used for highway fills and shoulders, and has been reprocessed and leveled for equipment parks and industrial building sites. A major concern about the use of coal waste is its potential for water pollution. Any utilization project must consider the composition of the refuse material, its size, and strength.

The Extraction Technology Branch of EPA performed an in-house study of coal waste material used as a base course for construction of a parking apron. Three test sections were installed to investigate the water quality of leachate from an all-refuse base, a refuse-fly ash base, and a refuse-fly ash-lime base. After five years the latter two areas exhibited a yearly average leachate pH value greater than 6.0, and iron and aluminum contents were less than 1 mg/ℓ and 15 mg/ℓ, respectively. The test section using an all-refuse base continues to leach acid water with high metals content. These preliminary results indicate that the refuse needs to be blended with an alkaline material before use. Though the water quality of leachate from the all-refuse area is relatively poor, only a very small quantity collected; thus drainage protection, encapsulation or surface sealing might serve to protect the environment from adverse impacts.

Considerable research and development efforts have gone into the utilization of coal mine waste and refuse in a slightly altered form. Through sintering or roasting processes, coal refuse can be converted to a fairly inert, durable, lightweight aggregate for use as a medium for vegetation or construction material. When coal waste is subjected to high temperatures, on the order of 1300°C, sulfur tends to gasify, volatile materials are driven off, and carbonates and organics are decomposed. Other constituents are either reduced or oxidized leaving a stable product which is sterile, chemically inert, and full of fine holes. Studies at the University of Kentucky (27) have demonstrated that this type of product from coal refuse serves well as a growing medium for containerized plants and does not release chemicals which are toxic to the plants.

The material described above is similar to the incinerated refuse produced at burning refuse and gob piles. For years, incinerated refuse or "Red Dog" has been excavated for use as a highway fill material and skid-preventive road coating. This naturally occurring material may vary considerably in physical and chemical composition, however, from one refuse dump to another.

Other studies at the University of Kentucky were undertaken to investigate the use of sintered coal refuse in portland cement concrete, bituminous concrete, and concrete blocks (28). This lightweight aggregate performed well in all three applications. Its strength was slightly lower than that of control specimens constructed with limestone aggregate, but this is more than offset by the benefits of the lower unit weight of the sintered coal refuse mixtures. One benefit noted in the use of coal refuse for sintered aggregate production was the extra heating value from coal left in the refuse by the cleaning process.

Coal refuse material is also being used in the United States as a resource for additional yields of coal and for a source of other raw materials. Advanced processing methods and equipment have made possible the reclamation of thousands of tons of coal from old waste piles and slurry ponds. In the area of fine coal recovery, tremendous resources are available from disposal sites where fine coal was placed as an unuseable waste. Anthracite coal waste banks are especially suited for reprocessing because of the relative cleanliness of the fuel from an environmental standpoint and the economic value of the product. Immediately proceeding the mine backfilling work at Scranton, Pennsylvania, the refuse piles were reprocessed for a second yield of steam coal.

Aside from coal itself, aluminum seems to be the next most economically recoverable component of coal waste. The U.S. Bureau of Mines has found that ash from most any type of coal refuse contains about 24% Al_2O_3 (29). Recovery of this aluminum, and some additional heat value, may help to offset the ash handling and air pollution problems associated with burning coal refuse. Other elements, such as iron, calcium, magnesium, manganese, titanium, etc., are also found in coal waste. These too, to varying degrees, may be recoverable, based on future supply and demand and on advanced recovery processes.

REFERENCES

(1) Public Law 9587, Surface Mining Control and Reclamation Act of 1977, August 3, 1977.

(2) Reprint from the *1975 Bureau of Mines Minerals Yearbook, Coal—Bituminous and Lignite,* U.S. Bureau of Mines, Washington DC.

(3) *1971 Minerals Yearbook, Volume 1—Metals, Minerals, and Fuels.* U.S. Bureau of Mines. Washington, DC, 1973, pp 329-395.

(4) Deurbrouck, A.W. "Coal Preparation 1973." Mining Congress Journal 60:65-67, February, 1974.

(5) Blackwood, T.R., et al, "Coal Refuse Piles, Abandoned Mines, and Outcrops," Draft Report Contract 68-02-1874, U.S. Environmental Protection Agency, Cincinnati, Ohio.

(6) Kosowski, Z.V. *Control of Mine Drainage from Coal Mine Mineral Wastes, Phase II.* EPA-R2-73-230. Office of Research and Monitoring, U.S. Environmental Protection Agency, Washington, DC, May 1973, 83 pp.

(7) Martin, J.F. "Quality of Effluents from Coal Refuse Piles," U.S. Environmental Protection Agency. Paper given at the First Symposium on Mine and Preparation Plant Refuse Disposal, Louisville, Kentucky, October 1974, 12 pp.

(8) Caruccio, F.T. *Estimating the Acid Potential of Coal Mine Refuse: The Ecology of Resource Degradation and Renewal,* Edited by Chadwick and Goodman, Blackweel Scientific Publications. pp 197-205.

(9) Caruccio, F.T., et al, *Paleoenvironment of Coal and its Relation to Drainage Quality.* EPA-600/7-77-067. Office of Research and Development, U.S. Environmental Protection Agency, Cincinnati, Ohio, June 1977, 108 pp.

(10) Schubert, J.P., R.D. Olsen, and S.D. Zellmer "Monitoring the Effects of Coal Refuse Disposal and Reclamation on Water Quality in Southwestern Illinois." Paper presented at the Fourth Joint Conference on Sensing of Environmental Pollutants, American Chemical Society, 1977, 8 pp.

(11) Libicki, J. "Impact of GOB and Power Plant Ash Disposal on Ground Water Quality and its Control." Paper given at the Third Symposium on Coal Preparation, Louisville, Kentucky, October 1977, 20 pp.

(12) Coalgate, J.L., D.J. Akers, and R.W. Frum *Gob Pile Stabilization Reclamation, and Utilization.* Office of Coal Research, DOI Report No. 75. Coal Research Bureau School of Mines, West Virginia University, Washington, DC, May 1973, 127 pp.

(13) Hummer, H.D. and L.J. Vogel "Refuse Removal and Disposal. pp 16.3-16.37 in Leonard and Mitchells, Eds., *Coal Preparation.* The American Institute of Mining, Metallurgical, and Petroleum Engineers, Inc., New York. 1968.

(14) Schramm, J.R. "Plant Colonization Studies on Black Wastes from Anthracite Mining in Pennsylvania." *Trans. Am. Phil. Soc.* 56:30-40, February 1966.

(15) Scott, R.B. *Sealing of Coal Refuse Piles.* Project Report—Mining Pollution Control Branch. National Environmental Research Center, U.S. Environmental Protection Agency, Cincinnati, Ohio. July 1973. 15 pp.

(16) *Methods and Costs of Coal Refuse Disposal and Reclamation.* U.S. Bureau of Mines, Pittsburgh, PA, Information Circular 8576. 36 pp.

(17) Wahler, W.A. *Coal Wastes: What Can be Done? Coal Mining and Processing.* pp 71-74, 102, 104, January 1974.

(18) E. D'Appolonia Consulting Engineers, Inc., *Engineering and Design Manual—Coal Refuse Disposal Facilities,* Mining Enforcement and Safety Administration. Washington, DC.

(19) E. D'Appolonia Consulting Engineers, Inc., *Coal Refuse Inspection Manual,* Mining Enforcement and Safety Administration. Washington, DC April 1975.

(20) Jankovsky, C.K., "Disposal of Coal Refuse Slurry Underground." *Mining Congress Journal,* September 1977.

(21) "Surface Mining Reclamation and Enforcement Provisions, U.S. Department of the Interior." *Federal Register,* Vol. 42, No. 239, December 13, 1977, Part II. pp 62639-62715.

(22) Truax-Tracer Coal Company, *Control of Mines Drainage from Coal Mine Minerals Wastes, Phase 1.* 14010 DDH. Office of Research and Monitoring, U.S. Environmental Protection Agency, Washington, DC, August 1971, 148 pp.

(23) Greene-Sullivan Demonstration Project, Draft Report of Grant No. S802593, U.S. Environmental Protection Agency, Cincinnati, Ohio November, 1977.

(24) Medrick, C. and A.F. Grandt "Lime Treatment Experiments—Gob Revegetation in Illinois." Paper presented at the Illinois Mining Institute.

(25) Snyder, G.A., F.A. Zuhl and E.F. Burch "Solidification of Fine Coal Refuse." *Mining Congress Journal,* December 1977.

(26) Allen, Alice S., and C.W. Anderson "Recent Developments in the Use of Mine Wastes for Subsidence Control." Paper presented before the Fourth Mineral Waste Utilization Symposium. Chicago, Illinois, May 1974, 9 pp.

(27) Buxton, J.W., D.E. Knavel and G. Yost "Sintered Coal Refuse as a Growing Medium for Plants." Paper presented at the Third Kentucky Coal Refuse Disposal and Utilization Seminar, Lexington, Kentucky, May 1977, 4 pp.

(28) Rose, J.G. "Sintered Coal Refuse as a Construction Aggregate." Paper presented at the Third Kentucky Coal Refuse Disposal and Utilization Seminar, Lexington, Kentucky, May 1977, 6 pp.

(29) Kealy, D., et al "Those Waste Banks Could Be Sources for Fuel, Alumina." *Coal Mining and Processing,* August 1976.

Successful Revegetation of Coal-Mined Lands in the United States

by Willie R. Curtis
M.Sc. Hydrologist and Project Leader
Northeastern Forest Experiment Station Forest Service
U.S. Dept. Agriculture, 370 Reed Road, Broomall, PA 19008

INTRODUCTION

Surface mining for coal in the United States began in Illinois just over 100 years ago. Natural revegetation of many of these areas probably began soon after mining ceased, because natural forces continually operate to vegetate disturbed land. For example, many of the 10- to 30-year-old coal surface mines in Alabama have been naturally reforested with loblolly pine and Virginia pine and now produce more commercial softwood timber than the surrounding unmined forest land (Lyle et al, 1976).

Although some surface mines have been revegetated by natural means, the vegetation process can usually be hastened by planting. Among the earliest documented attempts at reclamation is the seeding of sweet clover on spoil-bank ridges by the Wayne Coal Company in Ohio as early as 1918 (Kohnke, 1950). Tree planting followed in 1920, and several species of trees were planted on coal mine spoils in Illinois during that year. Reforestation of spoil banks was started in Pennsylvania as early as 1923. Then in 1929 research was started in Indiana to determine the adaptability of various tree species to conditions on spoil banks.

RESEARCH

A formal research program on the forestation of strip-mined land was started in 1937 by the U.S. Forest Service through its Central States Forest Experiment Station at Columbus, Ohio (Lane, 1968). Research was conducted in cooperation with the coal mining industry, state agencies, and agricultural experiment stations. The results of this early research were summarized by Limstrom (1960). It involved mostly the evaluation of survival and growth of many different tree species. The value of planting "nurse trees" such as black locust and European alder in mixture with other tree species received much attention. Limstrom's reasons and recommendations for planting the black locust are still valid. Successful forestation has been accomplished on thousands of acres of surface-mined land through the use of species and planting procedures recommended by the research foresters. This early research still serves as the basis for many revegetation guides in use today.

Before the 1960's, most of the research and consequently most of the reclamation centered on tree species; there was little concern for erosion and sedimentation. Anyway, most of the sediment from erosion on the ungraded spoil piles was trapped in the valleys between these piles. Herbaceous cover was considered a hindrance to the establishment and growth of planted tree seedlings.

Beginning in the early 1960's more and more emphasis has been given to the quick establishment of herbaceous cover for erosion control and for aesthetic purposes. This new emphasis coincided with greatly expanded surface mining activity in the Appalachian region. Here, water runoff and erosion produce severe sedimentation problems. Subsequently, the use of herbaceous vegetation for ground cover and erosion control increased to the extent that for the past 10 years or so relatively few trees have been planted. Now it seems that the move is again toward planting of trees for more permanent cover. At the same time there is need for the quick cover of herbaceous species for erosion control. This has led researchers to concentrate on simultaneous establishment of herbaceous cover and woody plants. Diversified plantings of grasses, legumes, shrubs, and tree species are still needed to provide for a variety of future uses of the land. Reforestation is anticipated as a result of natural plant succession in much of the eastern United States, especially where an herbaceous cover is established for erosion control without further management.

As early as the mid-1940's some research on the use of herbaceous species in reclamation was reported (Tyner and Smith, 1945). Subsequent trials and other research have been conducted by the Forest Service, Soil Conservation Service, Agricultural Research Service, and various state experiment stations. The results of these experiments have been incorporated into revegetation guides published by Federal, State, and private agencies.

Many plantings have been made by industry-related groups such as reclamation associations and coal producers' associations, and by many individual mining companies. These plantings may also be used in evaluating reclamation success.

Almost everyone who engages in surface mine revegetation research begins with experiments related to the adaptability of species and varieties to spoil conditions at hand. Results often vary widely from area to area, even for the same species. If the spoils were described more specifically perhaps we could begin to understand the reasons for this. For example, spoil with a pH of 3.5 to 4.0 derived from one geologic formation may be no more toxic to plants than spoil with a pH of 4.5 derived from a different formation. Spoil pH is often the only analysis reported and the implication is that particular plants will grow in all spoils of similar pH.

SOME FACTORS AFFECTING SUCCESSFUL REVEGETATION EFFORTS

Several of the more important factors that affect the successful establishment of vegetation on surface mine spoils will be discussed.

Seedbed Preparation

That a suitable seedbed is necessary for the rapid and successful establishment of seeded vegetation on mine spoils was shown by Vogel (1974). On plots that were rototilled before seeding, vegetative cover was established faster and was more dense at the end of the first and second growing seasons than on plots that were not tilled. Tillage was more beneficial for seedings made between April 1 and October 15. The main advantage of tillage is that it breaks up the crust on the surface of the spoils and increases the number of microsites favorable for germination of seed and growth of the plants. Seeding on freshly graded spoils would provide the same advantages as tillage at lower cost.

Amendments

The amendments to be discussed are lime, fertilizer, and mulch.

Lime: The chemical properties that most often influence the establishment and growth of vegetation on strip-mine spoils are related to chemical reaction and imbalances in plant nutrients. Acidity or alkalinity is a very useful criterion for predicting the capacity of spoil to support vegetation. Its intensity is expressed as pH, a measure of the concentration of hydrogen ions present. Most spoils in the eastern United States are in the acid range (below pH 7.0) while most spoil in the western states is alkaline.

Some plant species are more tolerant than others of acid conditions. Nearly all plant species will grow in spoils with a pH above 5.5. A more limited number will grow when pH is between 4.5 and 5.5. Fewer species can tolerate spoil below 4.5 and only a few will grow where pH is below 4.0. In the West, species tolerant of alkaline conditions must be selected.

There are other problems in establishing vegetation on acid spoils. Toxic levels of aluminum and manganese are soluble at low pH's and become harmful to plants. Roots of many plants will not grow into materials with toxic levels of soluble aluminum. Thus plant growth is stunted because the roots do not take up nutrients.

Some things can be done to ameliorate acid spoils. Amendments such as organic matter, fly ash, and topsoil may be added, but lime is the most used. Topsoiling is considered the most economical way of treating alkaline western spoils.

The most common practice now is to apply enough lime to raise pH to 5.5. At this pH, manganese and aluminum become less soluble and are no longer available in a form toxic to plants. Also phosphates and other nutrients become more readily available to plants.

Fertilizer: Most surface mine spoils are deficient in essential plant nutrients. Nitrogen is almost universally lacking; phosphorus is limited in most spoils but is adequate in some. Potassium generally is not limiting for the establishment of herbaceous vegetation.

A deficiency in nitrogen severely limits establishment and growth of herbaceous species, especially grass. Thus, the addition of nitrogen fertilizer is a must for successfully reclaiming mine spoils.

Further applications of nitrogen fertilizer may be avoided by planting or seeding nitrogen-fixing plant species. Legumes are used for this purpose, although the long-lived legumes usually take more than 1 year to get established. This means that annual legumes or grasses must be used for the initial quick cover needed to control erosion.

Mulches: Various mulches can be used during the critical period of seedling establishment to improve the microclimate and help conserve soil moisture. Among the many materials that have been tested for use on surface mine spoils are pulp fiber, grain straw, hay, wood chips, wood bark, leaves, and some chemical binders.

Generally wood pulp fiber is mixed with seed and fertilizer and applied with a hydroseeder. The amounts of this mulch usually used (900 to 1,100 kg/ha) are inadequate to significantly reduce erosion. Also, it is doubtful that they have a significant effect on moisture conservation.

Straw at 2,200 to 3,400 kg/ha, distributed evenly over the surface after seeding, can be a factor in the successful establishment of vegetation on spoils. However, it is often necessary to hold the straw in place with a binder or to run over it with a disc or crimper. Hay falls into the same category as straw. Both may retain seed which can result in a volunteer stand of vegetation. However, seed of some weeds may be undesirable.

Wood chips are good because they stay in place on the spoil. However, efforts to establish vegetation with wood chips as mulch have been disappointing. Shredded tree bark applied at the rate of 94 m^3/ha has been found to be quite effective in conserving moisture, retarding runoff and erosion, and providing microclimatic conditions favorable to plant establishment and growth. Leaves, too, have been found to be a valuable mulch. They can be applied with a blower or with a regular farm manure spreader. Leaves must be disced into the spoil to hold them in place.

Naturizer is a trade name of a mixture of composted municipal waste and sewage sludge. Initial tests indicate that reclaiming surface mine sites may be a beneficial way to dispose of these waste products. Using trade, firm or corporation names in this publication is for the information and convenience of the reader. Such use does not constitute an official endorsement or approval by the U.S. Dept. of Agriculture or Forest Service of any product or service to the exclusion of others that may be suitable.

Many chemical binding agents have been developed for erosion control. When used alone these chemicals cannot serve all the same purposes as mulch; i.e., they cannot provide moisture conservation, stabilize the soil, and moderate surface temperature.

Sewage sludge and effluent have been used successfully to vegetate some problem spoils. Much research is underway at present into the use of these and other waste materials in reclamation.

Topsoiling is now required on nearly all surface mine disturbance in the United States. The so-called "topsoil" is removed from the area about to be mined and stockpiled for later spreading on the graded site. Almost all the unconsolidated overburden may be classed as topsoil. In many cases the importance of topsoiling cannot be disputed, but in some cases another stratum in the overburden may have more desirable characteristics for vegetating than the topsoil. In these cases the best material for plant growth should be placed on the surface.

Time of Seeding

For a long time most seedings of mine spoil in the Appalachian region were made in the spring or fall, with spring generally being preferred. Many people believed that even in a humid region it was impossible to vegetate mine spoils by direct seeding in the summer. Vogel (1974) has shown that selected mixtures of temporary and long-lived herbaceous species can be used anytime between March 1 and October 15 in eastern Kentucky. Results of his study have been applied with success to other areas in the Appalachian region. Best initial cover was achieved during early spring and late fall with Balbo rye or annual ryegrass. Pearl millet, Sudan grass-sorghum hybrid, and weeping love grass provided quickest cover from mid-May to mid-July. Kentucky-31 fescue and *sericea lespedeza* became the primary cover on the areas after two or three growing seasons.

In humid regions, barren areas should be seeded anytime mining is completed between March 1 and October 15. The seed mix can be changed to fit climatic conditions throughout this period.

Seeding a temporary species for quick cover with the permanent species is desirable in most situations. The permanent species generally becomes established and replaces the temporary ones.

In any seeding, the species selected often is less important than adherence to good vegetation practices; i.e., preparation of a suitable seedbed, application of sufficient fertilizer, and, when necessary, application of lime and mulch.

In the Southwest it is important to seed when moisture conditions are best. For example, fourwing saltbush should be planted only when the probability of summer thunderstorms exceeds 50% and existing soil water stress is less than 2 atmospheres of tension.

Climate

The amount and distribution throughout the year of rainfall are strong factors in the establishment of vegetation. In the East rainfall is not usually a problem, but in the West and Southwest it may be the limiting factor in plant establishment. Supplemental water is recommended for areas that receive less than 200 mm of precipitation.

Species Selection

Species of vegetation for use on surface mine spoils should be selected according to the time of year, type of spoil, purpose of seeding or planting, and the expected future use of the land.

One of the first considerations is to establish a quick cover for erosion control. This generally means a mixture of herbaceous annuals and perennials. Some woody species may be desirable. When temporary quick-cover species are mixed with permanent species, there is a chance that the temporary cover will prevent or retard the development of a satisfactory stand of the permanent cover. It is important to limit the amount of seed of the annual or quick-cover species to prevent this.

Many experiments have been conducted to determine the adaptability of various herbaceous and woody species for vegetating strip-mine spoils. Some species have been selected for their tolerance to acid conditions; others have been selected specifically for use on alkaline spoils. Many species of grasses are well suited to surface mine reclamation because they are adapted to a wide range of climatic and spoil conditions. Some species are desirable because germination is usually rapid and growth is quite good during the first growing season.

Many plant species have been tried with varying degrees of success in the United States. For each of the major coal producing regions there are preferred plant species in each of four major categories: grasses, forbs, shrubs, and trees. The grasses and forbs may be further classified as temporary, semipermanent, or permanent. Temporary species are used to give quick cover for site protection; semipermanent species are reasonably persistent perennials; permanent species are those expected to persist for many years.

Annual grasses can be used for quick cover and as companion crops with slower-developing perennials. Some species are better adapted to spring seeding, others should be seeded in the summer, and still others in the fall.

Leguminous forbs are considered essential in ground-cover mixtures. Associated plants generally benefit from the nitrogen fixed by the legumes. Annual legumes germinate quickly and provide good cover within a few months. Perennial leguminous forbs often are more difficult to establish; many have hard seed coats or other specific dormancy requirements that delay germination. The availability of soil phosphorus is generally more critical for the establishment of legumes than for grasses. Seeding rates of companion grass species should be carefully regulated to prevent excessive competition and possible failure of the legumes. Each species of legumes must be inoculated with specific strains of *Rhizobium* bacteria to stimulate nodulation and to assure nitrogen fixing.

Shrubs for reclamation plants have not been emphasized although they have significant value for site protection, wildlife food and cover, and aesthetics. Possibly in the future more emphasis will be placed on the use of shrubs.

Trees have generally been the mainstay of reclamation efforts. A few tree species can be directly seeded but most are normally established by planting seedlings. Many species are adapted to surface mine spoils, but species native to the area are usually recommended. Unfortunately, high-value tree species are usually the most difficult to establish on spoils. Grass and legume ground covers generally reduce tree growth, at least in the early stages. In later years legumes may enhance growth. Black locust, autumn olive, and bristly locust are nitrogen fixers and can be beneficial as nurse trees for other tree species.

CURRENT PRACTICES

Revegetation practices in the U.S. are determined or influenced mainly by legal requirements. The seeding of grasses and herbaceous legumes to establish a cover for erosion control and for aesthetics is generally recommended. Thus, watershed protection automatically becomes the primary concern in land use. Once the herbaceous vegetation is well established it can be used for pasture and hay because species good for forage usually are used for vegetating mine spoils, for example, smooth bromegrass and alfalfa in Illinois and tall fescue and annual lespedeza in southern Appalachia. Livestock grazing is becoming common on vegetated strip mines even in the mountainous areas of Appalachia. Unfortunately, the tendency is to overgraze or start grazing before the vegetation is well established, which hinders the development of adequate cover.

The land use in effect at time of mining often influences the choice of revegetation practices. For example, prime agricultural land in the flatlands of Illinois is reclaimed for immediate return to row crop production.

On the other hand, forested land is usually not replanted to forest species. Some states require that woody species be planted in addition to the herbaceous species on steep slopes and other sites not intended for agricultural use. Normally the planting of woody species is optional for the landowner or mining company. The legal requirements in some states unintentionally discourage tree planting because they permit release of vegetation bonds as soon as the requirements for the herbaceous vegetative cover are met.

Black locust is still the most commonly used tree species. It is especially popular for use on steep slopes because it can be established by direct seeding in mixture with herbaceous species. Locust seedlings respond to fertilizer, especially phosphorus, as herbaceous legumes do, and first-year seedling heights of 30 cm or more are not uncommon. Unpopular "doghair" stands of locust sometimes develop where seeding rates are excessive. European black alder and autumn olive are other nitrogen-fixing woody plants that are used as nurse plants or site improvers and for wildlife habitat.

Red pine, white pine, and hybrid poplar have been successful in the northern Appalachian coal field. European white birch has been the most successful tree species on extremely acid spoils in Pennsylvania. Virginia pine has been widely planted in the middle and southern Appalachian region, and loblolly pine is often planted or direct-seeded in southern Appalachia. Most species of pine are tolerant of acid spoils, especially in association with the acid-tolerant mycorrhizal fungus *Pisolithus tinctorius* (Marx 1975).

Tree planting now is rare in the Midwest even though reforestation studies showed that high-value hardwoods can be produced on spoils. This is due in part to political and social pressures that favor agricultural uses of all surface-mined lands. Also, hand planting of tree seedlings costs more and entails more labor-related problems than seeding of herbaceous species. Another reason for choosing agricultural uses over forestry is the greater and quicker economic return from the reclaimed land. Yet some of the 30- to 50-year-old hardwood stands, especially of black walnut, in Indiana and Illinois demonstrate the economic potential of growing forests on mined lands.

There is a need to establish a quick cover for erosion control at the same time that trees are planted, but the herbaceous vegetation may adversely affect survival and growth of the planted trees. Some people recommend seeding the herbaceous species one year and planting trees the next. Others suggest using herbicides to kill patches of herbaceous vegetation where tree seedlings will be planted.

A more practical method is to plant trees and herbaceous species at the same time. The herbaceous cover should consist mainly of legumes because the growth of trees usually is enhanced by legumes but is suppressed by grass (Vogel 1973). When seeding must be done is late spring and summer, and herbaceous cover can be established with summer annuals. These will die in the fall and provide a mulch over winter. Trees and perennial herbaceous species can be planted the following spring.

On areas that can be traversed with farm machinery, early competition of herbaceous species with trees can be avoided by planting the herbaceous species and trees in alternate strips. To further assure their establishment and early growth, tree seedlings should be inoculated with the appropriate mycorrhizal fungi (Marx 1975).

Revegetated surface mines can provide excellent wildlife habitat and water-based recreation, as is shown on many old strip mine sites. Present vegetation practices in the East often provide poor wildlife habitat because they use mainly a few species of grasses and legumes such as tall fescue and *sericea lespedeza* that provide little food or diversity of habitat.

SUMMARY

Several species of hardwoods and pines planted 30 years ago have been reasonably successful on spoils in both the Midwest and Appalachian coal fields. Especially encouraging is the success of high-value hardwoods such as black walnut, red oak, yellow poplar, and white pine. Many of the older strip mined lands are providing excellent wildlife habitat and water-based recreational uses.

Currently, most surface mines are being vegetated with herbaceous species to provide quick erosion control. Tree planting has been discouraged by economic, legal, and social pressures. Establishment of both trees and herbaceous cover is feasible especially by concurrent planting of tree seedlings and unaggressive herbaceous legumes.

REFERENCES

Kohnke, H. 1950 "The reclamation of coal mine spoils." In A.G. Normal (ed.) *Advances in Agronomy*, Vol. II p 317-349. Academic Press, New York.

Lane, R.D. 1968 "Forest Service reclamation research," *Min. Congr. J.* 54: 38-42.

Limstrom, G.A. 1960 *Forestation of strip mined land in the central states.* U.S. Department Agriculture, Agric. Handb. 166. U.S. Government Printing Office, Washington, D.C.

Lyle, E.S., Jr., D.T. Janes, D.R. Hicks and D.H. Weingartner 1976 "Some vegetation and soil characteristics of coal surface mines in Alabama." *Proc. 4th Symp. Surf. Min. Reclam. Oct. 19-21, Louisville, KY*, Natl. Coal Assoc., Washington, D.C. p 140-152.

Marx, D.H. 1975 "Mycorrhizae and establishment of trees on strip-mined land." *Ohio J. Sci.* 75: 288-297.

Tyner, E.H. and R.M. Smith 1945 "The reclamation of the strip-mined coal lands of West Virginia with forage species." *Soil Sci. Sco. Am. Proc.* 10: 429-436.

Vogel, W.G. 1973 "The effect of herbaceous vegetation on survival and growth of trees planted on coal-mine spoils." p 197-207. *Proc. Res. Appl. Technol. Symp. Min. Land Reclam. Mar. 7-8, Pittsburgh, PA* Bitum. Coal Res., Inc., Monroeville, PA.

Vogel, W.G. 1974 "All-season seeding of herbaceous vegetation for cover on Appalachian strip-mine spoils." p 175-186. *Proc. 2nd Res. Appl. Technol. Symp. Min. Land Reclam. Oct. 22-24, Louisville, KY* Natl. Coal Assoc., Washington, DC.

COAL INFORMATION SOURCES AND DATA BASES 1980

by Carolyn C. Bloch

Determining the proper source of coal information is often more perplexing than perusing that information whence retrieved.

This directory comprises a listing of international, national and state agencies, departments, offices and cooperating information sources, such as libraries and universities that deal in some capacity with coal data, whether it be legal and regulatory, assistance programs and technology. This information can serve as an easy, economical tool for scientists, managers and engineers concerned with all phases of coal studies and projects, provided they have the know-how to obtain it. This time-saving directory of sources offers the assistance they need.

The following cites main source categories and **examples of some** specific sources.

1. **GENERAL SOURCES**
   Department of Commerce
   NTIS
   Department of Energy
   Technical Information Center
   NASA
   Scientific and Technical Inf. System
   Smithsonian Science Inf. Exchange
   Electric Power Research Institute
2. **LIBRARIES**
   Department of Energy
   Library
   DOE Laboratory Libraries
   Department of Interior
   Office of Library and Inf. Services
   Library of Congress
   Science and Technology Division
3. **PUBLICATIONS AND INFORMATION**
   Department of Interior
   Bureau of Mines
   Government Printing Office
   Office of Technology Assessment
   National Academy of Sciences
4. **COAL POLICY, BUDGET, AND PROCUREMENT INFORMATION**
5. **COAL-RELATED LEGISLATIVE COMMITTEES**
6. **COAL RESEARCH AND TECHNOLOGY**
   Department of Energy
   Asst. Sec. for Resource Applications
   Asst. Sec. for Fossil Energy
   Technical Information Center
   Department of Interior
   Bureau of Mines
   U.S. Geological Survey
   Department of Commerce
   Asst. Sec. for Science and Technol.
   National Bureau of Standards
   NASA
   Office of Information
7. **PATENTS, INVENTIONS AND TECHNOLOGY TRANSFER SOURCES**
   Department of Commerce
   U.S. Patent and Trademark Office
   NTIS
   Department of Energy
   DOE Patents for Licensing
8. **COAL ASSISTANCE PROGRAMS**
   Department of Energy
   Coal Loan Guaranty Program
   Department of Interior
   Office of Surface Mining
   Reclamation and Enforcement
9. **INFORMATION ON COAL RESOURCES, SUPPLIES AND TRANSP.**
10. **SOURCES FOR ENVIRONMENTAL AND SAFETY INFORMATION**
    Environmental Protection Agency
    Department of Labor
    Mine Safety and Health Admin.
    Dept. of Health and Human Services
    NIOSH
11. **SURFACE MINING AND LEASING INF.**
    Department of Energy
    Asst. Sec. for Resource Applications
    Department of Interior
    Bureau of Land Management
12. **SOURCES FOR INTERNATIONAL DATA**
    Department of Commerce
    NTIS
    Department of Energy
    Department of Interior
    Branch of Foreign Data
    State Department
13. **UNIVERSITY SOURCES**
14. **PRINCIPAL STATE COAL ORGANIZATIONS**
15. **ASSOCIATIONS WITH COAL-RELATED INFORMATION**

ISBN 0-8155-0830-1 **128 pages**

Other Noyes Publications

LIGNITE TECHNOLOGY 1980

Edited by Perry Nowacki

Energy Technology Review No. 53
Chemical Technology Review No. 146

This book discusses the present and future use of lignite for electric power generation, and the problems associated with the use of low-rank coals.

Lignite and other subbituminous coals represent a large domestic, yet relatively unused source of low-cost energy. The abundant deposits are located largely in the Northern Great Plains and Rock Mountain regions, as well as in Texas, and are near the surface.

Lignite is a low-sulfur coal, but produces copious amounts of alkaline fly ash, and ash fouling of boiler tubes is the most serious operating problem when burning it. Yet lignite has high permeability and reactivity which properties make it a superior feedstock for gasification and liquefaction.

A partial and condensed table of contents with **chapter headings and examples of some subtitles** follows here.

ISBN 0-8155-0783-6 **228 pages**

Other Noyes Publications

COAL LIQUEFACTION PROCESSES 1979

by Perry Nowacki

Chemical Technology Review No. 131
Energy Technology Review No. 45

With the precarious position of petroleum supplies in the world market and the unfavorable balance of payments caused by overdependence on foreign sources, it may be expedient to exploit coal further as a source of liquid fuels. In the United States coal is the most abundant natural energy resource. Through its conversion to liquid fuels, it could supply a portion of the energy required for transportation, heating, etc.

Coal can furnish a complete range of liquid fuels, as well as chemical feedstocks. To achieve maximum yields, however, judicious adjustment of temperature, pressure, time and catalyst, as elaborated in this book, is essential. The text reviews the histories of important domestic and foreign coal liquefaction processes—pyrolysis, solvent extraction, and catalytic and indirect liquefaction. It details the chemistry of the processes and their inherent problems, also discussing economics, developmental status, and environmental impacts.

Chapter headings and **examples of some** important subtitles follow below:

ISBN 0-8155-0756-9 **339 pages**

Other Noyes Publications

COAL RESOURCES, CHARACTERISTICS AND OWNERSHIP IN THE U.S.A. 1978

Edited by Robert Noyes

Coal currently occupies less than eighteen percent of the nation's energy market. This is despite the fact that coal reserves are adequate for at least a century at present consumption rates, while the reserves of the alternative fossil fuels, petroleum and natural gas, are probably measurable only in decades. The probable course of the petroleum supply picture is further beclouded by economic and political factors, making any kind of probability forecast next to impossible.

In the late 1960s, the abrupt rise of environmentalism and the ensuing passage of legislation, setting standards of air quality, imposed direct restrictions on the sulfur content of coal used for the generation of heat. The scarcity of coals with low sulfur content in the Eastern U.S. and the lack of an effective and economic method of sulfur removal have helped to limit the redevelopment of the nation's most abundant and available energy source. The variations in properties of coal have given rise to geographical patterns of coal mining operations for specific industrial purposes.

This book presents an accurate picture of U.S. coal reserves, the nature and composition of the coal and of the land and minerals ownership. It is based on federally funded studies and is thus a valuable tool in the all-out attack on the energy crisis. A partial and condensed table of contents follows here.

ISBN 0-8155-0698-8

346 pages

Other Noyes Publications

ENERGY SAVING BY INCREASING BOILER EFFICIENCY 1979

Edited by Lee Yaverbaum

Energy Technology Review No. 40

The boiler plant manager or operator will find this definitive text furnishes the precise assistance he needs to implement efficiency improvement techniques. With the sharp boiler maintenance described he can realize fuel savings of from 10 to 30%.

To keep efficiency high each device in the system must be adjusted for optimum performance. This book presents the necessary information regarding adjustments, modifications, equipment and maintenance that should be considered, from the simplest adjustment of fuel/air ratio (which can effect remarkable fuel savings of several percentage points) to the more complex auxiliary equipment. It contains material on installing combustion controls, minimizing heat losses, adjusting combustion conditions, and removing impurities.

Not to be disregarded are the data on alternate fuels. The testing so essential to better boiler functioning is also discussed in detail, and in the last chapter the economic effects of the suggested boiler changes are evaluated.

Following is a condensed table of contents with **examples of some** important subtitles.

ISBN 0-8155-0745-3 **226 pages**

Other Noyes Publications

OIL SHALE TECHNICAL DATA HANDBOOK

Edited by Perry Nowacki

Energy Technology Review No. 63
Chemical Technology Review No. 182

A ready reference for the evaluation of appropriate technology for oil shale development, this book will be extremely useful to engineering and managerial personnel. The book defines the oil resource and lists the properties of shale and the oil and gas derived from it. Recovery technologies potentially compatible with the resource are described. Associated processes, such as mining, materials handling, beneficiation, upgrading, wastewater treatment, and spent shale disposal, have also been included. Emphasis has been placed on documenting available design information such as maximum module size, operating conditions, yields, utility requirements, etc.

The following is a partial table of contents, illustrating chapter headings and some important subtitles.

ISBN 0-8155-0835-2 (1981) **309 pages**

Other Noyes Publications

FLUE GAS CLEANING WASTES DISPOSAL AND UTILIZATION

Edited by D.L. Khoury

Pollution Technology Review No. 77
Energy Technology Review No. 65

Flue gas desulfurization (FGD) wastes and fly ash, together, are generally referred to as flue gas cleaning (FGC) wastes. Modern fossil-fueled boilers (utility and industrial) produce large quantities of FGC wastes. Those boilers employing conventional coal combustion present a broad spectrum of potential environmental problems, which, due to regulatory constraints pertaining to air and water pollution control, will require focusing on the environmental management of FGC wastes.

This study, prepared under the direction of Arthur D. Little, Inc., assembles and reviews data from ongoing activities relating to pollution control technology for conventional coal-fired combustion sources. Generation of FGC wastes; disposal options including current practice, R&D and field studies; and utilization practice, including technical and economic assessment of current practice and R&D studies, are covered. Emphasis is placed on wastes produced by commercially demonstrated technologies and, where data are available, those technologies in advanced developmental stages.

A **partial table of contents** for this book is given below.

ISBN 0-8155-0847-6 (1981) **646 pages**